T0331873

FUNDAMENTAL CONCEPTS IN MODERN ANALYSIS

An Introduction to Nonlinear Analysis

Second Edition

FUNDAMENTAL CONCEPTS IN MODERN ANALYSIS

An Introduction to Nonlinear Analysis

Second Edition

Vagn Lundsgaard Hansen

Technical University of Denmark, Denmark

Revised with **Poul G. Hjorth**

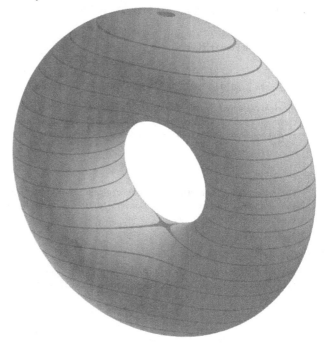

 **World Scientific**

NEW JERSEY · LONDON · SINGAPORE · BEIJING · SHANGHAI · HONG KONG · TAIPEI · CHENNAI · TOKYO

Published by

World Scientific Publishing Co. Pte. Ltd.

5 Toh Tuck Link, Singapore 596224

USA office: 27 Warren Street, Suite 401-402, Hackensack, NJ 07601

UK office: 57 Shelton Street, Covent Garden, London WC2H 9HE

Library of Congress Cataloging-in-Publication Data

Names: Hansen, Vagn Lundsgaard, author.

Title: Fundamental concepts in modern analysis : an introduction to nonlinear analysis /
  Vagn Lundsgaard Hansen, Technical University of Denmark, Denmark ;
  revised with Poul G. Hjorth.

Description: Second edition. | New Jersey : World Scientific, [2020] |
  Originally published: 1999. | Includes bibliographical references and index.

Identifiers: LCCN 2019046832 | ISBN 9789811209406 (hardcover)

Subjects: LCSH: Mathematical analysis.

Classification: LCC QA300 .H353 2020 | DDC 515--dc23

LC record available at https://lccn.loc.gov/2019046832

British Library Cataloguing-in-Publication Data

A catalogue record for this book is available from the British Library.

Revised with Poul G. Hjorth

For any available supplementary material, please visit
https://www.worldscientific.com/worldscibooks/10.1142/11529#t=suppl

Printed in Singapore

To my students

Preface to the Second Edition

The second edition opens with a new chapter (Chapter 1) describing a geometrical approach to the system of real numbers. The focus is on the limit operations of quantities measured by numbers, which are crucial in mathematical analysis. In the following three chapters (Chapters 2-4), the basic material on topology in the first edition, has been reorganized and revised with Poul G. Hjorth.

The fundamental concept of compactness in metric and topological spaces is treated in Chapter 3, and in more detail than in the first edition. In the theory of metric spaces, the notion of a sequentially compact space is defined by a convergence property of sequences in the space. In the theory of topological spaces, the more general notion of a compact space is defined through a refinement property of coverings of the space with open sets. In the second edition, the two notions of compactness are developed with independent proofs of the main results. Thereby the material on compactness is apt for applications also in functional analysis, where the notion of sequentially compactness prevails.

The material on differentiability in normed vector spaces in the first edition, has also been reorganized. Chapter 5 concentrates on the first derivative of a differentiable mapping, and a new section on partial derivatives has been added. Higher order derivatives are treated in Chapter 6, where new material has been added to make a more complete account. The additions include proofs of symmetry of higher order derivatives and of Taylor's formula.

In Chapter 7, the theory developed in the preceding chapters is applied to the foundations of differentiable manifolds with a view towards global analysis and differential geometry and topology. In Chapter 8 we offer an elementary introduction to singularity theory in finite dimensions, and in

Chapter 9 to Morse theory in infinite dimension. The material in Chapters 7-9 is, except for minor revisions, identical to the material in the last three chapters of the first edition.

Most of the figures in the book have been redrawn and more descriptive captions have been added to the figures.

The exercise material has been reorganized from a collection of problem sets at the end of the book to a section at the end of each chapter with exercises and further results. Several new exercises have been added.

In the layout of the book, the text has been marked with various symbols to clarify the structure. The completion of a Proof has been marked with a $\square$, the end of an Example with a ◀, and the end of a Remark with a ◁.

I am grateful to my colleagues Jens Gravesen, Steen Markvorsen, Michael Pedersen, Morten Brøns, Andreas Aabrandt and Kristoffer Jon Albers for constructive remarks, technical assistance and helpful comments over the years.

In a recent class based on material from the first edition of the book and conducted by Hjorth, a number of students have supplied valuable comments and minor corrections to the text. In particular Hjorth and I would like to thank Tobias Boklund, Lukas Kluge, Lasse Mohr Mikkelsen, Asger Limkilde and Aksel Kaastrup Rasmussen.

Technical University of Denmark, August 2019

Vagn Lundsgaard Hansen

Preface to the First Edition

Many advanced mathematical disciplines, such as global differential geometry, the calculus of variations, dynamical systems and the theory of Lie groups, have a common foundation in general topology and analysis in normed vector spaces. The purpose of this book is to introduce students to basic parts of this foundation and to give them a firm basis for further studies of mathematics.

This book derives from a course at the advanced undergraduate or beginning graduate level offered to engineering students at the Technical University of Denmark. The intention of the course is to give the mathematically inclined and interested engineering students an opportunity to go into some depth with fundamental mathematical notions from analysis that are important not only from a mathematical point of view but also occur frequently in theoretical parts of the engineering sciences, and to introduce them to proofs in mathematics and to mathematical reasoning. It is my hope that the book will also appeal to university students in mathematics and in the physical sciences.

The book opens with a study of fundamental concepts from general topology: metric spaces, topological spaces, compactness, connectedness, function spaces. Then follows a study of fundamental concepts in analysis: normed vector spaces, differentiability in normed vector spaces, and the Inverse Function Theorem in Banach spaces. The theory developed is applied to lay the foundations of differentiable manifolds with a view towards global analysis and differential geometry. In the last two chapters we offer elementary introductions to singularity theory in finite dimensions, respectively Morse theory in infinite dimension.

Major parts of the book are a translation and revision of the lecture notes "Grundbegreber i den Moderne Analyse" for the above-mentioned course, published in 1986 by the Department of Mathematics, Technical University of Denmark. The English translation of the first three chapters has been prepared with the very efficient help of Dan Erik Krarup Sørensen. The figures were drawn by Beth Beyerholm.

I am grateful to several people for valuable comments on the material in the book. In particular, I am indebted to the students who tested the material in practice. Among them, Jonas Bjerg, Peter Gross, Lars Gæde, Christian Henriksen, Jan Kristensen, Jens Christian Larsen, Anders Høst-Madsen, Thomas Randrup, Henrik Obbekær Rasmussen, Peter Røgen and Dan Erik Krarup Sørensen deserve particular mentioning for detailed comments. Jennifer Brockbank suggested many improvements in the translation of the first chapters.

My late colleague Niels Vigand Pedersen was a most valued discussion partner at the early stages of the Danish book.

It is a particular joy to thank my good colleague Poul Hjorth who has lectured on the material in the book and has contributed many valuable remarks. As a special favour, he has read most of the text and has suggested several improvements in the language. In this connection, I am also very grateful to Robert Sinclair.

Lyngby, February 1999   Vagn Lundsgaard Hansen

Preliminary Notions

There are certain standard notations and terminologies used throughout mathematics. In this explanatory note we list some of these.

Logical symbols

$\forall$	for all	$\wedge$	and
$\exists$	there exists	$\vee$	or
$:$	such that	$\subseteq$	subset
		$\subset$	proper subset
$\implies$	implies	$\in$	belongs to
$\iff$	if and only if	$\notin$	does not belong to

Sets of numbers

$\mathbb{N}$	the natural numbers	$\mathbb{R}$	the real numbers
$\mathbb{Z}$	the integers	$\mathbb{R}^+$	positive numbers
$\mathbb{Q}$	the rational numbers	$\mathbb{R}_0^+$	non-negative numbers
$\mathbb{I}$	the irrational numbers	$\mathbb{C}$	the complex numbers

Notions from set theory

A set S can be declared by listing the elements. For example,

$$S = \{x \mid x \text{ has property } \mathcal{P}\}$$

denotes a set S of elements x characterized by a given property $\mathcal{P}$. The property $\mathcal{P}$ will often be expressed in terms of logical symbols.

Below we list some basic sets and constructions with sets.

$\emptyset$	the empty set
$A \cup B$	the union of sets A and B, i.e. $\{x \mid x \in A \vee x \in B\}$
$A \cap B$	the intersection of sets A and B, i.e. $\{x \mid x \in A \wedge x \in B\}$
$A \sqcup B$	the union of disjoint sets A and B, i.e. $A \cap B = \emptyset$

$\bigcup_{\alpha \in I} A_\alpha$ union of sets A_α indexed by α in an index set I

$\bigcap_{\alpha \in I} A_\alpha$ intersection of sets A_α indexed by α in an index set I

$A \times B$ the product set of A and B, i.e. $\{(x, y) \mid x \in A,\ y \in B\}$

$S \setminus A$ the set difference, i.e. $\{x \in S \mid x \notin A\}$

Notions related to mappings

$f : A \to B$ a mapping of A into B

$f(A)$ the *image* of f, i.e. $f(A) = \{b \in B \mid \exists a \in A : f(a) = b\}$

$f^{-1}(C)$ the *preimage* of subset $C \subseteq B$ under the mapping
$f : A \to B$, i.e. $f^{-1}(C) = \{a \in A \mid f(a) \in C\}$

Relations and equivalence classes

A *relation* $\sim$ in a set S is a subset $R \subseteq S \times S$, i.e. a distinguished set
of ordered pairs of points $x, y \in S$. We write $x \sim y$ if $(x, y) \in R$.

The relation $\sim$ is called an *equivalence relation* if

(i) $x \sim x$ ($\sim$ is reflexive)

(ii) $x \sim y \implies y \sim x$ ($\sim$ is symmetric)

(iii) $(x \sim y) \wedge (y \sim z) \implies x \sim z$ ($\sim$ is transitive) .

If $\sim$ is an equivalence relation in S, then S can be partitioned into a
corresponding system of disjoint subsets, so-called *equivalence classes* S_α,
indexed by $\alpha \in I$, and defined by

$$x, y \in S_\alpha \iff x \sim y .$$

If on the other hand,

$$S = \bigsqcup_{\alpha \in I} S_\alpha ,$$

then we can define a relation $\sim$ in S by

$$x \sim y \iff x, y \in S_\alpha \text{ for some } \alpha \in I .$$

Clearly $\sim$ is an equivalence relation in S.

An equivalence relation in a set S and a partition of S into a disjoint
union of equivalence classes amounts in other words to the same thing.

If $\sim$ is an equivalence relation in S, the set of equivalence classes is
denoted by $\tilde{S} = S/\!\sim$ and $\pi : S \to \tilde{S}$ denotes the mapping which to an
element $x \in S$ associates its equivalence class $\pi(x) \in \tilde{S}$.

Contents

Chapter 1

It Began with the Numbers

The development of mathematical analysis is intimately related to developing a system of numbers where in harmony with arithmetical operations like addition, multiplication and division you can also perform appropriate limit operations of quantities measured by the numbers.

Numbers originated in the early needs of mankind for counting and for measuring in relation to both quantities and spatial objects. A particularly interesting early mathematical artefact is the Ishango bone found in 1960 on the shores of Lake Edward on the border of Uganda and Zaire. The bone is named after a small settlement living at this location in prehistoric times and it is generally supposed to be about 11,000 years old. There is evidence that the Ishango man has carved the bone according to some kind of pattern. The carvings could indicate that some arithmetic was done. Other observations suggest that the bone could have been a lunar calendar, but it all remains speculations. There is evidence of mathematical activities in Africa more than 30,000 years ago, and after the Ishango bone was found in 1960, it has generally been accepted that mathematics in ancient Egypt in relation to the pyramids and surveying have an African background. From Egypt mathematics found its way to Mesopotamia, where the mathematical source material is known from about 1800 BC.

Neither in the above-mentioned early cultures nor in ancient China, with known mathematical sources from about 300 BC, is there any evidence of systematic formal mathematical theories or proofs of mathematical results. Such activities began with the Greeks around 600 BC.

The Greeks discovered to their dismay the existence of irrational quantities, which could not be measured by fractions of whole numbers, and hence turned to develop a geometric theory of proportions. With this theory they could measure irrational quantities such as the area of a circle by approximating it with regular polygons - the beginnings to limiting processes. But there was still a long way to go to develop an abstract number system in which such limiting processes could be formalized.

It took humankind more than 2,000 years after the Greek contributions before the real number system as we know it today was finally developed shortly before 1900. We begin this chapter by sketching a geometric approach to the real number system.

1.1 A geometric approach to the real numbers

The real number system is a highly abstract structure, the understanding of which can be eased by linking the numbers to points on an oriented line - a number axis. It has to be admitted at once that one of the deep basic difficulties in the foundations of mathematics, consists in formally linking the real numbers to points on a (mathematical) line, which in itself is a highly abstract construct. A geometric approach offers a good intuitive feeling, however, and provides a path to more rapid progress in the early stages without being stuck with deep philosophical and set theoretical questions. And a geometric approach can even help to illuminate the profound nature of the philosophical problems in the foundations of mathematics.

Choose an oriented axis: a line with a preferred sense of direction; cf. Figure 1.1. The choice of the axis is arbitrary but once chosen, it is kept fixed. Furthermore, we choose a fixed subdivision of the oriented axis into intervals of equal lengths.

1.1.1 *Marking the rational numbers*

We can mark the *integers* (whole numbers) $\mathbb{Z}$,

$$\ldots, -2, -1, 0, 1, 2, \ldots,$$

along the division points, by choosing one of the division points as 0 and marking the positive integers in the positive direction according to the chosen orientation of the axis, and the negative integers in the opposite direction from 0. The positive integers, called the *natural numbers* $\mathbb{N}$, have been employed by humans in an intuitive and non-conceptual manner, even in the oldest cultures; some cultures did not go beyond 2, though. Much later, the negative integers were introduced by Hindu mathematicians to represent 'deficits'; the first use of negative numbers is often ascribed to Brahmagupta about 628, but it goes back to about 400 AD. It was also around that time the Hindus began to use the number 'zero' as a usual number; earlier the Egyptians and the Greeks (sources from about 300 BC) had used 'zero' only as a 'place-holder' to indicate the absence of a number.

If we subdivide each of the intervals of equal length on the oriented axis marked by the integers, in q subintervals of equal length, we get a set of division points along which we can mark all *fractions* with a *denominator* q and an arbitrary integer p as *numerator*, the numbers p/q. By letting q run through all the natural numbers we can thereby mark all fractions, representing the so-called *rational numbers* $\mathbb{Q}$, along the oriented axis.

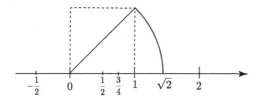

Fig. 1.1   On the real number axis, the number $\sqrt{2}$ can clearly be located by a simple geometrical construction. The number $\sqrt{2}$ is, however, *not* a rational number.

1.1.2   *Catching the real numbers*

We now realize that there are points on the axis that have not yet been included: there are 'holes' in the axis. For example, the Greeks discovered that the diagonal of the unit square is a quantity that cannot be represented by a rational number. If we lay down this length from 0 we arrive at a new point, $\sqrt{2}$; cf. Figure 1.1. Before long we realize that there are many more 'holes' in the axis than points corresponding to rational numbers: examples include $\sqrt[3]{5}$ and π.

We now introduce the *real numbers* $\mathbb{R}$ as the magnitudes represented by the lengths of intervals with one of the endpoints at 0 and the other endpoint at an arbitrary point on the given oriented axis; the magnitudes are counted with sign corresponding to the orientation of the axis. In this way, the real numbers are identified with the points on the oriented axis, which accordingly is referred to as a *number axis*: in particular, we get $\sqrt{2}$ as above. The real numbers not represented by rational magnitudes are called *irrational numbers*. The set of irrational numbers is denoted by $\mathbb{I}$.

If we want to describe the real numbers completely from the rational numbers, we can do it expediently by the following procedure. We imagine that we catch the real numbers in so-called *nested intervals*. By a *nested interval sequence* we understand a decreasing sequence of closed intervals

$$[a_1, b_1] \supseteq [a_2, b_2] \supseteq \cdots \supseteq [a_n, b_n] \supseteq \ldots,$$

in which the length of the interval $[a_n, b_n]$ approaches 0, for increasing n. We can then introduce the real numbers as such 'limit points' of nested interval sequences, in which we use only rational numbers as endpoints of the intervals. After this has been done, every nested interval sequence catches exactly one real number, the only point in common to all the intervals. We can take this property, the *principle of nested intervals*, as a basic property that distinguishes the real numbers from the rational numbers.

Together with the well-known arithmetical properties and ordering relation of numbers, the real number system is completely characterized by the following fundamental property, illustrated in Figure 1.2, page 7. In other words, there is a 'unique' mathematical structure having these properties.

Property 1.1.1 (The principle of nested intervals).
Let $[a_1, b_1] \supseteq [a_2, b_2] \supseteq \cdots \supseteq [a_n, b_n] \supseteq \cdots$ be a decreasing sequence of closed and bounded intervals of $\mathbb{R}$. Then:

(i) *The intersection $\bigcap_{n=1}^{\infty} [a_n, b_n]$ is a closed and bounded interval of $\mathbb{R}$.*

(ii) *If the length of the interval $|b_n - a_n|$ approaches 0 for increasing n then the intersection $\bigcap_{n=1}^{\infty} [a_n, b_n]$ contains exactly one real number.*

A nested interval sequence $[a_1, b_1] \supseteq [a_2, b_2] \supseteq \cdots \supseteq [a_n, b_n]$ in which $|b_n - a_n| \to 0$ for $n \to \infty$, we call, for short, an *interval trap*.

1.1.3   *Digesting the real numbers*

One of the important discoveries in the foundations of mathematics in the nineteenth century was that a tenable basis can be found for the number system by this or some similar method. Definitive constructions of the real numbers by purely logical arithmetical constructions (without appeal to intuition) were not given until the latter half of the nineteenth century, when Karl Weierstrass (1815–1897), Charles Méray (1835–1911), Richard Dedekind (1831–1916) and Georg Cantor (1845–1918), almost simultaneously and independently, each presented constructions. It should be mentioned that it was a particularly difficult problem to relate the arithmetical constructions of the real numbers to the points on a number axis.

As we shall prove later in section 1.5, the set of rational numbers can be counted, whereas the set of all real numbers is uncountable. With a number axis as the setting for the real number system, one may well understand and appreciate the profound nature of the question whether there are sets of real numbers of 'size' between the set of rational numbers and the set of all real numbers, known as *the continuum hypothesis*.

Using the identification of the set of real numbers with the points on a (mathematical) number axis, one can also discuss a fundamental philosophical problem in the mathematical description of the physical world. The problem has to do with the question about the existence of indivisible elements already discussed by the ancient Greek philosophers - in particular, Democritus (c. 460–400 BC). This is still an open question, and so it remains unknown whether a material physical line can be subdivided into arbitrarily small pieces. This is not a problem in the abstract mathematical world, where an interval can be subdivided *ad infinitum* into arbitrarily small subintervals. It may come as a surprise that there are already fundamental philosophical problems in the use of mathematical models in the description of the physical world at this basic level.

1.2 Supremum and infimum

There are other ways than the *principle of nested intervals* of expressing that the real number system is complete in the sense that there are no 'holes' in a real number axis.

One of these alternative characterizations of the real number systems is very directly related to the ordering relation in the set of real numbers $\mathbb{R}$, and is of fundamental importance for many constructions in mathematical analysis. We start out by fixing some terminology in connection with general ordering relations.

1.2.1 *Ordering relations*

An *ordering* $\leq$ of elements in a set S is a relation, where $x \leq y$ if the ordered pair (x, y) of elements $x, y \in S$ belongs to the subset R of the product set $S \times S$ defining the relation, and which satisfies the following conditions:

(i) $x \leq x$ ($\leq$ is reflexive)

(ii) $(x \leq y) \wedge (y \leq x) \implies x = y$ ($\leq$ is anti-symmetric)

(iii) $(x \leq y) \wedge (y \leq z) \implies x \leq z$ ($\leq$ is transitive) .

A set S together with an ordering relation is called an *ordered set*. In this book, the only ordering relation we shall make use of in practice is the usual ordering $\leq$ in the set of real numbers $\mathbb{R}$.

Let A be a subset in the ordered set S with ordering relation $\leq$.

An element $x_{\max} \in A$ is called a *maximal element* in A if $x \leq x_{\max}$ for all $x \in A$. Similarly, an element $x_{\min} \in A$ is called a *minimal element* in A if $x_{\min} \leq x$ for all $x \in A$.

An element $s \in S$ is called an *upper bound* for A if $x \leq s$ for all $x \in A$. If A is *bounded from above*, an upper bound $s^* \in S$ is called a *least upper bound*, or a *supremum*, of A, if $s^* \leq s$ for all other upper bounds $s \in S$.

An element $t \in S$ is called a *lower bound* for A if $t \leq x$ for all $x \in A$. If A is *bounded from below*, a lower bound $t^* \in S$ is called a *greatest lower bound*, or an *infimum*, of A, if $t \leq t^*$ for all other lower bounds $t \in S$.

1.2.2 *Existence of supremum and infimum*

We have chosen *the principle of nested intervals*, as the basic property that distinguishes the real numbers from the rational numbers. This property is equivalent to the existence of a supremum (an infimum) for every non-empty subset A of $\mathbb{R}$ bounded from above (bounded from below). Alternatively, either of these very important properties of the system of real numbers can therefore be taken as the completeness property of the real numbers, which together with the well-known arithmetical properties and the ordering relation characterizes the real number system.

Property 1.2.1 (Existence of Supremum).

Every non-empty subset A of $\mathbb{R}$ which is bounded from above has a smallest upper bound, which is called the least *upper bound of A, or* supremum *of A, and is denoted by* $\sup A$.

Property 1.2.2 (Existence of Infimum).

Every non-empty subset A of $\mathbb{R}$ which is bounded from below has a largest lower bound, which is called the greatest *lower bound of A, or* infimum *of A, and is denoted by* $\inf A$.

It is easy to deduce the existence of supremum (and infimum) from the principle of nested intervals. We leave this to the reader.

We prove now conversely that the principle of nested intervals (Property 1.1.1) follows from the existence of supremum. Note that the existence of supremum automatically implies the existence of infimum.

Proof. First we prove (i). Since the sequence (a_n) is bounded above, e.g. by b_1, the number $a = \sup a_n$ exists. Analogously, the number $b = \inf b_n$

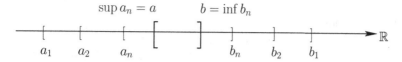

Fig. 1.2   Principle of nested intervals. Beginning with $[a_1, b_1]$, each subsequent interval is nested inside the previous one. If the interval length decreases to 0, a unique real number exists which belongs to all the intervals.

exists, since the sequence (b_n) is bounded below, e.g. by a_1. Because every b_n is an upper bound for (a_n), we have that $a \leq b_n$ for all $n \in \mathbb{N}$. But then a is a lower bound for (b_n), from which it follows that $a \leq b$; cf. Figure 1.2. So what is left is to show that $[a, b] = \cap_{n=1}^{\infty}[a_n, b_n]$. The inclusion $\subseteq$ is clear, since $a_n \leq a \leq b \leq b_n$ for all $n \in \mathbb{N}$. The inclusion $\supseteq$ follows by observing that a number c satisfying $a_n \leq c \leq b_n$ for all $n \in \mathbb{N}$ is an upper bound for (a_n) and a lower bound for (b_n) and therefore has to satisfy $a \leq c \leq b$.

Next we prove (ii). Since $[a, b] \subseteq [a_n, b_n]$ for all $n \in \mathbb{N}$, it is clear that $a = b$ when $|b_n - a_n| \to 0$ for $n \to \infty$. In this situation we therefore have

$$\bigcap_{n=1}^{\infty} [a_n, b_n] = \{a\} \ .$$

This completes the proof. $\square$

1.3   On the nature of proof in mathematics

Proofs in mathematics serve at least two important purposes: verification and explanation. From the point of view of the coherence of a mathematical theory, verification of statements is the more important. From the point of view of learning, the explanation aspect must have a high priority. It was the Greeks who introduced proofs into mathematics: in fact, the Greek word for proof, or demonstration, also carries the meaning of explanation. I advocate presenting proofs that have a strong component of explanation.

1.3.1   *Division with remainder*

As a very simple example, consider the following explanation of the rule for division with remainder. Let n and m be arbitrary positive integers, where $n < m$. Choose a number axis and walk in steps of length n along the axis in the positive direction starting at 0. After a uniquely determined number of steps q, you arrive at the unique point from which the next step will bring you past m. Then, clearly, there is a unique number r (the *remainder*) in the interval $0 \le r < n$ such that $m = nq + r$. This is the rule for division with remainder.

Using the rule for division with remainder, it is easy to prove that if, for an arbitrarily given pair of non-zero integers p, q, we consider all integers of the form $m = up + vq$, for integers u, v, and let d be the smallest strictly positive integer of this form, then d is the greatest common divisor of p and q. Accordingly, the greatest common divisor d of p and q satisfies the equation $rp + sq = d$, for some pair of integers r, s (Exercise 1.11). This basic result is the starting point for a journey into elementary number theory.

1.3.2   *The isoperimetric problem*

As a more advanced example, I shall present a topic pointing the way to advanced mathematical theories such as optimization theory and the calculus of variations. The topic is the *isoperimetric problem* for geometrical figures in the plane bounded by closed curves (without self-intersections). The curves can be polygonal curves, i.e. n-gons, for each integer $n = 3, 4, \ldots$, or, more generally, curves which can be ascribed a finite length (rectifiable Jordan curves). For each class of such curves we can formulate

The isoperimetric problem: *Among all closed curves with a fixed length (in the given class), find the one that encloses the maximum area.*

The formulation of the problem already raises some questions. Does there exist a solution to the problem? Is it unique? The formulation of the problem indicates that the answer to both of the questions is affirmative. This is indeed true, but it is highly non-trivial. More details about this are given in Chapter 9, Section 9.4, page 265.

Many of the problems in the general case can be illustrated by the class of triangles. First we need a result about isosceles triangles.

Theorem 1.3.1. *Among all triangles with a given base and a given perimeter, the isosceles triangle has the largest area.*

The theorem follows by observing that for all triangles with a given perimeter and a given base, the vertex opposite the base in each of these triangles lies on an ellipse with the endpoints of the base as focal points. Hence the height in such a triangle is largest at the vertex of the ellipse. This explains that the area is largest possible exactly when the triangle is isosceles over the base.

We can then handle the isoperimetric problem for the class of triangles.

Theorem 1.3.2. *Among all triangles with a prescribed perimeter, the equilateral triangle has the largest area.*

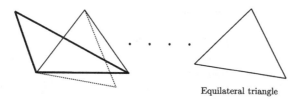

Equilateral triangle

Fig. 1.3   The isoperimetric problem for triangles: Given a fixed perimeter (sum of the sides), which shape of triangle has the largest area?

The proof (explanation) of this theorem (fact) can be tackled as follows. Start with an arbitrary triangle with a prescribed perimeter. By an iterative process, we can construct a sequence of triangles with increasing area, but preserving the perimeter, by making the triangle isosceles over each one of the sides in turn; cf. Figure 1.3. This (infinite) sequence of triangles in the limit approaches the equilateral triangle with the prescribed perimeter. By continuity of the area of a triangle on its shape (Heron's formula), it follows that the equilateral triangle is the triangle enclosing the largest area among all triangles with the prescribed perimeter.

For the case of quadrilaterals one can solve the isoperimetric problem without involving limit processes and continuity of the area function. The solution is that the square contains the largest area among the quadrilaterals with a prescribed perimeter. Surprisingly, it is easier to solve the isoperimetric problem for quadrilaterals than for triangles. But this seems to be very special for $n = 4$. For general $n \geq 3$, methods from analysis of functions of several variables seem to be needed. The solution is as follows.

Theorem 1.3.3. *Among all n-gons with a prescribed perimeter, the regular n-gon encloses the largest area.*

For the general isoperimetric problem about closed planar Jordan curves, the circle is the solution.

Theorem 1.3.4. *Among all closed rectifiable Jordan curves in the plane with a prescribed length, the circle encloses the largest area.*

A proof of the general isoperimetric problem can be based on approximation of rectifiable Jordan curves with closed polygonal curves (n-gons) and exploiting results on continuous real-valued functions (area functions) in

$2n$ real variables defined on closed and bounded subsets in $2n$-dimensional real number space.

1.3.3 *A surprising limit process*

In the proof of the isoperimetric problem for triangles we used a limit argument, which we could justify by appealing to the fact that the area of a triangle depends continuously on the shape of the triangle.

That such limit process needs to be handled carefully can be demonstrated by the following example.

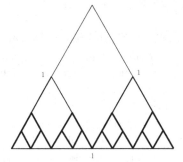

Fig. 1.4   A surprising limit process: All the polygonal detours between the bottom left and the bottom right have length 2. Yet their limit, the direct path, has length 1.

Consider an equilateral triangle in the plane with side length 1 and choose a fixed base line in the triangle; cf. Figure 1.4. The two other sides forms a polygonal curve over the base line of length 2. Divide each of the two edges in this polygonal curve at the middle and draw the connecting lines from the midpoints of the edges to the midpoint of the base line in the triangle. Thereby we get a polygonal curve with four edges of equal length over the base line, and again with total length 2. Now divide in a similar manner each of the four edges in the just constructed polygonal curve at the middle and connect by lines to points on the base line to obtain a polygonal curve with eight edges of equal length and still with total length 2. Stepwise continuing this process we get in the nth step a polygonal curve with 2^{n+1} edges of equal length over the base line and still with total length 2. And what is strange here? The strange thing is that the sequence of polygonal curves constructed by the above procedure approaches the base line in the original triangle, which has length 1. In the limit the length of the sequence of polygonal curves therefore drops from 2 to 1. The explanation is, that the length of a curve does not depend continuously on the curve alone, also variations in the tangent have influence.

1.4 Aesthetics and the search for simplicity

Aesthetics hold an important place in mathematics. Somehow mathematical conjectures that are desirable in a given context and can be embodied naturally into a theory are likely to be true. When elegant and novel proofs are given for such conjectures, most mathematicians will probably agree that it is a testimony to aesthetics in mathematics.

1.4.1 *An appealing proof*

In his renowned book "A mathematicians apology" of 1940, the famous number theorist G.H. Hardy includes a discussion of aesthetics in mathematics. After some thoughts he chooses to present Euclid's proof that there are infinitely many prime numbers, as an example of an aesthetically appealing proof in mathematics. Euclid's proof is one of the first proofs in mathematics that uses the method of *reductio ad absurdum*, or *proof by contradiction*.

Theorem 1.4.1. *There are infinitely many prime numbers.*

In essence, Euclid's proof goes as follows. Suppose that $2, 3, 5, \ldots, P$ is the complete list of primes. Under this hypothesis, consider the number N obtained by multiplying all these primes together, and then adding 1:
$$N = 2 \cdot 3 \cdot 5 \cdot \ldots \cdot P + 1 .$$
The number N cannot be divisible by 2, for then also the difference of N and $2 \cdot 3 \cdot 5 \cdot \ldots \cdot P$ would be divisible by 2; but the difference is 1, which is not divisible by 2. In the same way, we see that N cannot be divisible by 3, or by 5, or by any of the primes up to and including P. It follows that N is a prime, or is divisible by some prime different from any of the primes $2, 3, 5, \ldots, P$. In both cases, we get a prime greater than P. This contradicts our hypothesis that $2, 3, 5, \ldots, P$ was the complete list of primes, and hence this hypothesis is false.

From time to time there is some interest in the news media in the standing record of the largest known prime number. Due to Euclid's theorem this competition will never come to an end. On the other hand it is a famous unsolved problem in number theory as to whether there are infinitely many *pairs* of prime numbers with any given even gap between them (so-called twin primes if the fixed gap is 2).

1.4.2   *The importance of simplicity*

In mathematics it is important to strive for simple and transparent arguments, since this is the best way to ensure the correctness of results and thus, the coherence of mathematical theories. Since coherence is vital to mathematics it is accepted as an integrated part of mathematical research to publish alternative proofs of already established original results. The goal is to find the "proof from THE BOOK", as Paul Erdős (1913–1996) has termed brilliant proofs, which 'feel right' and appear to be 'perfect'.

1.5   Cardinality of sets of real numbers

An arbitrary set A is said to be *countable*, if the elements in the set can be indexed either by a finite set of natural numbers (a finite set) or by the full set of natural numbers $\mathbb{N}$ (an infinite countable set). In other words, A is infinite countable if the elements in A can be assigned numbers in such a way that $\{a_1, a_2, \ldots, a_n, \ldots\}$ is the complete list of elements in A.

The definitions applies in particular to sets of real numbers.

The set of rational numbers $\mathbb{Q}$ is obviously not a finite set, but it is (infinitely) countable.

Theorem 1.5.1. *The set of rational numbers $\mathbb{Q}$ is countable.*

Proof. The set of all ordered pairs of integers (p, q) can be lined up in a spiral, winding its way out from $(0, 0)$ in a rectangular coordinate system in the Euclidean plane; cf. Figure 1.5. Following the spiral, the rational numbers can be ascribed numbers by forming the fraction $\frac{p}{q}$ at each integral pair (p, q) where it makes sense and leaving out in the process those integral pairs (p, q), where the rational number $\frac{p}{q}$ has already been listed. This way we get a listing of the rational numbers: $1, 0, -1, -2, 2, \frac{1}{2}, -\frac{1}{2}, \ldots$. $\square$

In contrast to this, we can use the principle of nested intervals (interval traps) to prove that the full set of real numbers is not countable.

Theorem 1.5.2. *The set of real numbers $\mathbb{R}$ is not countable.*

Proof. The proof is by contradiction. Assume therefore that $\mathbb{R}$ is an (infinite) countable set, and that $\{r_1, r_2, \ldots, r_n, \ldots\}$ is the complete list of all real numbers. To obtain a contradiction, we proceed as follows. Choose a closed interval $[a_1, b_1]$ in $\mathbb{R}$ which does not contain r_1. Then partition the interval $[a_1, b_1]$ evenly into three subintervals. At least one of these subintervals does not contain r_2. Choose such a subinterval and denote it by $[a_2, b_2]$.

By successively partitioning intervals into three subintervals as indicated, we can construct an interval trap

$$[a_1, b_1] \supseteq [a_2, b_2] \supseteq \cdots \supseteq [a_n, b_n] \supseteq \cdots ,$$

in which the interval $[a_n, b_n]$ does not contain the real number r_n. The interval trap determines a real number c, such that

$$\{c\} = \bigcap_{n=1}^{\infty} [a_n, b_n] .$$

Clearly $c \neq r_n$ for all $n \in \mathbb{N}$. This is in contradiction with the assumption that $\{r_1, r_2, \ldots, r_n, \ldots\}$ was the complete list of real numbers. $\square$

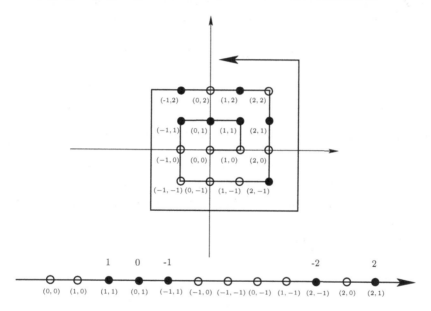

Fig. 1.5 How to count the set of rational numbers.

The German mathematician Georg Cantor (1845–1918) defined an equivalence relation between sets known as cardinality: Two sets A and B are said to have the same *cardinality*, if a one-to-one correspondence between their elements can be established by a bijective mapping $f : A \to B$.

The equivalence class of a set A with respect to cardinality defines a so-called *cardinal number*, denoted by $|A|$. An ordering relation $\leq$ of cardinal numbers of sets A and B is introduced by the definition: $|A| \leq |B|$ if a one-to-one correspondence can be established between the elements in A and a subset of B by an injective mapping $f : A \to B$.

The cardinal number of a finite set A is just the number of elements in A. Sets with the cardinality of the natural numbers $\mathbb{N}$ are exactly the infinite countable sets. The set of real numbers $\mathbb{R}$ is not countable and since $\mathbb{N}$ is a subset of $\mathbb{R}$, the cardinality of $\mathbb{N}$ (countability) is strictly less than the cardinality of $\mathbb{R}$, which is called cardinality of the *continuum*.

It is an undecidable question in mathematics whether there are sets of real numbers of cardinality between the set of rational numbers (countable cardinality) and the set of all real numbers (continuum cardinality). By work of the Austrian mathematician Kurt Gödel in 1931 and the American mathematician Paul Cohen in 1963, it has in fact turned out that both possibilities are consistent with the generally accepted foundations of set theory. The question is known as *the continuum hypothesis*.

Exercises and Further Results

Exercise 1.1. Let A, B and C be sets. Suppose $A \subseteq B \subseteq C$ and $C \subseteq A$. Prove that $A = B = C$.

Exercise 1.2. Let A, B and C be sets. Prove that if $A \subseteq B \subseteq C$, then $C \setminus B \subseteq C \setminus A$.

Exercise 1.3. Determine for each of the following sets of real numbers whether $\sup, \inf, \min$ and $\max$ exists, and in case of existence, determine the value:

    (a) $\{x \in \mathbb{R} \mid x \in \,]0,1[\}$
    (b) $\{x \in \mathbb{R} \mid x \in \,]0,1]\}$
    (c) $\{x \in \mathbb{R} \mid x \in \,[0,1]\}$
    (d) $\{\log(n) \mid n \in \mathbb{N}\}$
    (e) $\{y \in \mathbb{R} \mid y = \arctan(x),\, x \in \mathbb{R}\}$
    (f) $\{y \in \mathbb{R} \mid y = \arctan(x),\, x \in [0,\infty[\}$
    (g) $\{y \in \mathbb{R} \mid y = \arctan(x),\, x \in \,]0,\infty[\}$
    (h) $\{|x| + |y| \mid (x,y) \in \mathbb{R}^2,\, 1 < x^2 + y^2 \leq 4,\, y > 0\}$
    (i) $\{x \in \mathbb{R} \mid \cos(x) = 1\}$.

Exercise 1.4. Prove that for a subset A of $\mathbb{R}$, each one of the existence statements in Property 1.2.1 and Property 1.2.2 follows from the other.

Exercise 1.5. Deduce the existence of a supremum from the principle of nested intervals.

Exercise 1.6. Let (x_n) be a bounded sequence of real numbers. Define two new sequences (a_n) and (b_n) by

$$a_n = \inf\{x_n, x_{n+1}, \ldots, x_{n+k}, \ldots\}$$
$$b_n = \sup\{x_n, x_{n+1}, \ldots, x_{n+k}, \ldots\}\ .$$

1) Prove that (a_n) is a monotonically increasing sequence and that (b_n) is a monotonically decreasing sequence and that $a_n \leq b_n$ for all $n \in \mathbb{N}$.

2) Prove that both sequences (a_n) and (b_n) are convergent sequences and determine the limits $a = \lim_{n\to\infty} a_n$ and $b = \lim_{n\to\infty} b_n$.

The limits a and b are called respectively the *limit inferior* and the *limit superior* of the original sequence (x_n).

3) Prove that (x_n) contains a convergent subsequence with limit a and a convergent subsequence with limit b. Prove also that a and b is respectively the *lower limit* and the *upper limit* for convergent subsequences of (x_n).

4) Prove that (x_n) is convergent if and only if $a = b$.

5) Work out in detail the above considerations for the sequence (x_n), where $x_n = (-1)^n \frac{1+n}{n}$ for $n \in \mathbb{N}$.

Exercise 1.7. Let $x \in \mathbb{R}$ be an arbitrary real number such that $x \geq -1$. Prove by induction Bernoulli's inequality

$$(1+x)^n \geq 1 + nx \quad \text{for all} \quad n \in \mathbb{N}\ .$$

Exercise 1.8. Prove by induction that $2^n > n^2$ for $n > 4$.

Exercise 1.9. Prove by induction the *Binomial Theorem*:

$$(a+b)^n = \sum_{k=0}^{n} \binom{n}{k} a^{n-k} b^k\ .$$

Exercise 1.10. Can a set be one of its own elements? Some sets are certainly not one of their own elements. For let X be the set consisting of all sets which is not one of their own elements, i.e.,

$$X = \{E \,|\, E \notin E\}.$$

Discuss the question: Is $X \in X$?

This question is known as *Russel's Paradox,* and is due to the English philosopher and mathematician Bertrand Russel (1872–1970). The paradox created a considerable amount of discussion around the year 1900, and led to a very thorough examination of the axiomatic basis for set theory (the so-called Zermelo-Fraenkel-axioms).

Exercise 1.11. Make use of the rule for division with remainder to prove that for an arbitrarily given pair of non-zero integers p, q, there exists a pair of integers r, s such that the greatest common divisor d of p and q satisfies the equation $rp + sq = d$.

Exercise 1.12. The *power set* $\mathcal{P}(G)$ of an arbitrary set G is the set consisting of all the subsets of G, including the empty set $\emptyset$ and G itself.

Write down the power set $\mathcal{P}(G)$ for a set $G = \{a, b\}$ with two elements, and a set $G = \{a, b, c\}$ with three elements.

Determine the cardinality of $\mathcal{P}(G)$ for a set G with n elements.

Exercise 1.13. Prove that the open interval $]0, 1[$ has the same cardinality as the real line.

Exercise 1.14. Prove that the set of polynomials with integer coefficients is a countable set.

Exercise 1.15. A real number is said to be an *algebraic number* if it is root in a real polynomial with integer coefficients. Prove that the set of algebraic numbers form a countable subset of $\mathbb{R}$.

Exercise 1.16. Prove that if A is a countable set and B is a countable set, then $A \times B$ is a countable set.

Exercise 1.17. Prove that if A, B, C, and D are arbitrary sets for which $|A| = |B|$ and $|C| = |D|$ then $|A \times C| = |B \times D|$.

Exercise 1.18. Prove that $\mathbb{N}^2 = \mathbb{N} \times \mathbb{N}$ is countable. Prove by induction that $\mathbb{N}^n$ is countable for all $n \in \mathbb{N}$.

Exercise 1.19. Prove that if $|A| \le |\mathbb{N}|$ and $|\mathbb{N}| \le |A|$ then $|A| = |\mathbb{N}|$.

Exercise 1.20. Prove the following facts about the countability of sets:
  1) The sets of even numbers and odd numbers are both countable sets.
  2) The disjoint union of two countable sets is countable.
  3) The union of finitely many countable sets is a countable set.
  4) A countable union of countable sets is again a countable set.

Exercise 1.21. Prove that the set S of real sequences (x_n), where each x_n is either 0 or 1, is not a countable set.

Chapter 2

Basic Concepts in Topology

Concepts like convergence of sequences of points and continuity of functions are fundamental in the subject of mathematical analysis. A more profound study of these concepts, among other things, takes place in the mathematical discipline called general, or point-set, *topology*.

In this chapter we introduce basic concepts for the study of mutual relations between points in topological spaces and continuity of mappings.

In the next chapter we introduce more concepts from general topology in a search for generalizations to topological spaces of the following fundamental theorems from classical analysis:

Theorem A. *A continuous real-valued function $f : [a, b] \to \mathbb{R}$ defined in a closed and bounded interval $[a, b]$ is bounded.*

Theorem B. *A continuous real-valued function $f : [a, b] \to \mathbb{R}$ defined in a closed and bounded interval $[a, b]$ attains every value between $f(a)$ and $f(b)$.*

Theorem A and Theorem B lead respectively to compact sets and connected sets in topological spaces.

2.1 The classical setting for continuity

Informally speaking, a real-valued function $f = f(x) : \mathbb{R} \to \mathbb{R}$ of one real variable $x \in \mathbb{R}$ is said to be continuous if small variations in x only cause small variations in $f(x)$. If we think of x as the input and $f(x)$ as the corresponding output, then this heuristic definition of continuity expresses that small variations in the input only give small variations in the output. This definition, however, is too imprecise to work with. Indeed, consider e.g. $f(x) = 10^{79}x$. This is a linear function, thus in particular a continuous function, but with some justification one can assert that even small variations in x cause large variations in $f(x)$. Hence we need a more precise definition which better encapsulates the idea that we have control over the output of a continuous function.

Definition 2.1.1. Let $U \subseteq \mathbb{R}$ be a subset of $\mathbb{R}$, and let $f : U \to \mathbb{R}$ be a real-valued function. We say that $f : U \to \mathbb{R}$ is *continuous at a point* $x_0 \in U$, provided that

$$\forall \varepsilon > 0 \ \exists \delta > 0 \ \forall x \in U : |x - x_0| < \delta \ \Rightarrow \ |f(x) - f(x_0)| < \varepsilon .$$

If $f : U \to \mathbb{R}$ is continuous at every point $x_0 \in U$, we say that f is *continuous*.

Remark 2.1.2. In the above definition we can think of $\varepsilon > 0$ as an admissible margin of error in the output and of $\delta > 0$ as an associated margin of tolerance in the input. Continuity of $f : U \to \mathbb{R}$ at a point $x_0 \in U$ then means that for any given admissible margin of error $\varepsilon > 0$ in the output, we can find a corresponding admissible margin of tolerance $\delta > 0$ in the input such that, as long as x stays within the tolerance $\delta > 0$ of x_0, the actual value $f(x)$ stays within the admissible margin of error $\varepsilon > 0$ of $f(x_0)$. ◁

Continuity of $f : U \to \mathbb{R}$ means that f is continuous at every point $x_0 \in U$. Thus, in Definition 2.1.1, a $\delta > 0$ corresponding to a given $\varepsilon > 0$ in general depends on the point $x_0 \in U$. By using quantifiers, the definition of continuity of $f : U \to \mathbb{R}$ may be written

$$\forall x_0 \in U \ \forall \varepsilon > 0 \ \exists \delta > 0 \ \forall x \in U : \ |x - x_0| < \delta \ \Rightarrow \ |f(x) - f(x_0)| < \varepsilon .$$

This should not be confused with

$$\forall \varepsilon > 0 \ \exists \delta > 0 \ \forall x, y \in U : \ |x - y| < \delta \ \Rightarrow \ |f(x) - f(y)| < \varepsilon .$$

In the second case, the tolerance $\delta > 0$ does not depend on $y \ (= x_0)$, whereas it does in the first case. In fact, in the second case we define that $f : U \to \mathbb{R}$ is *uniformly continuous*.

Example 2.1.3. Consider $f(x) = x^2$ for $x \in \mathbb{R}$. For all $x, y \in \mathbb{R}$ we have

$$|f(x) - f(y)| = \left|x^2 - y^2\right| = |x + y| \cdot |x - y| .$$

For any $\varepsilon > 0$ and any fixed $y \in \mathbb{R}$, we can find $\delta > 0$ such that $|f(x) - f(y)| < \varepsilon$ when $|x - y| < \delta$. Thus the function f is continuous. However, we cannot choose $\delta > 0$ independently of y, since by choosing x and y sufficiently large, we can obtain that $|f(x) - f(y)| > \varepsilon$ even though $|x - y| < \delta$. Hence, the function f is not uniformly continuous. ◀

Remark 2.1.4. Replacing the strict inequalities $< \delta$ and $< \varepsilon$ by weak inequalities $\leq \delta$ and $\leq \varepsilon$ does not affect the definitions. ◁

Let $\mathbb{R}^n$ denote the space consisting of n-tuples $x = (x_1, \ldots, x_n)$ of real numbers, and let $f_1, \ldots, f_k : \mathbb{R}^n \to \mathbb{R}$ be k real-valued functions of n variables. As usual, we can collect these into a mapping

$$f = (f_1, \ldots, f_k) : \mathbb{R}^n \to \mathbb{R}^k,$$

where $f(x) = (f_1(x), \ldots, f_k(x))$, which we can elaborate to

$$f(x_1, \ldots, x_n) = (f_1(x_1, \ldots, x_n), \ldots, f_k(x_1, \ldots, x_n)).$$

If $U \subseteq \mathbb{R}^n$ is a subset of $\mathbb{R}^n$, and the functions $f_1, \ldots, f_k$ are defined only in U, we obtain a mapping

$$f = (f_1, \ldots, f_k) : U \to \mathbb{R}^k.$$

The above definition of continuity of the function $f : \mathbb{R} \to \mathbb{R}$, or more generally, $f : U \to \mathbb{R}$ defined in a subset $U \subseteq \mathbb{R}$, can easily be generalized to concern also mappings $f : \mathbb{R}^n \to \mathbb{R}^k$, or more generally, $f : U \to \mathbb{R}^k$ defined in a subset $U \subseteq \mathbb{R}^n$. We only have to replace $|x - x_0|$, respectively $|f(x) - f(x_0)|$, by the Euclidean distances in $\mathbb{R}^n$, respectively $\mathbb{R}^k$, formally defined in Example 2.2.3 below; cf. Figure 2.1 for $n = k = 2$.

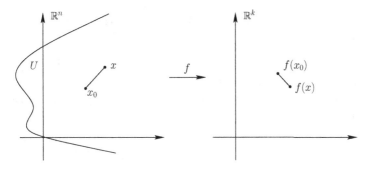

Fig. 2.1   The basic set-up for checking continuity. The function f maps $\mathbb{R}^n$ to $\mathbb{R}^k$. Can one bring the image points $f(x)$ and $f(x_0)$ arbitrarily close together by bringing the points x and x_0 sufficiently close together?

2.2   Continuity of mappings in metric spaces

It is useful to introduce a notion of continuity of mappings in more abstract situations, where the input for the mappings are not merely n-tuples of real numbers, but perhaps a collection of functions or a collection of differential equations. Hence we want to generalize the definition of continuity so that it can be applied to mappings $f : X \to Y$, where X and Y are more general spaces than just $\mathbb{R}^n$ and $\mathbb{R}^k$ or subsets of these spaces. We can perform an immediate generalization, which can be used in many situations, when X and Y are equipped with the notion of a distance.

A distance function (or a metric) in a set M is a function $d : M \times M \to \mathbb{R}$ which to any pair of points $x, y \in M$ associates a real number $d(x, y)$, called the *distance* from x to y. Furthermore, to get a reasonable notion of distance, it has proved fruitful to require that the following three conditions are satisfied:

MET 1 (positive definite)
$$d(x, y) \geq 0, \text{ for all } x, y \in M, \quad d(x,y) = 0 \iff x = y.$$

MET 2 (symmetry)
$$d(x, y) = d(y, x), \text{ for all } x, y \in M.$$

MET 3 (the triangle inequality)
$$d(x, z) \leq d(x, y) + d(y, z), \text{ for all } x, y, z \in M.$$

Definition 2.2.1. A *metric* in a set M is a function $d : M \times M \to \mathbb{R}$ satisfying the conditions MET 1, MET 2, MET 3.

A pair (M, d) consisting of a set M together with a specific metric d on M is called a *metric space*.

Remark 2.2.2. The same set M can be equipped with several different metrics, in which case the corresponding metric spaces are counted as different.

When the context leaves no doubt as to which metric is being considered, usually the metric is not mentioned explicitly. ◁

The idea of metric spaces was introduced by the French mathematician Maurice Fréchet (1878–1973) in his doctoral thesis of 1906.

A few examples of metric spaces are in order.

Example 2.2.3. Consider the set of real n-tuples $M = \mathbb{R}^n$.

For points $x = (x_1, \dots, x_n)$ and $y = (y_1, \dots, y_n)$ in $\mathbb{R}^n$ we set
$$d(x, y) = \sqrt{\sum_{i=1}^{n}(x_i - y_i)^2} \; .$$

Clearly MET 1 and MET 2 are satisfied. To prove that MET 3 is satisfied we need some preparations.

The space $\mathbb{R}^n$ is an n-dimensional real vector space with the usual coordinatewise definition of addition and multiplication by scalars. The points in $\mathbb{R}^n$ are then identified with vectors. We define the *inner product* by

$$\langle x, y \rangle = \sum_{i=1}^{n} x_i y_i ,$$

and the associated *norm* by

$$\|x\| = \sqrt{\sum_{i=1}^{n} x_i^2} .$$

Then we have the *Cauchy-Schwarz inequality:*

$$|\langle x, y \rangle| \le \|x\| \|y\| \quad \text{for all} \quad x, y \in \mathbb{R}^n .$$

Proof. Let $x, y \in \mathbb{R}^n$ be arbitrarily chosen, but fixed vectors in $\mathbb{R}^n$. For every real number $t \in \mathbb{R}$ we have the inequality

$$0 \le \langle x + ty, x + ty \rangle = \|y\|^2 t^2 + 2\langle x, y \rangle t + \|x\|^2 .$$

For $y \neq 0$, this describes a polynomial of degree 2 in t (a parabola) with at most one zero. Hence the discriminant of the polynomial must satisfy

$$4\langle x, y \rangle^2 - 4\|x\|^2 \|y\|^2 \le 0 .$$

From this we get the inequality $|\langle x, y \rangle| \le \|x\| \|y\|$ as asserted. For $y = 0$, the inequality is trivially satisfied. $\square$

The Cauchy-Schwarz inequality implies the *triangle inequality:*

$$\|x + y\| \le \|x\| + \|y\|.$$

Proof. Using the Cauchy-Schwarz inequality, the small computation

$$\begin{aligned}
\|x + y\|^2 = \langle x + y, x + y \rangle &= \|x\|^2 + \|y\|^2 + 2\langle x, y \rangle \\
&\le \|x\|^2 + \|y\|^2 + 2\|x\| \cdot \|y\| \\
&= (\|x\| + \|y\|)^2
\end{aligned}$$

reveals the triangle inequality. $\square$

By noting that

$$d(x, y) = \|x - y\|,$$

the condition MET 3 now follows immediately:

$$\begin{aligned}
d(x, z) = \|x - z\| = \|x - y + y - z\| \\
\le \|x - y\| + \|y - z\| = d(x, y) + d(y, z).
\end{aligned}$$

Thereby we have verified that $\mathbb{R}^n$ equipped with the distance function d is a metric space. This metric space is called n-dimensional *Euclidean space*, and the metric d, *the Euclidean metric* in $\mathbb{R}^n$.

For $n = 1$ we get the usual concept of distance on a line by identifying the line with $\mathbb{R}$ represented by a real number axis. For $n = 2, 3$ we rediscover the usual concept of distance in a plane, respectively 3-dimensional space, when these are identified with $\mathbb{R}^2$, respectively $\mathbb{R}^3$, by choosing an orthonormal coordinate system in the set. ◀

Before the next example, we recall the basic property of existence of supremum of subsets of $\mathbb{R}$ bounded from above, introduced in Section 1.2, page 6.

Property (Supremum). *Every non-empty subset A of $\mathbb{R}$ which is bounded from above has a smallest upper bound, which is called the* least *upper bound of A, or* supremum *of A, and is denoted by* $\sup A$.

From the existence of a supremum we easily get:

Property (Infimum). *Every non-empty subset A of $\mathbb{R}$ which is bounded from below has a largest lower bound, which is called the* greatest *lower bound of A, or* infimum *of A, and is denoted by* $\inf A$.

Example 2.2.4. For K a non-empty, but otherwise arbitrary, fixed set, let M be the set of all bounded real-valued functions $f : K \to \mathbb{R}$ defined in K.

For $f, g \in M$ put

$$d(f, g) = \sup_{x \in K} |f(x) - g(x)|$$
$$= \sup \left\{ |f(x) - g(x)| \mid x \in K \right\}.$$

Note that since

$$A = \left\{ |f(x) - g(x)| \mid x \in K \right\}$$

is bounded from above, the number $d(f, g) = \sup A$ exists. Note also that we have indicated different ways of writing $\sup A$ in the present example.

MET 1 follows since $0 \leq |f(x) - g(x)| \leq \sup A$ for all $x \in K$. MET 2 is also easily proved, since $|f(x) - g(x)| = |g(x) - f(x)|$ for all $x \in K$.

To prove that MET 3 is satisfied we proceed as follows.

Let f, g, h be three functions in M. For every $x \in K$, we have

$$
\begin{aligned}
|f(x) - h(x)| &= |f(x) - g(x) + g(x) - h(x)| \\
&\leq |f(x) - g(x)| + |g(x) - h(x)| \\
&\leq \sup_{x \in K} |f(x) - g(x)| + \sup_{x \in K} |g(x) - h(x)| \\
&= d(f, g) + d(g, h).
\end{aligned}
$$

From this we see that $d(f, g) + d(g, h)$ is an upper bound for

$$
\left\{ |f(x) - h(x)| \mid x \in K \right\}.
$$

Since any upper bound is greater than or equal to the least upper bound, we conclude that

$$
d(f, h) = \sup_{x \in K} |f(x) - h(x)| \leq d(f, g) + d(g, h).
$$

Hence d is a metric in M.

If we define addition of functions in M and multiplication of functions by real numbers using the obvious pointwise defined operations (see Example 4.1.4), M gets the structure of a real vector space. This vector space has finite dimension if K contains only a finite number of points. If K contains infinitely many points there will be no system of finitely many functions spanning M and hence the vector space M is infinite dimensional in this case. ◄

Example 2.2.5. Let M be any set. Set

$$
d(x, y) = \begin{cases} 1 & \text{for } x \neq y \\ 0 & \text{for } x = y. \end{cases}
$$

It is easy to see that d is a metric in M. Called *the discrete metric* in M. ◄

Example 2.2.6. Let (M, d) be a metric space, and let $A \subseteq M$ be a subset of M. Then A inherits a metric from (M, d) called *the induced metric*. More formally: If $d : M \times M \to \mathbb{R}$ is the metric in M, we get the induced metric in A by taking the restriction of d to $A \times A$. ◄

We finish this section with the definition of the notion of continuity of a mapping in the setting of metric spaces.

Definition 2.2.7. Let (X, d_X) and (Y, d_Y) be metric spaces. We say that the mapping $f : X \to Y$ is *continuous at a point* $x_0 \in X$, if

$$
\forall \varepsilon > 0 \ \exists \delta > 0 \ \forall x \in X : d_X(x, x_0) < \delta \quad \Rightarrow \quad d_Y(f(x), f(x_0)) < \varepsilon.
$$

We say that the mapping $f : X \to Y$ is *continuous* if f is continuous at every point $x_0 \in X$.

2.3   The topology of a metric space

A closer study of continuity of mappings in the setting of metric spaces reveals that it is not the specific metrics as such that are decisive for continuity to make sense but rather a class of subsets defined by the metrics leading to the concept of the underlying *topology* in a metric space.

First some definitions. Let (M, d) be an arbitrary metric space, let x_0 be a point in M, and let $r \in \mathbb{R}^+$ be a positive real number. Then the set

$$B_r(x_0) = \left\{ x \in M \mid d(x_0, x) < r \right\}$$

is called the *open ball* or the *open sphere* in M with *center* x_0 and *radius* r.

Using open balls, the continuity of a mapping $f : X \to Y$ between metric spaces (X, d_X) and (Y, d_Y) at a point $x_0 \in X$ (Definition 2.2.7) can now be formulated as follows (cf. Figure 2.2):

$$\forall \varepsilon > 0 \ \exists \delta > 0 : \ f\left(B_\delta(x_0)\right) \subseteq B_\varepsilon\left(f(x_0)\right),$$

or equivalently:

$$\forall \varepsilon > 0 \ \exists \delta > 0 : B_\delta(x_0) \subseteq f^{-1}\left(B_\varepsilon\left(f(x_0)\right)\right).$$

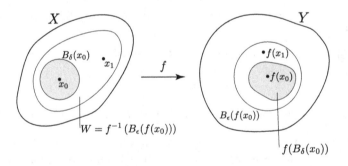

Fig. 2.2   Continuity of a mapping. The mapping $f : X \to Y$ is continuous at $x_0 \in X$, if for any given radius $\varepsilon > 0$, the inverse image $f^{-1}(B_\varepsilon(f(x_0)))$ of the open ball $B_\varepsilon(f(x_0))$ centred at $f(x_0) \in Y$ contains an open ball $B_\delta(x_0)$ centred at $x_0 \in X$ with radius $\delta > 0$.

Definition 2.3.1. Let (M, d) be a metric space. A subset W of M is called an *open set* in the metric space (M, d), if for every $x \in W$ there exists a $\delta > 0$ such that $B_\delta(x) \subseteq W$.

Remark 2.3.2. An open ball $B_r(x_0)$ in the metric space (M, d) is an open set in M, since for $x \in B_r(x_0)$ the triangle inequality shows that $B_\delta(x) \subseteq B_r(x_0)$, if $0 < \delta \leq r - d(x_0, x)$. ◁

Example 2.3.3. Open intervals in $\mathbb{R}$ are open sets.

In $\mathbb{R}^2$, the sets

$$\{(x, y) \mid -2 < x < 3, -1 < y < 1\} \quad \text{and} \quad \{(x, y) \mid x^2 + y^2 \neq 1\}$$

are examples of open sets.

More generally, a subset of $\mathbb{R}^n$ described by a finite number of strict inequalities defined by continuous functions is open. ◄

Example 2.3.4. Let M be any set equipped with the discrete metric introduced in Example 2.2.5.

In this case, the open ball $B_r(x_0)$ in M with center $x_0 \in M$ and radius $r \leq 1$ is just the set $\{x_0\}$. For radius $1 < r$, the open ball $B_r(x_0) = M$.

Consequently any subset in M is an open set, when M is equipped with the discrete metric. ◄

Continuity of mappings can be described in a fruitful way using open sets.

Theorem 2.3.5. *Let* $f : X \to Y$ *be a mapping between metric spaces* (X, d_X) *and* (Y, d_Y). *Then* f *is continuous if and only if for every open set* V *in* Y, *the set* $f^{-1}(V)$ *is an open set in* X.

Proof. First assume that f is continuous. Let V be an arbitrary open set in Y. We have to show that $f^{-1}(V)$ is an open set in X. For that purpose let $x \in f^{-1}(V)$ be an arbitrary point in $f^{-1}(V)$ in case $f^{-1}(V) \neq \emptyset$. Since $f(x) \in V$, and V is open in Y, we can find an $\varepsilon > 0$ such that $B_\varepsilon(f(x)) \subseteq V$. Since f is continuous at x, for this $\varepsilon > 0$ we can find a $\delta > 0$, so that $B_\delta(x) \subseteq f^{-1}(B_\varepsilon(f(x)))$. But then $B_\delta(x) \subseteq f^{-1}(V)$, and since $x \in f^{-1}(V)$ was an arbitrarily chosen point, this shows that $f^{-1}(V)$ is an open set in X, possibly the empty set.

Now assume conversely, that all preimages of open sets in Y under f are open sets in X. We then have to show that f is continuous at every point $x \in X$. Let $x_0 \in X$ be an arbitrary point in X, and let $\varepsilon > 0$ be given. According to Remark 2.3.2, $B_\varepsilon(f(x_0))$ is an open set in Y, and hence by the assumption, $f^{-1}(B_\varepsilon(f(x_0)))$ is an open set in X. But then there exists a $\delta > 0$, such that $B_\delta(x_0) \subseteq f^{-1}(B_\varepsilon(f(x_0)))$ proving that f is continuous at the point $x_0 \in X$. $\square$

Theorem 2.3.5 shows that the family of open sets in a metric space plays a decisive role in the study of continuity of mappings between metric spaces. The family of open sets is called the *topology* in the metric space.

2.4 Topological spaces

The main features of the family of open sets in a metric space (M, d) are captured by three fundamental properties.

Theorem 2.4.1. *The family of open sets $\mathcal{T}$ in a metric space (M, d) has the following fundamental properties.*

TOP 1 *If $\{U_i \in \mathcal{T} \mid i \in I\}$ is an arbitrary system of subsets in M from $\mathcal{T}$, then the union $\cup\{U_i \in \mathcal{T} \mid i \in I\}$ of these subsets also belongs to $\mathcal{T}$.*

TOP 2 *If $U_1, \ldots, U_k$ is an arbitrary finite system of subsets in M from $\mathcal{T}$, then the intersection $\cap_{i=1}^{k} U_i$ of these subsets also belongs to $\mathcal{T}$.*

TOP 3 *The empty set $\emptyset$, and the set M itself belong to $\mathcal{T}$.*

The proof of Theorem 2.4.1 is left to the reader.

Now consider an arbitrary set M equipped with a family of distinguished subsets $\mathcal{T}$ satisfying the above properties TOP 1, TOP 2, TOP 3. Then $\mathcal{T}$ is called a *topology* on M, and the pair $(M, \mathcal{T})$ is called a *topological space*. In this general context, the subsets of M belonging to the family $\mathcal{T}$ is referred to as the family of *open sets* in the topological space.

For later reference we define the concepts formally.

Definition 2.4.2. Let M be a fixed set and let $\mathcal{T}$ be a family of subsets in M satisfying the conditions TOP 1, TOP 2, and TOP 3. Then $\mathcal{T}$ is called a *topology* on M. And the pair $(M, \mathcal{T})$ consisting of the set M together with a specific topology $\mathcal{T}$ on M is called a *topological space*. The specified subsets in $\mathcal{T}$ are called the *open sets* in the topological space.

Remark 2.4.3. Strictly speaking, a topological space is a pair $(M, \mathcal{T})$ consisting of a set M and a topology $\mathcal{T}$ on M. However, we shall most often just refer to M as a topological space, and not mention the family $\mathcal{T}$ explicitly. Accordingly, whenever we refer to a set M as a topological space in the future, we tacitly assume that there is a family of subsets $\mathcal{T}$ in M with the properties TOP 1, TOP 2 and TOP 3 - the open sets in the topology. ◁

Remark 2.4.4. A short formulation of the conditions TOP 1 and TOP 2 is that the family of subsets $\mathcal{T}$ of the set M is closed under arbitrary unions of sets (TOP 1) and finite intersections of sets (TOP 2) in $\mathcal{T}$.

The restriction to finiteness in TOP 2 is needed. For example, on the real axis $\mathbb{R}$ consider the system of open intervals $]-1/n, 1/n[$ for $n = 1, 2, \ldots$ The intersection of the intervals is the set containing only 0, and this is not an open set in $\mathbb{R}$. ◁

Remark 2.4.5. To test whether a family $\mathcal{T}$ of subsets of a set M satisfies the condition TOP 2, it is sufficient to inspect whether the intersection $U \cap V$ of two arbitrary subsets U and V in the family $\mathcal{T}$ also belongs to $\mathcal{T}$. ◁

We shall now list some examples of topological spaces.

Example 2.4.6. Any metric on a set M determines a topology on M. It consists of the family $\mathcal{T}$ of open sets corresponding to the metric. Every metric space is then a topological space.

Conversely, not every topological space admits a metric. In case the topology in a topological space actually can be described by a metric, we say that the topology is *metrizable*.

The topology on $\mathbb{R}^n$ coming from the Euclidean metric is called *the usual topology*, or *the Euclidean topology*, on $\mathbb{R}^n$. Unless explicitly specified, $\mathbb{R}^n$ is always equipped with this topology. ◀

Example 2.4.7. Let M be an arbitrary set. Then the set of *all* subsets of M will of course satisfy TOP 1, TOP 2, and TOP 3, and therefore it is a topology on M. This topology is called *the discrete topology, the largest topology*, or *the finest topology* on M.

The discrete topology on M arises from the discrete metric on M. This follows easily by observing that with respect to the discrete metric on M, any subset in M consisting of just one point in M is an open set. ◀

Example 2.4.8. If we go to the other extreme, we can take the family of subsets in M, consisting of only the empty set $\emptyset$ and the set M itself. This family of subsets in M forms a topology on M, which is called *the indiscrete topology, the smallest topology*, or *the coarsest topology* on M. When M contains at least two elements, this topology on M is not metrizable. ◀

Example 2.4.9. Let M be a topological space with the topology $\mathcal{T}$, and let A be an arbitrary subset in M (not necessarily from $\mathcal{T}$).

Consider the family of subsets in A which we get by taking the intersections of A and the subsets of M from the family $\mathcal{T}$; cf. Figure 2.3. We denote this family of subsets in A by $\mathcal{T}_A$, i.e.

$$\mathcal{T}_A = \{V = A \cap U \mid U \in \mathcal{T}\}.$$

It is easy to verify that $\mathcal{T}_A$ is a topology on A. This topology is called *the induced topology, the subspace topology*, or *the trace topology* on A.

If the topology $\mathcal{T}$ on M stems from a metric on M, then the induced topology $\mathcal{T}_A$ on A stems from the induced metric on A. ◄

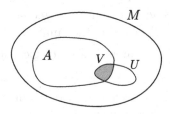

Fig. 2.3   The topology *induced* on a subset A from a surrounding topological space M consists of the subsets $V = A \cap U$ where U belongs to the topology on M.

Example 2.4.10. Let X and Y be topological spaces. We wish to define a topology on the product set $X \times Y$.

It is tempting to take sets of the form $U \times V$, where U is open in X and V is open in Y, to be the open sets in $X \times Y$. But if we do so, then TOP 1 can fail, since indeed a set obtained as the union $(U_1 \times V_1) \cup (U_2 \times V_2)$ does not necessarily have the right product form. (It is easy to construct simple counterexamples in $\mathbb{R}^2 = \mathbb{R} \times \mathbb{R}$. Consider for instance two rectangles.)

To avoid this difficulty, we consider the larger family of subsets in $X \times Y$, which arises by including arbitrary unions of sets of the form $U \times V$. Hence, we consider the family $\mathcal{T}_{X \times Y}$ of subsets in $X \times Y$ of the form

$$\bigcup_{i \in I} U_i \times V_i,$$

where U_i is an open set in X, V_i is an open set in Y and I is any index set.

It should come as no surprise that the family $\mathcal{T}_{X \times Y}$ is a topology on $X \times Y$. The conditions TOP 1 and TOP 3 are now both trivially satisfied. To prove that TOP 2 is satisfied, we just have to use that

$$\left(\bigcup_{i \in I} U_i \times V_i \right) \cap \left(\bigcup_{j \in J} U'_j \times V'_j \right) = \bigcup_{i \in I} \bigcup_{j \in J} (U_i \cap U'_j) \times (V_i \cap V'_j).$$

The topology $\mathcal{T}_{X \times Y}$ is called *the product topology* on $X \times Y$.

If the topologies on X and Y stem from metrics, then the product topology on $X \times Y$ also stems from a metric. ◄

Example 2.4.11. Let M be a topological space, and let $\tilde{M}$ be an arbitrary set. Furthermore, we are given a mapping $\pi : M \to \tilde{M}$. Usually π will be surjective, but this is not necessary for what follows.

Let $\mathcal{T}$ be the topology on M. Consider the matching family $\tilde{\mathcal{T}}$ of subsets V in $\tilde{M}$, for which the corresponding preimages $U = \pi^{-1}(V)$ under π belong to $\mathcal{T}$; cf. Figure 2.4. In other words we set

$$\tilde{\mathcal{T}} = \{V \subseteq \tilde{M} \mid U = \pi^{-1}(V) \in \mathcal{T}\}.$$

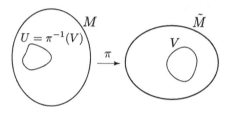

Fig. 2.4   Quotient topology. Given a topology $\mathcal{T}$ on a set M, and a mapping $\pi : M \to \tilde{M}$, then the sets V in $\tilde{M}$ such that the set $\pi^{-1}(V)$ belongs to $\mathcal{T}$ defines a topology on $\tilde{M}$.

By using the set theoretical formulae

$$\pi^{-1}\left(\bigcup_{i\in I} V_i\right) = \bigcup_{i\in I}\pi^{-1}(V_i) \quad \text{and} \quad \pi^{-1}\left(\bigcap_{i\in I} V_i\right) = \bigcap_{i\in I}\pi^{-1}(V_i),$$

it is easily verified that $\tilde{\mathcal{T}}$ is a topology on $\tilde{M}$.

The topology $\tilde{\mathcal{T}}$ on $\tilde{M}$ is called *the quotient topology* on $\tilde{M}$ induced by the mapping π. The name is due to the fact that $\tilde{M}$ often arises from M by a process of identification, under which points in $\tilde{M}$ emerge as equivalence classes of points in M.

Even if the topology on M stems from a metric, the quotient topology on $\tilde{M}$ does not necessarily do so, and even when it does, it is usually not in a natural way. ◄

The original motivation for introducing the notion of a topological space was to study the continuity of mappings in a more general setting than metric spaces. In view of Theorem 2.3.5, the only reasonable definition of continuity of a mapping between topological spaces is the following definition.

Definition 2.4.12. Let $f : X \to Y$ be a mapping between topological spaces X and Y. We say that f is *continuous* if, for every open set V in Y, the set $f^{-1}(V)$ is open in X.

What have we gained by this? First of all, we avoid referring to specific metrics. Secondly – and of course most importantly – some topological spaces can only be equipped with a metric in a very unnatural way. In fact, many important topological spaces, like certain function spaces, are not metrizable. Typically this occurs when the domain in the function space is not compact. We give an example in Theorem 8.5.1, page 248.

Another benefit is that we now have an elegant tool for handling theoretical questions in connection with continuous functions. For as Theorem 2.3.5 shows, Definition 2.4.12 is indeed the obvious generalization of the concept of continuity from the setting of metric spaces to the setting of topological spaces. Enjoy e.g. the following proof that the composition of two continuous mappings is itself continuous.

Theorem 2.4.13. *If $f : X \to Y$ and $g : Y \to Z$ are continuous mappings between topological spaces, then the* composite mapping $g \circ f : X \to Z$ *is also continuous.*

Proof. Let V be an open set in Z. We need to show that $(g \circ f)^{-1}(V)$ is an open set in X. With this in mind note that

$$(g \circ f)^{-1}(V) = f^{-1}\left(g^{-1}(V)\right).$$

Since g is continuous, $g^{-1}(V)$ is an open set in Y, and since f is continuous, $f^{-1}\left(g^{-1}(V)\right)$ is an open set in X. This completes the proof. $\square$

2.5 Local theory in topological spaces

2.5.1 *Neighbourhoods*

Let M be a topological space, and let $x \in M$ be a point in M.

Definition 2.5.1. An open subset U of M which contains x is called an *open neighbourhood* of x in M. More generally, a subset N of M which contains x *and* an open neighbourhood U of x, that is $x \in U \subseteq N$ (see Figure 2.5), is called a *neighbourhood* of x in M.

Remark 2.5.2. For a metric space (M, d), the system of open balls $B_\delta(x)$ for $\delta > 0$ is a particularly useful system of open neighbourhoods of $x \in M$. ◁

Example 2.5.3. On the real axis $\mathbb{R}$, the closed interval $[-1, 1]$ is a neighbourhood of 0, but not of -1 or 1. ◂

Lemma 2.5.4. *A subset W of M is open if and only if every point $x \in W$ has a neighbourhood N_x in M such that $N_x \subseteq W$.*

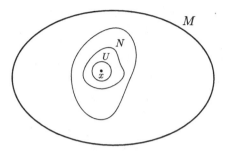

Fig. 2.5   For a point x in a topological space M, an open set U containing x is an *open neighbourhood* of x. Any set N which contains both x and an open neighbourhood of x is called *a neighbourhood* of x.

Proof. First, assume that W is open in M. Then W itself is an open neighbourhood N_x of every point $x \in W$.

Next, assume that every point $x \in W$ has a neighbourhood N_x with $N_x \subseteq W$. According to the definition of a neighbourhood, every point $x \in W$ then also has an open neighbourhood U_x such that $U_x \subseteq W$. Since $W = \cup_{x \in W} U_x$ it now follows by TOP 1 that W is open in M. $\square$

2.5.2   *Continuity at a point*

Lemma 2.5.4 reveals that neighbourhoods play the same role for topological spaces as open balls do for metric spaces. As an example we can define the continuity of a mapping at a point.

Definition 2.5.5. Let $f : X \to Y$ be a mapping between topological spaces. We say that f is *continuous at a point* $x_0 \in X$, if for every neighbourhood M of $f(x_0)$ in Y there exists a neighbourhood N of x_0 in X, such that $f(N) \subseteq M$.

It is now easy to prove the following theorem.

Theorem 2.5.6. *Let $f : X \to Y$ be a mapping between topological spaces. Then f is continuous if and only if f is continuous at every point $x \in X$.*

Proof. First, assume that f is continuous, and consider a point $x_0 \in X$. Let V be an arbitrary open neighbourhood of $f(x_0)$ in Y. Then since f is continuous, $U = f^{-1}(V)$ is an open neighbourhood of x_0 in X for which $f(U) \subseteq V \subseteq Y$. This proves that f is continuous at the arbitrarily chosen point $x_0 \in X$.

Now, assume conversely that f is continuous at every point $x \in X$, and let V be an arbitrarily chosen open set in Y. We have to prove that $f^{-1}(V)$ is an open set in X. According to Lemma 2.5.4, we only need to show that any point $x \in f^{-1}(V)$ has a neighbourhood N_x in X such that $N_x \subseteq f^{-1}(V)$, or equivalently, $f(N_x) \subseteq V$. However, this is obvious since f is continuous at x, and V is an open neighbourhood of $f(x)$ in Y. $\square$

With Theorem 2.5.6 we have completed the circle of ideas by getting back to the original definition of continuity in the setting of metric spaces.

2.5.3 *Basis for a topology*

We finish this section with a few remarks about the concept of a *basis* for a topology.

Definition 2.5.7. Let $(M, \mathcal{T})$ be a topological space.

A *fundamental system of neighbourhoods* (or a *local basis*) for a point $x \in M$ is a system of neighbourhoods $\{N_i \mid i \in I\}$ of $x \in M$ with the property that every neighbourhood N of $x \in M$ contains a neighbourhood from the system $\{N_i \mid i \in I\}$.

A *basis* for the topology $\mathcal{T}$ is a system of open sets $\{B_i \mid i \in I\}$ from $\mathcal{T}$ with the property that every open set U from $\mathcal{T}$ can be written as the union of open sets from the system $\{B_i \mid i \in I\}$.

A particularly important case is when the index set I in Definition 2.5.7 is either finite or the set of natural numbers $\mathbb{N}$. In the latter case we say that the corresponding systems of neighbourhoods or open sets are *countable*. In either situation, we say that the systems are *at most countable* or *numerable* (they can be assigned numbers).

Definition 2.5.8. A topological space $(M, \mathcal{T})$ satisfies the *first axiom of countability* if every point $x \in M$ has a numerable fundamental system of neighbourhoods.

A topological space $(M, \mathcal{T})$ satisfies the *second axiom of countability* if the topology $\mathcal{T}$ has a numerable basis.

Example 2.5.9. Let (M, d) be a metric space, and let $x \in M$ be an arbitrary point in M. It is clear that the system $\{B_{1/n}(x) \mid n \in \mathbb{N}\}$ of open balls is a numerable fundamental system of neighbourhoods for $x \in M$. So a metric space satisfies the first axiom of countability. Therefore, in particular, a topological space which does not satisfy the first axiom of countability cannot be metrizable. ◄

A topological space $(M, \mathcal{T})$ satisfying the second axiom of countability clearly also satisfies the first axiom of countability. The converse is not always the case. For instance, an infinite non-countable set (e.g. the real numbers $\mathbb{R}$, cf. Theorem 1.5.2) equipped with the discrete topology satisfies the first but not the second axiom of countability.

2.6 Points in relation to a subset

Let M be a topological space with the system of open sets $\mathcal{T}$, and let W be an arbitrary subset of M.

Definition 2.6.1. A point $x \in W$ is called an *interior point* of W if there exists an open neighbourhood U_x of x in M completely contained in W, i.e. $U_x \subseteq W$. The collection of interior points of W is called *the interior* of W, and is denoted by int W.

It may happen that int W is the empty set $\emptyset$. For example, the set of rational numbers $\mathbb{Q}$ and the set of irrational numbers $\mathbb{I} = \mathbb{R} \setminus \mathbb{Q}$ both have empty interiors in $\mathbb{R}$. The interior of a subset depends on the topological space in which it is considered as a subset. For instance, the interior of the closed interval $[a, b]$ considered as a subset of $\mathbb{R}$ is exactly the open interval $]a, b[$, whereas it has an empty interior considered as a subset of $\mathbb{R}^2$. In some sense, int W measures how 'fat' W is, considered as a subset of M.

Theorem 2.6.2. *The interior of a subset W in the topological space M has the following properties:*

(i)   int W *is an open set in M.*

(ii) *If U is an open set in M so that $U \subseteq W$, then $U \subseteq$ int W.*
      (i.e. int W is the 'largest' open set in M contained in W.)

(iii) *If U is an open set in M so that int $W \subseteq U \subseteq W$, then $U =$ int W.*

(iv) int W *is the union of all open sets U in M for which $U \subseteq W$.*

Proof. Proof of (i). If int $W = \emptyset$, the statement is trivially true. So assume that int $W \neq \emptyset$. According to the definition of an interior point, we can, for any $x \in$ int W, find an open neighbourhood U_x of x in M such that $U_x \subseteq W$. Since U_x is an open neighbourhood of any of its points, clearly $U_x \subseteq$ int W. This shows that int W is an open set in M by Lemma 2.5.4.

Proof of (ii). Let U be an open set in M so that $U \subseteq W$. Since U is an open set, it is an open neighbourhood U_x of any point $x \in U$ for which

$U_x = U \subseteq W$. But then $x \in$ int W, proving that $U \subseteq$ int W.

Proof of (iii). Follows immediately from (ii).

Proof of (iv). The union of all open sets U in M such that $U \subseteq W$ is itself an open set $\tilde{U}$ in M such that $\tilde{U} \subseteq W$. According to (i) int W is itself one of the open sets in this union, thus int $W \subseteq \tilde{U} \subseteq W$. But then according to (iii), we have that $\tilde{U} =$ int W. $\square$

Theorem 2.6.2 has the following immediate corollary.

Corollary 2.6.3. *A subset W of a topological space M is open if and only if $W =$ int W.*

Complementary to the concept of an interior point, we have the concept of an exterior point.

Definition 2.6.4. A point $x \in M$ is called an *exterior point* of the subset W in the topological space M, if x is an interior point in the set $M \setminus W$. The collection of exterior points of W is called *the exterior* of W in M.

We do not need any special symbol for the exterior of a set, since *the exterior of W in M* is exactly the interior of $M \setminus W$ in M, i.e. int $(M \setminus W)$.

Finally, we have a third type of points in relation to W considered as a subset of M.

Definition 2.6.5. A point $x \in M$ is called a *boundary point* of the subset W in the topological space M if it is neither an interior point nor an exterior point of W. In other words, a point $x \in M$ is a boundary point of W if every open neighbourhood U_x of x in M contains points both from W and from the complementary set $M \setminus W$. The collection of boundary points for W is called *the boundary* of W, and is denoted by ∂W.

Note that a boundary point of W is not necessarily contained in W. If for example, $W = \,]a, b]$ is a half-open interval in $\mathbb{R}$, then the boundary point a does not belong to W, while the boundary point b does belong to W.

Usually, the boundary of a set corresponds to the intuitive imaginations. But life can surprise. Consider for instance the subset $\mathbb{Q}$ of $\mathbb{R}$, which consists of the rational numbers. Here all points in $\mathbb{Q}$ are boundary points for $\mathbb{Q}$. And even worse. Actually every point in $\mathbb{R}$ is a boundary point of $\mathbb{Q}$. Hence a subset W of a topological space M can consist of nothing but boundary points, and even more surprisingly, it can even happen that $\partial W = M$.

The interior, the exterior, and the boundary of W give a partitioning of M into disjoint sets. We have

$$M = \text{int } W \sqcup \text{int } (M \setminus W) \sqcup \partial W,$$

where $\sqcup$ marks that the sets int W, int $(M \setminus W)$ and ∂W are pairwise disjoint.

In concrete situations some of the sets can be empty. We just gave an example for which int $W = \text{int } (M \setminus W) = \emptyset$.

2.7 Closed sets

Definition 2.7.1. A subset A of a topological space M is said to be *closed* if the complementary set $M \setminus A$ of A is an open set in M.

In general a subset of M need not to be neither open nor closed. On the other hand, it is also possible for a subset to be both open and closed, as it is the case for any subset in a set equipped with the discrete metric. One should not make the mistake of thinking that a closed set is just a set which is not an open set.

Theorem 2.7.2. *A subset A in the topological space M is closed if and only if A contains all its boundary points.*

Proof. If A is a closed set, the complementary set $M \setminus A$ is open and consists of nothing but exterior points of A. But then all boundary points of A have to be contained in A.

Conversely, if all boundary points of A are contained in A, then all points in $M \setminus A$ must be exterior points of A, so that $M \setminus A = \text{int } (M \setminus A)$. But then $M \setminus A$ is an open set according to Theorem 2.6.2, and hence A is closed. $\square$

The system of closed sets in a topological space M has the following 'complementary' properties to the system of open sets:

A1 If $\{A_i \mid i \in I\}$ is an arbitrary system of closed sets in M, then the intersection $\bigcap \{A_i \mid i \in I\}$ is also a closed set in M.

A2 If $A_1, \ldots, A_k$ is an arbitrary finite system of closed sets in M, then the union $\bigcup_{i=1}^{k} A_i$ is also a closed set in M.

A3 The empty set $\emptyset$ and the set M itself are closed sets in M.

The properties A1, A2, and A3 follow from the corresponding properties TOP 1, TOP 2, and TOP 3 by using the following formulae from set theory:

$$M \setminus \bigcap_{i \in I} A_i = \bigcup_{i \in I} (M \setminus A_i) \quad \text{and} \quad M \setminus \bigcup_{i \in I} A_i = \bigcap_{i \in I} (M \setminus A_i).$$

Under the formation of a complementary set there is complete duality between the concepts of open and closed sets. Indeed, for any subsets U and A in the topological space M it is obvious that:

$$U \quad \text{open} \quad \Longleftrightarrow \quad M \setminus U \quad \text{closed}$$

and

$$A \quad \text{closed} \quad \Longleftrightarrow \quad M \setminus A \quad \text{open.}$$

The main reason for introducing the system of closed sets in a topological space is that it is easier to use in certain situations. In this context the following result is of interest.

Theorem 2.7.3. *A mapping $f : X \to Y$ between topological spaces X and Y is continuous if and only if for every closed set A in Y, the set $f^{-1}(A)$ is closed in X.*

Proof. First, assume that $f : X \to Y$ is continuous, and let A be a closed set in Y. Then $Y \setminus A$ is an open set in Y, and since f is continuous, the set $X \setminus f^{-1}(A) = f^{-1}(Y \setminus A)$ is therefore an open set in X. But then $f^{-1}(A)$ is a closed set in X, as should be proved.

Next assume conversely that $f^{-1}(A)$ is a closed set in X for any closed set A in Y. Let V be any open set in Y. We have to show that $f^{-1}(V)$ is an open set in X. But because $X \setminus f^{-1}(V) = f^{-1}(Y \setminus V)$ this follows immediately, since $Y \setminus V$ is a closed set in Y, and therefore by the assumption that $f^{-1}(Y \setminus V)$ is a closed set in X. $\square$

2.8 The closure of a set

Let M be a topological space with the system of open sets $\mathcal{T}$.

Let W be an arbitrary subset of the topological space M. In Theorem 2.6.2 we have seen that W contains a 'largest' open set, namely int W. We shall now show that correspondingly, W is contained in a 'smallest' closed set $\overline{W}$, which we call the closure of W.

As a preparation we first introduce two types of points in relation to W considered as a subset of M.

Definition 2.8.1. A point $x \in M$ is called a *contact point* of W, provided that every open neighbourhood U_x of x in M contains at least one point from W. A point $x \in M$ is called an *accumulation point* of W, provided that every open neighbourhood U_x of x in M contains at least one point from W different from x.

Clearly interior points and boundary points of W are contact points of W. Conversely, it is also immediate from the definitions, that a contact point of W is either an interior point or a boundary point of W. In other words, *'contact point' is a common name for interior point and boundary point,* and therefore we have in fact not really introduced a new concept by this definition. In many cases, however, it is convenient to have the common notion.

A contact point of W which does not belong to W has to be an accumulation point. An *isolated point* in W is a contact point of W which is not an accumulation point of W. An isolated point in W is in other words a point $x \in W$, which has an open neighbourhood U_x in M, in which x is the only point from W.

Since interior points of a subset naturally belong to the subset, we can give Theorem 2.7.2 a more convenient formulation.

Theorem 2.8.2. *A subset A in the topological space M is closed if and only if A contains all its contact points (accumulation points).*

It is now clear what we have to do to in order to 'close' a subset.

Definition 2.8.3. The collection of all contact points of the subset W in the topological space M is called *the closure* of W and is denoted by $\overline{W}$.

It appears that $\overline{W}$ is obtained by adding to W those contact points for W which do not already belong to W.

From the description of contact points as either interior points or boundary points for W it follows immediately that

$$\overline{W} = \text{int } W \cup \partial W.$$

Hence one gets $\overline{W}$ by adding to W the boundary points of W which do not already belong to W.

Theorem 2.8.4. *The closure of a subset W in the topological space M has the following properties:*

(i) $\overline{W}$ *is a closed set.*

(ii)   *If A is a closed set in M so that $W \subseteq A$, then $\overline{W} \subseteq A$.*
       (i.e. $\overline{W}$ is the 'smallest' closed set in M which contains W.)

(iii) *If A is a closed set in M so that $W \subseteq A \subseteq \overline{W}$, then $A = \overline{W}$.*

(iv)   *$\overline{W}$ is the intersection of all closed sets A in M for which $W \subseteq A$.*

Proof. Proof of (i). By using

$$\overline{W} = \text{int } W \cup \partial W = M \setminus \text{int } (M \setminus W),$$

it follows immediately from Theorem 2.6.2 (i) that $\overline{W}$ is closed.

Proof of (ii). Let A be a closed set in M such that $W \subseteq A$. Let $x \in \overline{W}$ be an arbitrary point in $\overline{W}$. Since x is a contact point of W, any open neighbourhood of x in M will contain at least one point from W, thus also from A since $W \subseteq A$. But then x is a contact point for A, and since A is closed it follows that $x \in A$ by Theorem 2.8.2. This proves that $\overline{W} \subseteq A$.

Proof of (iii). Follows immediately from (ii).

Proof of (iv). The intersection of all closed sets A in M such that $W \subseteq A$ is itself a closed set $\tilde{A}$ in M such that $W \subseteq \tilde{A}$. According to (i), $\overline{W}$ is itself one of the sets over which the intersection is taken, hence $W \subseteq \tilde{A} \subseteq \overline{W}$. But then according to (iii) we have that $\tilde{A} = \overline{W}$. $\square$

Theorem 2.8.4 has the following immediate corollary.

Corollary 2.8.5. *A subset A in a topological space M is closed if and only if $A = \overline{A}$.*

We have the following simple results regarding the effect of taking the closure of the union, respectively the intersection, of two sets.

Theorem 2.8.6. *For subsets W_1 and W_2 in the topological space M the following two statements hold:*

$$(a) \quad \overline{W_1 \cup W_2} = \overline{W_1} \cup \overline{W_2} .$$

$$(b) \quad \overline{W_1 \cap W_2} \subseteq \overline{W_1} \cap \overline{W_2} .$$

Proof. Since $W_1 \cup W_2 \subseteq \overline{W_1} \cup \overline{W_2}$ and since $\overline{W_1} \cup \overline{W_2}$ is a closed set in M, it follows from Theorem 2.8.4 (ii) that,

$$\overline{W_1 \cup W_2} \subseteq \overline{W_1} \cup \overline{W_2}.$$

Since a contact point of either W_1 or W_2 is clearly a contact point of $W_1 \cup W_2$, we also get the opposite inclusion,

$$\overline{W_1} \cup \overline{W_2} \subseteq \overline{W_1 \cup W_2}.$$

This proves part (a).

A contact point for $W_1 \cap W_2$ is clearly a contact point of both W_1 and W_2, which proves part (b). $\square$

Remark 2.8.7. On the real axis $\mathbb{R}$ consider the subsets

$$W_1 = \{x \in \mathbb{R} \mid x < 0\} \quad \text{and} \quad W_2 = \{x \in \mathbb{R} \mid x > 0\}.$$

Then $\overline{W_1 \cap W_2} = W_1 \cap W_2 = \emptyset$, whereas $\overline{W_1} \cap \overline{W_2} = \{0\}$. Therefore $\overline{W_1 \cap W_2}$ is, in general, a proper subset of $\overline{W_1} \cap \overline{W_2}$ in Theorem 2.8.6 (b). ◁

In the mathematical theory of singularities of differentiable mappings, as well as in the qualitative theory of dynamical systems, one is interested in *generic* ('typical') properties of mappings or dynamical systems. Informally speaking, a property is said to be generic if it is valid for a dense subset of the objects under consideration (mappings or dynamical systems) when considered as points in an appropriate topological space.

Definition 2.8.8. A subset W in a topological space M is said to be *dense* in M if $\overline{W} = M$.

The concept of density can be given many equivalent formulations. We list some of them in the following theorem, the proof of which is left to the reader.

Theorem 2.8.9. *For a subset W in a topological space M the following statements are equivalent:*

(i)   *W is dense in M.*

(ii)   *Every point in M either belongs to W or is an accumulation point of W.*

(iii) *Every non-empty, open subset in M contains at least one point from W.*

(iv) *int $(M \setminus W) = \emptyset$.*

It is also of considerable interest to know whether a property of a mapping, or a dynamical system, is *stable* (robust under small perturbations). From a topological point of view, a property is stable if the objects with the desired property form an open set, when considered as points in an appropriate topological space.

Example 2.8.10. Let $\mathbb{Z}$ denote the integers and consider the subset $W = \mathbb{R} \setminus \mathbb{Z}$ of the real axis $\mathbb{R}$. Then W is an open and dense subset of $\mathbb{R}$. It is therefore both a generic and a stable property for a real number not to be an integer. The subset $\mathbb{I}$ of irrational numbers in $\mathbb{R}$ is dense but not open. Hence for a real number it is a generic property, but not a stable property, to be irrational. ◄

2.9 Limit points. Hausdorff spaces

The concept of a limit point for a mapping or a sequence, and the corresponding concept of convergence, can easily be introduced in the setting of topological spaces. We introduce these concepts simultaneously in metric spaces and in topological spaces, so that one can easily see the relation to the well-known situations from the Euclidean spaces.

Definition 2.9.1 (metric). Let $f : X \to Y$ be a mapping between metric spaces (X, d_X) and (Y, d_Y), and let $x_0 \in X$ and $y_0 \in Y$. We say that

$$f(x) \quad approaches \quad y_0 \quad \text{for} \quad x \quad approaching \quad x_0 ,$$

or that

$$f(x) \text{ has the } limit \ point \ y_0 \text{ for } x \ approaching \ x_0 ,$$

or that

$$f(x) \quad converges \ to \quad y_0 \quad \text{for} \quad x \quad converging \ to \quad x_0 ,$$

if:

$$\forall \varepsilon > 0 \ \exists \delta > 0 \ \forall x \in X \setminus \{x_0\} : \ d_X(x, x_0) < \delta \ \Rightarrow \ d_Y(f(x), y_0) < \varepsilon .$$

Definition 2.9.1 (topological). Let $f : X \to Y$ be a mapping between topological spaces X and Y, and let $x_0 \in X$ and $y_0 \in Y$. We say that

$$f(x) \quad approaches \quad y_0 \quad \text{for} \quad x \quad approaching \quad x_0 ,$$

or one of the other similar terminologies presented in the metric case, if:

For any open neighbourhood V_{y_0} of y_0 in Y there exists an open neighbourhood U_{x_0} of x_0 in X such that

$$f(U_{x_0} \setminus \{x_0\}) \subseteq V_{y_0}.$$

For sequences of points we have similar concepts.

Definition 2.9.2 (metric). Let (x_n), or more explicitly $x_1, x_2, \ldots, x_n, \ldots$, be a sequence of points in the metric space (M, d), and let $y_0 \in M$. We say that

$$x_n \quad \text{approaches} \quad y_0 \quad \text{for} \quad n \quad \text{going to} \quad \infty \,,$$

or that

$$x_n \text{ has the } \text{limit point } y_0 \quad \text{for} \quad n \quad \text{going to} \quad \infty \,,$$

or that

$$x_n \quad \text{converges to} \quad y_0 \quad \text{for} \quad n \quad \text{going to} \quad \infty \,,$$

if:

$$\forall \varepsilon > 0 \ \exists n_0 \in \mathbb{N} \ \forall n \in \mathbb{N} : \ n \geq n_0 \quad \Rightarrow \quad d(x_n, y_0) < \varepsilon \,.$$

Definition 2.9.2 (topological). Let (x_n), or more explicitly $x_1, x_2, \ldots$, $x_n, \ldots$, be a sequence of points in the topological space M, and let $y_0 \in M$. We say that

$$x_n \quad \text{approaches} \quad y_0 \quad \text{for} \quad n \quad \text{going to} \quad \infty \,,$$

or one of the other similar terminologies presented in the metric case, if:

For every open neighbourhood V_{y_0} of y_0 in M there exists an $n_0 \in \mathbb{N}$ such that

$$x_n \in V_{y_0} \quad \text{for} \quad n \geq n_0 \,.$$

Often, we just say that the sequence (x_n) in the topological space M is *convergent* with the *limit point* y_0, or that the sequence (x_n) *converges* to y_0.

In the two situations described in Definition 2.9.1, respectively Definition 2.9.2, we use the short notations:

$$f(x) \to y_0 \quad \text{for} \quad x \to x_0 \,, \quad \text{or} \quad \lim_{x \to x_0} f(x) = y_0 \,,$$

respectively

$$x_n \to y_0 \quad \text{for} \quad n \to \infty \,, \quad \text{or} \quad \lim_{n \to \infty} x_n = y_0 \,.$$

The definitions of continuity and convergence are closely related, and in the setting of metric spaces continuity of a mapping at a point can in fact be described by the convergence of sequences.

Theorem 2.9.3. *A mapping $f : X \to Y$ between metric spaces (X, d_X) and (Y, d_Y) is continuous at a point $x_0 \in X$ if and only if for every convergent sequence (x_n) of points in X with limit point x_0, the sequence $(f(x_n))$ of points in Y is convergent with limit point $f(x_0)$.*

Proof. First, assume that $f : X \to Y$ is continuous at $x_0 \in X$. Let (x_n) be a convergent sequence of points in X with limit point x_0. We have to show that $(f(x_n))$ is a convergent sequence of points in Y with limit point $f(x_0)$. For this, let an arbitrary $\varepsilon > 0$ be given. Since f is continuous at x_0, we can choose $\delta > 0$, such that for $x \in X$ with $d_X(x, x_0) < \delta$ we have $d_Y(f(x), f(x_0)) < \varepsilon$. Because the sequence (x_n) is convergent with limit point x_0, for $\delta > 0$ we can choose an $n_0 \in \mathbb{N}$ such that $d_X(x_n, x_0) < \delta$ for $n \geq n_0$. But then $d_Y(f(x_n), f(x_0)) < \varepsilon$, for $n \geq n_0$. This shows that $(f(x_n))$ converges to $f(x_0)$.

Now assume, conversely, that for every convergent sequence (x_n) of points in X with limit point x_0, we know that the sequence $(f(x_n))$ of points in Y converges towards $f(x_0)$. We have to prove that f is continuous at x_0. The proof is indirect. Assume for that purpose that f is not continuous at x_0. Then:

$$\exists \varepsilon_0 > 0 \ \ \forall \delta > 0 \ \ \exists x \in X : \ (d_X(x, x_0) < \delta) \wedge (d_Y(f(x), f(x_0)) \geq \varepsilon_0).$$

Consider such an $\varepsilon_0 > 0$, and choose for every $n \in \mathbb{N}$ a point $x_n \in X$ such that

$$(d_X(x_n, x_0) < 1/n) \quad \text{and} \quad (d_Y(f(x_n), f(x_0)) \geq \varepsilon_0).$$

Now, it is clear that (x_n) is a convergent sequence of points in X with limit point x_0, even though $(f(x_n))$ does not converge to $f(x_0)$. This yields a contradiction, which proves that f is continuous at x_0. $\square$

In a general topological space, a convergent sequence of points may have several limit points. Usually we want to avoid this, and therefore we need to require more of the topological space. In that connection a suitable separation axiom for the points in M was formulated in the book "Grundzüge der Mengenlehre" of 1914 by the German mathematician Felix Hausdorff (1868–1942), who is one of the pioneers in general topology.

Definition 2.9.4. A topological space M is called a *Hausdorff space*, if for every pair of distinct points x and y in M there exists a related pair of disjoint open neighbourhoods U of x and V of y.

Theorem 2.9.5. *A metric space (M, d) is a Hausdorff space.*

Proof. Let x and y be an arbitrary pair of distinct points in M. Then

$$r = d(x, y) > 0.$$

The triangle inequality now implies that the open sets $U = B_{r/2}(x)$ and $V = B_{r/2}(y)$ are disjoint open neighbourhoods of x and y, as required. $\square$

Theorem 2.9.6. *A sequence (x_n) of points in a Hausdorff space M has at most one limit point.*

Proof. The proof is by contradiction. So assume that the sequence (x_n) of points in the Hausdorff space M admits two distinct limit points x and y. Corresponding to x and y there is a pair of disjoint, open neighbourhoods U of x and V of y. Since (x_n) is convergent with limit point x, respectively y, we can choose $n_1 \in \mathbb{N}$, respectively $n_2 \in \mathbb{N}$, such that $x_n \in U$ for $n \geq n_1$, respectively $x_n \in V$ for $n \geq n_2$. For $n \geq \max(n_1, n_2)$ this yields a contradiction, since $x_n \in U \cap V$ and $U \cap V = \emptyset$. □

We finish this section with an example illustrating the idea of the Hausdorff axiom.

Example 2.9.7 (Line with an extra origo). Consider the set M defined by adding an extra origo 0^* to the real axis $\mathbb{R}$.

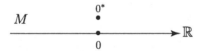

Fig. 2.6   A space consisting of the usual real line, plus an additional point 0^*.

We equip M with a topology, in which the open sets are the usual open sets in $\mathbb{R}$ together with those subsets U of M that contain 0^* and have the property that $(U \setminus \{0^*\}) \cup \{0\}$ is an ordinary open set in $\mathbb{R}$. It is easy to prove that this actually defines a topology on M.

The set M equipped with the above topology is not a Hausdorff space. Indeed, it is clear that we cannot separate the points 0 and 0^* in M with a pair of disjoint, open subsets. Furthermore, it is clear that the sequence $(1/n)$ admits both 0 and 0^* as limit points.

The topological space M is, in other words, not a Hausdorff space. It has, however, a weaker separation property, which is also interesting, namely the following property.

Separation Axiom T_1. *For each pair of distinct points x and y in M there exists an open neighbourhood U of x which does not contain y, and an open neighbourhood V of y which does not contain x.* ◄

A topological space satisfying the separation axiom T_1 is called a T_1-*space*. Correspondingly, the separation axiom behind the notion of a Hausdorff space is called T_2, and the space itself a T_2-*space*. The term T-space was introduced by the Russian mathematician Pavel S. Aleksandrov (1896–1982) and the German mathematician Heinz Hopf (1894–1971) in their book "Topologie I" of 1935. (T stands for the German word 'Trennung', which means separation.)

Exercises and Further Results

Exercise 2.1. Show that the real-valued function $f(x) = x^3$ is continuous, but not uniformly continuous, in $\mathbb{R}$.

Exercise 2.2. Prove that the real-valued function $f(x) = \sin(x)$ is uniformly continuous in $\mathbb{R}$.

Exercise 2.3. Let $f : [a, b] \to \mathbb{R}$ be a continuous function defined in the closed interval $[a, b]$. Prove that

$$\int_a^b |f(x)| \, dx = 0 \quad \Longleftrightarrow \quad f = 0.$$

Exercise 2.4. Prove the following formulae from set theory:

$$S \setminus \bigcap_{\alpha \in I} A_\alpha = \bigcup_{\alpha \in I} (S \setminus A_\alpha)$$
$$S \setminus \bigcup_{\alpha \in I} A_\alpha = \bigcap_{\alpha \in I} (S \setminus A_\alpha).$$

Exercise 2.5. Let $f, g : A \to \mathbb{R}$ be two real-valued functions defined in a subset A of $\mathbb{R}$, and let $x_0 \in A$ be a point in A.
    (1) Show that if f is continuous at x_0 and g is discontinuous at x_0, then $f + g$ is discontinuous at x_0.
    (2) Does the same conclusion hold for $f \cdot g$?

Exercise 2.6. Let $U, V \subseteq \mathbb{R}$ be subsets of $\mathbb{R}$ and let $f : U \to \mathbb{R}$ and $g : V \to \mathbb{R}$ be real-valued functions, such that $f(U) \subseteq V$. Then the *composite function* $g \circ f : U \to \mathbb{R}$ is defined by $(g \circ f)(x) = g(f(x))$ for all $x \in U$.
    Suppose that the function f is continuous at the point $x_0 \in U$ and that the function g is continuous at the point $f(x_0) \in V$. Show that if this is the case, then the composite function $g \circ f$ is continuous at $x_0 \in U$.

Exercise 2.7. Let $f : \mathbb{R} \to \mathbb{R}$ be a positive, real-valued function for which

$$f(x + y) = f(x)f(y) \quad \text{for all} \quad x, y \in \mathbb{R}.$$

(1) Show that $f(0) = 1$.

(2) Prove that there exists a unique real number $a > 0$ such that $f(n) = a^n$ for all integers $n \in \mathbb{Z}$.

(3) Show that f is continuous in all of $\mathbb{R}$ if f is continuous at 0.

(4) Do you know an example of such a function?

Exercise 2.8. Give an example of two subsets A and B in $\mathbb{R}$ with $A \cap B = \emptyset$, between which the distance

$$d(A, B) = \inf\{|x - y| \,|\, x \in A, y \in B\} = 0.$$

Exercise 2.9. On the 2-dimensional real number space $M = \mathbb{R}^2 = \mathbb{R} \times \mathbb{R}$, define three distance functions $d_1, d_2, d_\infty : M \times M \to \mathbb{R}$ as follows:

$$d_1((x_1, x_2), (y_1, y_2)) = |x_1 - y_1| + |x_2 - y_2|,$$
$$d_2((x_1, x_2), (y_1, y_2)) = \sqrt{(x_1 - y_1)^2 + (x_2 - y_2)^2},$$
$$d_\infty((x_1, x_2), (y_1, y_2)) = \max\{|x_1 - y_1|, |x_2 - y_2|\}.$$

(1) Show that each of the three functions defines a metric in M, and describe in each case the shape of the unit ball: $\{x \in M | d(0, x) \le 1\}$.

(2) Show the existence of positive real numbers a_1, b_1, a_2, b_2 for which

$$a_1 d_\infty(x, y) \le d_1(x, y) \le b_1 d_\infty(x, y),$$
$$a_2 d_\infty(x, y) \le d_2(x, y) \le b_2 d_\infty(x, y),$$

for all $x, y \in M$.

Exercise 2.10. Prove that in a metric space M with the discrete metric, every subset A is both open and closed.

Exercise 2.11. Let (M, d) be a metric space and let $A \subseteq M$. Show that a subset $U \subseteq A$ is open in the induced metric on A if and only if there exists an open set $V \subseteq M$ in M such that $U = A \cap V$.

Exercise 2.12. For an arbitrary pair of non-empty subsets A and B in a metric space (M, d), we define the distance between A and B by

$$d(A, B) = \inf\{d(x, y) \,|\, x \in A, y \in B\}.$$

Show that the closure of a subset in a metric space is exactly the set of points for which the distance to the subset is zero.

Exercise 2.13. Find the interior, the closure, the boundary and the isolated points for the following subsets in $\mathbb{R}^2$ with Euclidian metric:

   (1) $A = \mathbb{R}^2$,   (2) $A = \mathbb{Z} \times \mathbb{Z}$,   (3) $A = \mathbb{Q} \times \mathbb{Z}$.

Exercise 2.14. Let (X, d_X) and (Y, d_Y) be metric spaces.

   Define $d_{X \times Y} : (X \times Y) \times (X \times Y) \to \mathbb{R}_0^+$ by

$$d_{X \times Y}((x_1, y_1), (x_2, y_2)) = \max(d_X(x_1, x_2), d_Y(y_1, y_2)) .$$

1) Show that $d_{X \times Y}$ is a metric on $X \times Y$.

2) Show that the projections

$$p_X : X \times Y \to X, \quad p_X(x, y) = x$$

$$p_Y : X \times Y \to Y, \quad p_Y(x, y) = y$$

are continuous mappings.

Exercise 2.15. Let (M, d) be a metric space. For every pair of points $x, y \in M$, we set

$$\tilde{d}(x, y) = \frac{d(x, y)}{1 + d(x, y)} .$$

Show that $\tilde{d}$ is a metric on M with the property $0 \leq \tilde{d}(x, y) < 1$ for all $x, y \in M$.

   <u>Hint:</u> You may use that the function $\varphi : \mathbb{R}_0^+ \to \mathbb{R}_0^+$ defined by

$$\varphi(t) = \frac{t}{1 + t} \quad , \quad t \in \mathbb{R}_0^+$$

is increasing.

Exercise 2.16. Let K be an arbitrary set, and let (M, d) be a metric space in which $0 \leq d(x, y) \leq 1$ for all $x, y \in M$.

Let $F(K, M)$ denote the set of mappings $f : K \to M$.

Define $D : F(K, M) \times F(K, M) \to \mathbb{R}_0^+$ by

$$D(f, g) = \sup_{t \in K} d(f(t), g(t)) .$$

1) Show that D is a metric on $F(K, M)$.

2) Let $t_0 \in K$ be a fixed point in K and define

$$Ev_{t_0} : F(K, M) \to M \quad \text{by} \quad Ev_{t_0}(f) = f(t_0) .$$

Show that Ev_{t_0} is continuous. (Ev_{t_0} is called an *evaluation mapping*.)

Exercise 2.17. Let (M, d) be a metric space. For $x \in M$ and $r \in \mathbb{R}^+$, let $B_r(x)$ denote the open ball in M with centre x and radius r.

Show that the system of open balls in M has the following properties:

1) If $y \in B_r(x)$ then $x \in B_r(y)$.
2) If $y \in B_r(x)$ and $0 < s \leq r - d(x, y)$, then $B_s(y) \subseteq B_r(x)$.
3) If $d(x, y) \geq r + s$, where $x, y \in M$ and $r, s \in \mathbb{R}^+$, then $B_r(x)$ and $B_s(y)$ are mutually disjoint.

Exercise 2.18. Let (M, d) be a metric space. A subset K in M is called *bounded* in (M, d), if there exists a point $x \in M$ and an $r \in \mathbb{R}^+$ such that $K \subseteq B_r(x)$.

Examine the truth of each of the following three statements:

1) If two subsets K_1 and K_2 in M are bounded in (M, d), then their union $K_1 \cup K_2$ is also bounded in (M, d).

2) If $K \subseteq M$ is bounded in (M, d), then
$$K' = \bigcup_{x \in K} \{ y \in M \,|\, d(x, y) \leq 1 \}$$
is also bounded in (M, d).

3) If $K \subseteq M$ is bounded in (M, d), then
$$K'' = \bigcap_{x \in K} \{ y \in M \,|\, d(x, y) > 1 \}$$
is also bounded in (M, d).

Exercise 2.19. Prove that a subset A in a metric space M is a closed subset in M if and only if every sequence (x_n) in A, which converges in M, has limit point $x \in A$.

Exercise 2.20. Prove that every monotonically increasing (decreasing) sequence of real numbers is convergent if and only if it is bounded.

Exercise 2.21. Let (x_n) and (y_n) be convergent sequences of real numbers with limit points x_0 and y_0, respectively. Prove that the sequences $(x_n + y_n)$, $(x_n - y_n)$, $(x_n \cdot y_n)$, and (x_n/y_n) (when defined) are all convergent with limit points respectively, $x_0 + y_0$, $x_0 - y_0$, $x_0 \cdot y_0$, and x_0/y_0 (if $y_0 \neq 0$).

Exercise 2.22. Let $U \subseteq \mathbb{R}$ be a subset of $\mathbb{R}$ and let $f : U \to \mathbb{R}$ be a real-valued function. Let $x_0 \in U$ be an arbitrary point in U.

Prove that $f : U \to \mathbb{R}$ is continuous at x_0 if and only if for each convergent sequence (x_n) in U with limit point x_0, the image sequence $(f(x_n))$ is convergent with limit point $f(x_0)$.

Exercise 2.23. List all topologies that can be defined on a set $M = \{a, b\}$ containing only two elements a and b.

Exercise 2.24. Let $M = \{a, b, c, d, e\}$. Determine whether or not each of the following families of subsets in M is a topology on M:

$\tau_1 = \{M, \emptyset, \{a\}, \{a, b\}, \{a, c\}\}$,

$\tau_2 = \{M, \emptyset, \{a, b, c\}, \{a, b, d\}, \{a, b, c, d\}\}$,

$\tau_3 = \{M, \emptyset, \{a\}, \{a, b\}, \{a, c, d\}, \{a, b, c, d\}\}$.

Exercise 2.25. Let τ be the family of subsets in $\mathbb{R}$ consisting of $\mathbb{R}$, the empty set $\emptyset$, and all half infinite intervals $]q, \infty[$, where q is a rational number. Show that τ is not a topology on $\mathbb{R}$.

Exercise 2.26. Let $\tau = \{M, \emptyset, A, B\}$ be a family of four subsets in a set M, where A and B are distinct, non-empty subsets.

Which conditions must A and B satisfy for τ to be a topology on M?

Exercise 2.27. Let $\mathcal{T}$ be the system of subsets U in $\mathbb{R}$ which is of one of the following types:

Either (i) U does not contain 0,

or (ii) U does contain 0, and the complementary set $\mathbb{R} \setminus U$ is finite.

1) Show that $\mathcal{T}$ is a topology on $\mathbb{R}$.

2) Show that $\mathbb{R}$ with the topology $\mathcal{T}$ is a Hausdorff space.

3) Show that the topology $\mathcal{T}$ on $\mathbb{R}$ does not stem from a metric on $\mathbb{R}$, since there does not exist a countable system of open neighbourhoods of $0 \in \mathbb{R}$ in the topology $\mathcal{T}$ with the property that an arbitrary open neighbourhood of $0 \in \mathbb{R}$ contains a neighbourhood from the system.

Exercise 2.28. Let M be a topological space. For every pair of real-valued functions $f, g : M \to \mathbb{R}$, we can in the usual way define the functions $f + g$, $f - g$, $f \cdot g$, and (if $g(x) \neq 0$ for all $x \in M$) f/g.

1) Show that if f and g are continuous at a point $x_0 \in M$, then also $f + g$, $f - g$, $f \cdot g$, and (when it is defined) f/g are continuous at $x_0 \in M$. (Carry through the argument in at least one case.)

2) Assume that $f, g : M \to \mathbb{R}$ are continuous. Show that $U = \{x \in M | f(x) < g(x)\}$ is an open set in M.

3) Let $f_1, \ldots, f_k : M \to \mathbb{R}$ be contiuous real-valued functions and $a_1, \ldots, a_k \in \mathbb{R}$ real numbers. Show that $U = \{x \in M | f_i(x) < a_i, \ i = 1, \ldots, k\}$ is an open set in M.

Exercise 2.29. Let W_1 and W_2 be arbitrary subsets in the topological space M. Show that

1) $\text{int}(W_1 \cap W_2) = \text{int}W_1 \cap \text{int}W_2$.
2) $\text{int}(W_1 \cup W_2) \supseteq \text{int}W_1 \cup \text{int}W_2$.

Give an example that the equality sign in 2) does not apply in general.

Exercise 2.30. Let W_1 and W_2 be arbitrary subsets in the topological space M. Show that

1) $W_1 \subseteq W_2 \Rightarrow \overline{W_1} \subseteq \overline{W_2}$.
2) $\overline{W_1 \cup W_2} = \overline{W_1} \cup \overline{W_2}$.
3) $\overline{W_1 \cap W_2} \subseteq \overline{W_1} \cap \overline{W_2}$.
4) $\text{int}(W_1 \backslash W_2) \subseteq \text{int}W_1 \backslash \text{int}W_2$.
5) $\overline{W_1} \backslash \overline{W_2} \subseteq \overline{W_1 \backslash W_2}$.

Exercise 2.31. Let A be a non-empty, bounded subset of a real number axis $\mathbb{R}$. Prove that $\inf A$ and $\sup A$ are contact points of A in $\mathbb{R}$.

Exercise 2.32. Let M be a Hausdorff space, and let W be an arbitrary subset in M. (It is sufficient that M satisfies the separation property T_1.)

Show that every neighbourhood of an accumulation point $x \in M$ of W, contains infinitely many mutually different points from W.

Exercise 2.33. Let K be a subset of isolated points in a topological space M, and let (x_n) be an arbitrary sequence in K.

Prove that if (x_n) is convergent in K, then there exists a natural number $n_0 \geq 1$ such that $x_n = x_{n_0}$ for all $n \geq n_0$.

Exercise 2.34. Let M be a topological space with topology $\mathcal{T}$, and let $\pi : M \to \tilde{M}$ be a mapping into a set $\tilde{M}$. Let $\tilde{\mathcal{T}}$ be the quotient topology on $\tilde{M}$ induced from the topology $\mathcal{T}$ on M by the mapping π.

1) Let $\mathcal{T}'$ be a topology on $\tilde{M}$, such that $\pi : M \to \tilde{M}$ is continuous when M is considered with the topology $\mathcal{T}$ and $\tilde{M}$ with the topology $\mathcal{T}'$. Show that $\mathcal{T}' \subseteq \tilde{\mathcal{T}}$.

(The quotient topology $\tilde{\mathcal{T}}$ on $\tilde{M}$ is in other words the 'largest' topology on $\tilde{M}$ for which $\pi : M \to \tilde{M}$ is continuous.)

2) Show that when $\tilde{M}$ has the quotient topology determined by the mapping $\pi : M \to \tilde{M}$, then the following holds:

A mapping $f : \tilde{M} \to T$ into a topological space T is continuous if and only if the composite mapping $f \circ \pi : M \to T$ is continuous.

Exercise 2.35. Let M be a topological space with topology $\mathcal{T}$, and let A be an arbitrary subset in M. Equip A with the induced topology $\mathcal{T}_A$.

Show that a subset $B' \subseteq A$ is closed in A with the topology $\mathcal{T}_A$ if and only if there exists a closed subset $B \subseteq M$ in the topology $\mathcal{T}$ such that $B' = A \cap B$.

Exercise 2.36. Let $f : X \to Y$ be a mapping between topological spaces X and Y.

If $f : X \to Y$ maps a subset $X' \subseteq X$ in X into a subset $Y' \subseteq Y$ in Y, then f determines a mapping $f' : X' \to Y'$ defined by $f'(x) = f(x)$ for $x \in X'$.

When a subset of a topological space is considered as a topological space in the following, it is always with the induced topology.

1) Let $f' : X' \to Y'$ be a mapping determined by $f : X \to Y$ as above. Show that if f is continuous, then f' is continuous.

2) Let A_1 and A_2 be closed subsets in X such that $X = A_1 \cup A_2$. Let $f_1 : A_1 \to Y$ and $f_2 : A \to Y$ be the mappings determined by f, i.e. the restrictions of f to A_1 and A_2, respectively.

Show that if f_1 and f_2 are continuous, then f is continuous.

Exercise 2.37. Show that a topological space M is a T_1-space if and only if every subset in M containing exactly one point is a closed subset.

Exercise 2.38. Let $S = \mathbb{N} \cup \{0\}$ be the set of non-negative integers.

For every natural number $n \in \mathbb{N}$ we define a subset U_n in S by

$$U_n = \{n \cdot p \in S \mid p = 0, 1, 2, \dots\} \,.$$

1) Show that for all $n, m \in \mathbb{N}$, the intersection $U_n \cap U_m$ has the form U_k for a suitable $k \in \mathbb{N}$.

Consider the family $\mathcal{T}$ of subsets in S which consists of the empty set $\emptyset$ and all subsets U in S that can be written as a union of sets from $\{U_n | n \in \mathbb{N}\}$, i.e. $U = \cup_{\alpha \in A} U_{n_\alpha}$.

2) Show that $\mathcal{T}$ is a topology on S. (The system of subsets $\{U_n | n \in \mathbb{N}\}$ in S is called a *basis* for the topology $\mathcal{T}$.)

3) Show that the sequence $(x_n = n!)$ will converge to every point in the topological space $(S, \mathcal{T})$.

Chapter 3

Advanced Concepts in Topology

The advanced topological concepts of compactness and connectedness treated in this chapter are extremely important for applications in connection with properties of continuous functions.

We also introduce the important notion of completeness in metric spaces related to the convergence of sequences. In complete metric spaces certain natural candidates for convergent sequences (Cauchy sequences) actually do have limits. This property is decisive for many constructions in mathematical analysis.

3.1 Compact sets

We start out by stating two classical theorems concerning closed and bounded intervals in $\mathbb{R}$.

Theorem 3.1.1 (Bolzano-Weierstrass). *Every sequence (x_n) of points in a closed and bounded interval $[a, b]$ of $\mathbb{R}$ contains a convergent subsequence (x_{n_k}) with limit point x_0 in $[a, b]$.*

Theorem 3.1.2 (Heine-Borel). *Every covering of a closed and bounded interval $[a, b]$ of $\mathbb{R}$ by a system of open intervals $\{U_i \mid i \in I\}$ contains a finite subcovering.*

Theorem 3.1.2 asserts: If $\{U_i \mid i \in I\}$ is any system of open intervals in $\mathbb{R}$ such that $[a, b] \subseteq \cup_{i \in I} U_i$, then we can extract finitely many intervals from the system, say $U_{i_1}, U_{i_2}, \ldots, U_{i_n}$, such that $[a, b] \subseteq \cup_{k=1}^{n} U_{i_k}$.

Theorem 3.1.1 was anticipated by the Bohemian priest and mathematician Bernhard Bolzano (1781–1848) in 1817 and fully developed by the German mathematician Karl Weierstrass (1815–1897) in the 1860s.

The covering property in Theorem 3.1.2 was used by the German mathematician Eduard Heine (1821–1881) in 1872 in a study of uniform continuity and fully recognized as an important property by the French mathematician Emile Borel (1871–1956) in 1895 for countable open coverings, and by his pupil, the French mathematician Henri Lebesgue (1875–1941), in 1904 for general open coverings.

Proofs of the above theorems can be based on *the principle of nested intervals* (Property 1.1.1, page 4).

Proof of Theorem 3.1.1. Let (x_n) be a sequence of points in the closed and bounded interval $[a, b]$ in $\mathbb{R}$. By successively partitioning the interval at the middle we can construct an interval trap

$$[a, b] = [a_1, b_1] \supseteq [a_2, b_2] \supseteq \cdots \supseteq [a_m, b_m] \supseteq \cdots ,$$

in which every interval $[a_m, b_m]$ contains elements from the sequence (x_n) for infinitely many indices n. The interval trap determines a real number x_0 in $[a, b]$, such that

$$\{x_0\} = \bigcap_{m=1}^{\infty} [a_m, b_m] .$$

Any open neighbourhood U of x_0 contains all the intervals $[a_m, b_m]$ from a certain step m_0, and therefore contains elements from the sequence (x_n) for infinitely many indexes n. By successively choosing an element x_{n_k} from the sequence (x_n) in the open interval U_k of length $2/k$ centered at x_0 for each $k \in \mathbb{N}$, we can now extract a subsequence (x_{n_k}) with limit point x_0, thereby completing the proof. $\square$

Proof of Theorem 3.1.2. The proof is by contradiction. So assume that there exists a system of open intervals $\{U_i \mid i \in I\}$ which covers the closed and bounded interval $[a, b]$ in $\mathbb{R}$, but which does not contain a finite subcovering of $[a, b]$. By successively partitioning the interval at the middle we can construct an interval trap

$$[a, b] = [a_1, b_1] \supseteq [a_2, b_2] \supseteq \cdots \supseteq [a_m, b_m] \supseteq \cdots ,$$

in which none of the intervals $[a_m, b_m]$ can be covered by finitely many intervals from the system $\{U_i \mid i \in I\}$. The interval trap determines a real number c, such that

$$\{c\} = \bigcap_{m=1}^{\infty} [a_m, b_m] .$$

Since $c \in [a, b]$, there is an interval U_{i_0} from the system $\{U_i \mid i \in I\}$, such that $c \in U_{i_0}$. Since U_{i_0} is an open interval, the intervals $[a_m, b_m]$ are contained in U_{i_0} from a certain step m_0. This contradicts the construction of the intervals $[a_m, b_m]$. Our assumption led to a contradiction, and consequently from any covering of $[a, b]$ with open intervals we can extract a finite subcovering. $\square$

Using either Theorem 3.1.1 or Theorem 3.1.2 we can easily prove the following fundamental result.

Theorem 3.1.3. *Let $[a, b]$ be a closed and bounded interval of $\mathbb{R}$, and let $f : [a, b] \to \mathbb{R}$ be a continuous function. Then f is bounded. In other words, there exists a constant k, such that $|f(x)| \leq k$ for all $x \in [a, b]$.*

Proof by using Theorem 3.1.1. The proof is by contradiction. Hence we assume that f is not bounded. Then clearly we can find a sequence $x_1, x_2, \ldots, x_n, \ldots$ of points in $[a, b]$ with $|f(x_n)| > n$ for all $n \in \mathbb{N}$. According to Theorem 3.1.1, the sequence (x_n) contains a convergent subsequence (x_{n_k}) with limit point x_0 in $[a, b]$. Since f is continuous at x_0, there exists a $\delta > 0$, such that $|f(x)| \leq |f(x_0)| + 1$ for $|x - x_0| < \delta$. Now, choose an integer m such that $m > |f(x_0)| + 1$. Since x_0 is the limit point of the sequence (x_{n_k}) there exists an element x_{n_k} in the sequence with index $n_k > m$ such that $|x_{n_k} - x_0| < \delta$. But this yields a contradiction, since $n_k > m$ implies that $|f(x_{n_k})| > n_k > m$, and $|x_{n_k} - x_0| < \delta$ implies that $|f(x_{n_k})| \leq |f(x_0)| + 1 < m$. Hence our assumption led to a contradiction, and therefore f must be bounded. $\square$

Proof by using Theorem 3.1.2. Since f is continuous we can, for every $x \in [a, b]$, choose a $\delta_x > 0$ such that $|f(y)| \leq |f(x)| + 1$ for $|y - x| < \delta_x$. Let U_x be the open interval

$$U_x = \,]x - \delta_x, x + \delta_x[\,.$$

The system of open intervals $\{U_x \mid x \in [a, b]\}$ covers $[a, b]$. By Theorem 3.1.2 finitely many of these intervals, say $U_{x_1}, U_{x_2}, \ldots, U_{x_n}$, already cover $[a, b]$. Now, put

$$k = \max\{|f(x_1)|, |f(x_2)|, \ldots, |f(x_n)|\} + 1\,.$$

Since $U_{x_1}, U_{x_2}, \ldots, U_{x_n}$ cover $[a, b]$, there exists for every $x \in [a, b]$ at least one x_i, such that $x \in U_{x_i}$, or equivalently $|x - x_i| < \delta_{x_i}$. But then we have immediately that

$$|f(x)| \leq |f(x_i)| + 1 \leq k.$$

Thus f is bounded. $\square$

The results stated in Theorem 3.1.1 and Theorem 3.1.2 are essential, and they motivate the introduction of certain concepts that have proved to be of fundamental importance in mathematical analysis.

With the theorem of Bolzano and Weierstrass (Theorem 3.1.1) as inspiration we introduce the concept of 'sequentially compact subsets' in a metric space.

Definition 3.1.4. A subset K in a metric space M is called *sequentially compact* if every sequence (x_n) of points in K contains a convergent subsequence (x_{n_k}) with limit point x_0 in K.

We notice that, according to Theorem 3.1.1, any closed and bounded interval $[a, b]$ is a sequentially compact subset in $\mathbb{R}$.

Inspired by the Theorem of Heine and Borel (Theorem 3.1.2), we introduce the concept of a 'compact subset' in a topological space.

First we need a definition.

Definition 3.1.5. Let K be a subset of the topological space M.

By an *open covering* of K we understand a system $\{U_i \mid i \in I\}$ of open sets in M such that $K \subseteq \bigcup_{i \in I} U_i$.

By a *subcovering* of an open covering $\{U_i \mid i \in I\}$ of K we understand a subfamily of the open sets $\{U_i \mid i \in I\}$ which itself covers K.

An important issue is whether one can extract a finite subcovering from an open covering. This leads to the notion of compactness.

Definition 3.1.6. Let M be a topological space. A subset K in M is called *compact* if every open covering $\{U_i \mid i \in I\}$ of K contains a finite subcovering.

We note that, according to Theorem 3.1.2, any closed and bounded interval $[a, b]$ is a compact subset in $\mathbb{R}$.

3.2   Compact sets in Euclidean spaces

The main result of this section states that the compact subsets of $\mathbb{R}^n$ are precisely the subsets which are both closed and bounded. This way we obtain a perfect generalization of Theorem 3.1.2.

We approach this characterization of the compact subsets of $\mathbb{R}^n$ via two theorems which remain valid in more general topological spaces.

Theorem 3.2.1. *In a Hausdorff space M every compact subset K is closed.*

Proof. Let K be a compact subset in the Hausdorff space M, and consider the complement $M \setminus K$. We have to show that $M \setminus K$ is an open set in M. If $M \setminus K = \emptyset$ there is nothing to prove. So we assume that $M \setminus K \neq \emptyset$ and consider an arbitrary point $x \in M \setminus K$. Since M is a Hausdorff space, we can, for every point $y \in K$, choose a pair of disjoint open sets U_y and V_y in M, such that $x \in U_y$ and $y \in V_y$. The system of open sets $\{V_y \mid y \in K\}$ is an open covering of K. From this covering we can extract a finite subcovering, say $V_{y_1}, V_{y_2}, \ldots, V_{y_n}$, of K, since K is compact. Since $U_{y_i} \cap V_{y_i} = \emptyset$ for every $i = 1, \ldots, n$, the set $U = U_{y_1} \cap \cdots \cap U_{y_n}$ is an open neighbourhood of x in M, such that $U \subseteq M \setminus K$. According to Lemma 2.5.4, this shows that $M \setminus K$ is an open set in M, and hence that K is a closed set in M. $\square$

Definition 3.2.2. A subset K in a metric space (M, d) is called *bounded* if it is completely contained in an open ball in M.

Theorem 3.2.3. *In a metric space (M, d) every compact subset K is bounded.*

Proof. Let K be a compact subset in the metric space (M, d), and let $x_0 \in M$ be a fixed point in M. Consider the system $\{B_n(x_0) \mid n \in \mathbb{N}\}$ of open balls in M. Since every point in M is contained in these balls for n sufficiently large, it is clear that the system is an open covering of K. Since K is compact, finitely many of these balls will cover K, and hence obviously K is contained in that ball, among these finitely many balls, which has the largest radius. This proves that K is bounded. $\square$

Theorems 3.2.1 and 3.2.3 show that in an arbitrary metric space every compact subset is closed and bounded. If the n-dimensional real vector space $\mathbb{R}^n$ is considered with the usual topology induced by the Euclidean metric, then the converse is also true. Thereby we obtain the following main result.

Theorem 3.2.4 (Heine-Borel). *For a subset K in the n-dimensional Euclidean space $\mathbb{R}^n$, the following statements are equivalent:*

(1) *K is compact: Every open covering of K contains a finite subcovering.*

(2) *K is closed and bounded.*

Like Theorem 3.1.2, Theorem 3.2.4 is due to Heine and Borel, with a refinement by Lebesque. The power of the theorem lies mainly in the abstract ideas developed, and as earlier mentioned, the theorem crystallized in its final form only shortly after 1900.

Proof of Theorem 3.2.4. The implication (1)$\Rightarrow$(2) has been proved in Theorems 3.2.1 and 3.2.3.

It remains to prove (2)$\Rightarrow$(1). Assume therefore that K is a closed and bounded subset in $\mathbb{R}^n$. We shall prove that K is compact. The proof is by contradiction. Hence assume that there exists an open covering $\{U_i \mid i \in I\}$ of K from which we cannot extract a finite subcovering. As we shall see this leads to a contradiction.

Since K is bounded, we may choose an n-dimensional box

$$C^1 = [a_1^1, b_1^1] \times [a_2^1, b_2^1] \times \cdots \times [a_n^1, b_n^1]$$

such that $K \subseteq C^1$. We now partition each of the intervals $[a_i^1, b_i^1]$ at the middle, which yields 2^n new small boxes. Among these small boxes, there must be at least one, which we choose and denote by C^2, for which $C^2 \cap K$ cannot be covered by finitely many open sets from the system $\{U_i \mid i \in I\}$; cf. Figure 3.1.

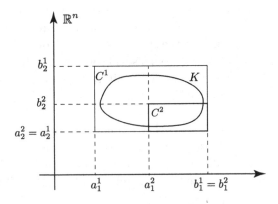

Fig. 3.1 The Heine-Borel theorem. Proving that for a closed and bounded set K in $\mathbb{R}^n$ any open covering must contain a finite subcovering.

By successively bisecting the intervals, we can construct a decreasing nested sequence of n-dimensional boxes

$$C^1 \supseteq C^2 \supseteq \cdots \supseteq C^k \supseteq \cdots,$$

in which it holds for every $k \in \mathbb{N}$ that $C^k \cap K$ cannot be covered by finitely many open sets from the system $\{U_i \mid i \in I\}$.

If we write the box C^k in the form

$$C^k = [a_1^k, b_1^k] \times [a_2^k, b_2^k] \times \cdots \times [a_n^k, b_n^k] ,$$

it is clear that for every $i = 1, \ldots n$ we get an interval trap

$$[a_i^1, b_i^1] \supseteq [a_i^2, b_i^2] \supseteq \cdots \supseteq [a_i^k, b_i^k] \supseteq \cdots .$$

This interval trap determines a real number c_i.

Now, consider the point $c = (c_1, \ldots, c_n) \in \mathbb{R}^n$. It is clear that

$$\bigcap_{k=1}^{\infty} C^k = \{c\}.$$

Every open neighbourhood of c in $\mathbb{R}^n$ therefore contains all the boxes C^k from a certain step k_0, thus in particular it contains infinitely many points from K. Indeed, if this was not the case, $C^k \cap K$ could have been covered by finitely many open sets from the system $\{U_i \mid i \in I\}$. Hence c is an accumulation point of K. Therefore $c \in K$ by Theorem 2.8.2, since K is closed.

Since $c \in K$ and the system $\{U_i \mid i \in I\}$ covers K, there must be an $i_0 \in I$ such that $c \in U_{i_0}$. From a certain step k_0, all the boxes C^k will be completely contained in the open neighbourhood U_{i_0} of c. But this contradicts that $C^k \cap K$ could not be covered by finitely many open sets from the system $\{U_i \mid i \in I\}$. This completes the proof. $\square$

Example 3.2.5. Let M be an arbitrary infinite set, and equip M with the discrete metric. Then every subset of M is both open and closed. Furthermore, it is bounded, since it is contained in any open ball of say radius 2. In particular, we note that every subset of M is both closed and bounded. However, an infinite subset A in M is not compact. For example $\{U_a = \{a\} \mid a \in A\}$ is an open covering of A which does not contain a finite subcovering. Therefore, in general, a closed and bounded subset of a metric space is not compact. This property is a particular property of the Euclidean space $\mathbb{R}^n$. ◄

3.3 Infinite subsets of compact sets

In this section we shall study the relationship between the concepts compact and sequentially compact. The main result is, that in metric spaces these concepts coincide.

First, however, we prove that compactness implies a variation of sequentially compactness valid in general topological spaces.

Theorem 3.3.1. *If K is a compact subset in a topological space M, then every infinite subset A of K has at least one accumulation point in K.*

Proof. The proof is by contradiction. Assume that A is an infinite subset of K without accumulation points in K. Then we can choose an open neighbourhood U_x around every point $x \in K$, such that U_x contains at most one point from A; namely the point x itself if $x \in A$. The open covering $\{U_x \mid x \in K\}$ does not contain a finite subcovering, since A is infinite. This contradicts the hypothesis that K is compact. Hence every infinite subset A in K has at least one accumulation point in K. $\square$

Then we are ready for the main theorem.

Theorem 3.3.2. *For a subset K in a metric space (M, d) the following statements are equivalent:*

(1) *K is compact: Every open covering of K contains a finite subcovering.*

(2) *K is sequentially compact: Every sequence (x_n) of points in K contains a convergent subsequence (x_{n_k}) with limit point x_0 in K.*

Proof. The implication (1)$\Rightarrow$(2) is an easy consequence of Theorem 3.3.1. The proof goes as follows. Let (x_n) be an arbitrary sequence in K. If the set $\{x_n\}$ is a finite set in K, at least one element in the set must be repeated for arbitrarily large indexes n and hence it is easy to extract a convergent subsequence (x_{n_k}) with this element in K as a limit point. If the set $\{x_n\}$ is an infinite set in K, it has at least one accumulation point in K, say $x_0 \in K$, according to Theorem 3.3.1. By successively choosing an element x_{n_k} from the sequence (x_n) in the open ball $B_{1/k}(x_0)$ for each $k \in \mathbb{N}$, we can now extract a subsequence (x_{n_k}) with limit point x_0, thereby completing the proof of the implication (1)$\Rightarrow$(2).

It remains to prove (2)$\Rightarrow$(1). Let therefore $\{U_i \mid i \in I\}$ be an arbitrary open covering of K. We shall prove that we can extract a finite subcovering from this covering. The case $K = \emptyset$ is trivial, so assume that $K \neq \emptyset$.

Assertion 3.3.3. *There exists a positive real number $r \in \mathbb{R}^+$, such that for every point $x \in K$ the ball $B_r(x)$ is contained in one of the sets U_i.*

Proof of Assertion 3.3.3. The proof is by contradiction. Assume therefore that for every $r \in \mathbb{R}^+$ there exists a point $x \in K$ such that $B_r(x)$ is not contained in any of the sets U_i. In particular, then, for every $n \in \mathbb{N}$, we can choose $x_n \in K$, such that $B_{1/n}(x_n)$ is not contained in any of the sets

U_i. From the sequence (x_n) in K, we can extract a convergent subsequence (x_{n_k}) with limit point $x_0 \in K$. Since $\{U_i \mid i \in I\}$ covers K, we can first choose $i_0 \in I$ such that $x_0 \in U_{i_0}$, and the set U_{i_0} being open in M, we can next choose $r_0 \in \mathbb{R}^+$ such that $B_{r_0}(x_0) \subseteq U_{i_0}$. Since x_0 is the limit point of the sequence (x_{n_k}) we can choose $m = n_k \in \mathbb{N}$, such that $x_m \in B_{r_0/2}(x_0)$ and $1/m < r_0/2$. But then, by the triangle inequality,

$$B_{1/m}(x_m) \subseteq B_{r_0}(x_0) \subseteq U_{i_0}.$$

This contradicts the assumption that $B_{1/m}(x_m)$ is not contained in any of the sets U_i, and hence proves Assertion 3.3.3. $\square$

Now, choose a positive real number $r \in \mathbb{R}^+$ in accordance with Assertion 3.3.3, and let $y_1 \in K$ be an arbitrary point in K. If $K \subseteq B_r(y_1)$, then K is covered by one of the sets from $\{U_i \mid i \in I\}$, and we are finished. Otherwise, we choose a point $y_2 \in K \backslash B_r(y_1)$. If $K \subseteq B_r(y_1) \cup B_r(y_2)$, then K is covered by two of the sets from $\{U_i \mid i \in I\}$, and again we are finished. Otherwise, we choose a point $y_3 \in K \setminus (B_r(y_1) \cup B_r(y_2))$, etc.

Assertion 3.3.4. *The process described above is finite.*

Proof Assertion 3.3.4. If the process was not finite, we would get a sequence (y_n) in K, with $d(y_n, y_m) \geq r$ for $n \neq m$. Such a sequence (y_n) does not contain any convergent subsequences, which contradicts (2). This proves Assertion 3.3.4. $\square$

According to Assertion 3.3.4, we can choose finitely many points $y_1, \ldots, y_p$ in K such that

$$K \subseteq B_r(y_1) \cup \cdots \cup B_r(y_p).$$

This proves that finitely many of the sets from the system $\{U_i \mid i \in I\}$ cover K, and hence the proof of Theorem 3.3.2 is complete. $\square$

Until now we have only considered compact subsets of a topological space. If the whole space itself is compact, then we call it a *compact topological space*. In such a space the following very useful statement holds.

Theorem 3.3.5. *Let M be a compact topological space. Then every closed subset K in M is compact.*

Proof. Let $\{U_i \mid i \in I\}$ be an arbitrary open covering of K. Since K is closed, the family of subsets $\{U_i \mid i \in I\} \cup \{M \setminus K\}$ is an open covering of M. By the compactness of M this covering contains a finite subcovering of M. Such a finite covering of M must comprise finitely many subsets from $\{U_i \mid i \in I\}$ covering K. This proves that K is compact. $\square$

3.4 Sequentially compact sets in Euclidean spaces

Since the notion of sequential compactness prevail in functional analysis and since it may be an easier concept to comprehend than compactness, we list the fundamental results about sequentially compact sets in Euclidean spaces. By Theorem 3.3.2, the results are consequences of the corresponding results for compact spaces, but for the convenience of the reader we also provide alternative proofs arguing directly with the notion of sequential compactness.

The main result of this section is that the sequentially compact subsets of the Euclidean spaces $\mathbb{R}^n$ are precisely the subsets which are both closed and bounded. This way we obtain a perfect generalization of Theorem 3.1.1.

We approach this characterization of the sequentially compact subsets of $\mathbb{R}^n$ via two theorems which are valid in arbitrary metric spaces.

Theorem 3.4.1. *In a metric space (M, d) every sequentially compact subset K is closed.*

Proof. Let K be a sequentially compact subset in the metric space (M, d). We shall prove that K is a closed set in M by proving that it contains all of its contact points; cf. Theorem 2.8.2. For that purpose, let $x \in \overline{K}$ be an arbitrary contact point of K. Since x is a contact point of K, we can choose a sequence (x_n) in K such that $d(x_n, x) < 1/n$ for all $n \in \mathbb{N}$. Since K is sequentially compact, we can extract a convergent subsequence (x_{n_k}) of the sequence (x_n) with limit point $x_0 \in K$. The subsequence (x_{n_k}) also has limit point x by the construction of the sequence (x_n). Since limit points are unique in a metric space we conclude that $x = x_0$, and hence that $x \in K$, which completes the proof that K is a closed subset of M. $\square$

Theorem 3.4.2. *In a metric space (M, d) every sequentially compact subset K is bounded.*

Proof. Let K be a sequentially compact subset in the metric space (M, d). We shall prove that K is a bounded subset in M. The proof is by contradiction. Hence assume that K is not bounded. Let $y_0 \in M$ be an arbitrarily chosen but then fixed point in M. We assume that K is not bounded, and hence we can choose a sequence (x_n) in K such that $d(y_0, x_n) > n$ for all $n \in \mathbb{N}$. Since K is sequentially compact, we can extract a convergent subsequence (x_{n_k}) of the sequence (x_n) with limit point $x_0 \in K$. Choose $k_0 \in \mathbb{N}$ such that $d(x_0, x_{n_k}) < 1$ for $k \geq k_0$. Hence for $k \geq k_0$, we get by the triangle inequality,

$$d(y_0, x_{n_k}) \leq d(y_0, x_0) + d(x_0, x_{n_k}) < d(y_0, x_0) + 1 .$$

For k sufficiently large we then have

$$n_k < d(y_0, x_{n_k}) < d(y_0, x_0) + 1 < n_k ,$$

which gives a contradiction. Hence K must be a bounded subset in M. $\square$

Theorems 3.4.1 and 3.4.2 show that in an arbitrary metric space every sequentially compact subset is closed and bounded. If the n-dimensional real vector space $\mathbb{R}^n$ is considered as a metric space with the Euclidean metric, then the converse is also true. Thereby we obtain the following main result.

Theorem 3.4.3 (Bolzano-Weierstrass). *For a subset K in the n-dimensional Euclidean space $\mathbb{R}^n$, the following statements are equivalent:*

(1) *K is sequentially compact: Every sequence (x_n) in K contains a convergent subsequence (x_{n_k}) with limit point x_0 in K.*

(2) *K is closed and bounded.*

Like Theorem 3.1.1, Theorem 3.4.3 is due to Weierstrass with a forerunner by Bolzano.

Proof of Theorem 3.4.3. The implication $(1) \Rightarrow (2)$ has been proved in Theorems 3.4.1 and 3.4.2.

It remains to prove $(2) \Rightarrow (1)$. Assume therefore that K is a closed and bounded subset in $\mathbb{R}^n$.

Let (x_p) be an arbitrary sequence in K. In order to prove that K is sequentially compact, we have to show that we can extract a convergent subsequence (x_{p_q}) from (x_p) with limit point $x_0 \in K$.

Since K is bounded, we may choose an n-dimensional box

$$C^1 = [a_1^1, b_1^1] \times [a_2^1, b_2^1] \times \cdots \times [a_n^1, b_n^1]$$

such that $K \subseteq C^1$. We now partition each of the intervals $[a_i^1, b_i^1]$ at the middle, which yields 2^n new small boxes. Among these small boxes, there must be at least one, which we choose and denote by C^2, for which $C^2 \cap K$ contains elements x_p for infinitely many indexes; cf. Figure 3.1, p. 56.

By successively bisecting the intervals, we can construct a decreasing nested sequence of n-dimensional boxes

$$C^1 \supseteq C^2 \supseteq \cdots \supseteq C^k \supseteq \cdots ,$$

in which it holds for every $k \in \mathbb{N}$ that $C^k \cap K$ contains elements x_p for infinitely many indexes.

If we write the box C^k in the form

$$C^k = [a_1^k, b_1^k] \times [a_2^k, b_2^k] \times \cdots \times [a_n^k, b_n^k] \,,$$

it is clear that for every $i = 1, \ldots n$ we get an interval trap

$$[a_i^1, b_i^1] \supseteq [a_i^2, b_i^2] \supseteq \cdots \supseteq [a_i^k, b_i^k] \supseteq \cdots \,.$$

This interval trap determines a real number c_i.

Now, consider the point $x_0 = (c_1, \ldots, c_n) \in \mathbb{R}^n$. It is clear that

$$\bigcap_{k=1}^{\infty} C^k = \{x_0\}.$$

Every open neighbourhood of x_0 in $\mathbb{R}^n$ therefore contains all the boxes C^k from a certain step k_0, thus in particular it contains points from the sequence (x_p) in K for infinitely many indexes. Hence x_0 is a contact point of K, and since K is closed, it follows that $x_0 \in K$ by Theorem 2.8.2.

By considering the system of open balls with center x_0 and radii $1/q$ for $q \in \mathbb{N}$, we can successively extract a subsequence (x_{p_q}) from (x_p) with limit point $x_0 \in K$. This completes the proof. $\square$

Example 3.4.4. Let M be an arbitrary infinite set, and equip M with the discrete metric. Then every subset of M is both open and closed. Furthermore, it is bounded, since it is contained in any open ball of say radius 2. In particular, we note that every subset of M is both closed and bounded. However, an infinite subset A in M is not sequentially compact, since sequences of mutually different elements in A are never convergent in A. Therefore, in general, a closed and bounded subset of a metric space is not sequentially compact. This is special for the Euclidean space $\mathbb{R}^n$. ◀

3.5 Completeness of metric spaces

Completeness of metric spaces is - like sequential compactness - an important topological notion related to the convergence of sequences. In complete metric spaces certain natural candidates for convergent sequences (Cauchy sequences) actually do have limits. This property is decisive for many constructions in mathematical analysis.

The French mathematician Augustin-Louis Cauchy (1789–1857), who is one of the pioneers in mathematical analysis, has formulated a convergence principle for sequences of real numbers, which has inspired the following definition.

Definition 3.5.1. A sequence (x_n) of points in (M, d) is called a *Cauchy sequence*, or a *fundamental sequence*, if for every $\varepsilon > 0$, there exists an $n_0 \in \mathbb{N}$ such that $d(x_n, x_m) < \varepsilon$ for all $n, m \geq n_0$. Or, using quantifiers:

$$\forall \varepsilon > 0 \ \exists n_0 \in \mathbb{N} \ \forall n, m \in \mathbb{N} : \qquad n, m \geq n_0 \quad \Rightarrow \quad d(x_n, x_m) < \varepsilon.$$

An equivalent, though in certain contexts more convenient, formulation of the concept of a Cauchy sequence can be expressed as follows using quantifiers:

$$\forall \varepsilon > 0 \ \exists n_0 \in \mathbb{N} \ \forall n, k \in \mathbb{N} : \qquad n \geq n_0 \quad \Rightarrow \quad d(x_n, x_{n+k}) < \varepsilon.$$

Theorem 3.5.2. *Every convergent sequence* (x_n) *in* (M, d) *is a Cauchy sequence.*

Proof. Assume that (x_n) is a convergent sequence in (M, d) with limit point $x_0 \in M$. Then we have:

$$\forall \varepsilon > 0 \ \exists n_0 \in \mathbb{N} \ \forall n \in \mathbb{N} : \qquad n \geq n_0 \quad \Rightarrow \quad d(x_n, x_0) < \frac{\varepsilon}{2}.$$

Since $d(x_n, x_m) \leq d(x_n, x_0) + d(x_0, x_m)$ according to the triangle inequality, this immediately yields:

$$\forall \varepsilon > 0 \ \exists n_0 \in \mathbb{N} \ \forall n, m \in \mathbb{N} : \qquad n, m \geq n_0 \quad \Rightarrow \quad d(x_n, x_m) < \varepsilon.$$

This shows that (x_n) is a Cauchy sequence. $\square$

The converse to Theorem 3.5.2 is not true. It depends on the metric space whether *all* Cauchy sequences in the space are convergent.

Example 3.5.3. Consider an arbitrary metric space (M', d') containing a convergent sequence (x_n) of pairwise different elements with limit point y_0. Remove the point y_0 from M' to form the subspace $M = M' \setminus \{y_0\}$. Equip M with the metric d induced from d'. Then (x_n) is a Cauchy sequence in the metric space (M, d), which is not convergent to a point in M. ◄

Metric spaces in which all Cauchy sequences are convergent are, however, sufficiently abundant and interesting that we introduce the following definition.

Definition 3.5.4. A metric space (M, d) is said to be *complete* if every Cauchy sequence in (M, d) is convergent.

Although Example 3.5.3 may seem artificial, it is the typical way that a metric space can fail to be complete.

An important result from classical analysis can then be expressed as follows.

Theorem 3.5.5 (The Cauchy condition for sequences). *The set $\mathbb{R}$ of real numbers with the usual metric is a complete metric space.*

Proof. Let (x_n) be an arbitrary Cauchy sequence in $\mathbb{R}$. Choose $n_0 \in \mathbb{N}$ such that $|x_n - x_m| < 1$ for $n, m \geq n_0$. Set

$$a = \max \left\{ |x_1|, \ldots, |x_{n_0-1}|, |x_{n_0}| + 1 \right\}.$$

Then $x_n \in [-a, a]$ for all $n \in \mathbb{N}$.

By Theorem 3.1.1, the sequence (x_n), considered as a sequence in $[-a, a]$, contains a subsequence (x_{n_k}) with limit point $x_0 \in [-a, a]$. Since (x_n) is a Cauchy sequence clearly the full sequence (x_n) must also be convergent with limit point x_0, thereby completing the proof. $\square$

By a similar argument we can prove the following theorem.

Theorem 3.5.6. *A sequentially compact metric space (M, d) is complete.*

Proof. Let (x_n) be an arbitrary Cauchy sequence in (M, d). Since (M, d) is sequentially compact, (x_n) contains a subsequence (x_{n_k}) with limit point $x_0 \in M$. Being a Cauchy sequence, clearly the full sequence (x_n) is also convergent with limit point $x_0 \in M$, thereby completing the proof. $\square$

The next result shows that completeness is inherited by closed subsets.

Theorem 3.5.7. *Let (M, d) be a complete metric space, and let A be a closed subset in (M, d). Then A with the metric induced from (M, d) is a complete metric space.*

Proof. Let (x_n) be an arbitrary Cauchy sequence in A. Then (x_n) is also a Cauchy sequence in (M, d), and since (M, d) is a complete metric space, (x_n) has a limit point $x_0 \in M$. Since A is closed in M, the limit point $x_0 \in A$. Thus (x_n) has a limit point in A, as should be proved. $\square$

As an illustration of the use of completeness we finish this section by stating and proving an important fixed point theorem for contracting mappings in complete metric spaces. In its final form, the theorem was formulated by Banach.

Definition 3.5.8. A mapping $T : M \to M$ in a metric space (M, d) is called a *contraction* if there is a real number λ with $0 \leq \lambda < 1$, for which

$$d(Tx, Ty) \leq \lambda d(x, y) \qquad \text{for all} \quad x, y \in M.$$

The number λ is called a *contraction factor*.

Remark 3.5.9. To simplify notation one often writes Tx for the image of a point $x \in M$ under a contraction $T : M \to M$ instead of the usual $T(x)$. ◁

Clearly a contraction $T : M \to M$ is continuous.

Together with the contraction $T : M \to M$ we shall consider its *iterates*. By this we understand the family of mappings $T^n : M \to M$ defined inductively by setting $T^2 = T \circ T$, and in general $T^n = T \circ T^{n-1}$. Note that if T is a contraction with contraction factor $\lambda \in [0, 1[$, then T^n is a contraction with contraction factor $\lambda^n \in [0, 1[$, since for all $x, y \in M$ we have the inequalities

$$d\left(T^n x, T^n y\right) = d\left(T(T^{n-1}x), T(T^{n-1}y)\right)$$
$$\leq \lambda d(T^{n-1}x, T^{n-1}y)$$
$$\vdots$$
$$\leq \lambda^n d(x, y).$$

A *fixed point* for $T : M \to M$ is a point $x_0 \in M$ satisfying $Tx_0 = x_0$.

Now we are ready to prove the fixed point theorem for contractions in complete metric spaces.

Theorem 3.5.10 (The Banach Fixed Point Theorem). *Every contraction* $T : M \to M$ *in a complete metric space* (M, d) *has exactly one fixed point* $x_0 \in M$.

The fixed point $x_0 \in M$ *is arrived at in the limit of iteration by* T *of any point* $x \in M$, *i.e.*

$$T^n x \to x_0 \quad for \quad n \to \infty.$$

Proof. First we prove uniqueness. Assume for that purpose that x_0 and x_1 are fixed points for T. Then we have

$$0 \leq d(x_0, x_1) = d(Tx_0, Tx_1) \leq \lambda d(x_0, x_1).$$

Since $\lambda \in [0, 1[$, this can only be satisfied when $d(x_0, x_1) = 0$, and hence $x_0 = x_1$. Therefore the contraction T has at most one fixed point.

Now to existence. Let $x \in M$ be an arbitrarily chosen point in M and consider the sequence $(T^n x)$ in M.

For any $n \in \mathbb{N}$, we have

$$d(T^n x, T^{n+1} x) = d\left(T^n x, T^n(Tx)\right) \leq \lambda^n d(x, Tx),$$

since T^n is a contraction with contraction factor $\lambda^n \in [0, 1[$.

Making repeated use of the triangle inequality followed by the above inequality, we then get, for arbitrary $n, k \in \mathbb{N}$:

$$d(T^n x, T^{n+k} x) \leq d(T^n x, T^{n+1} x) + d(T^{n+1} x, T^{n+k} x)$$

$$\vdots$$

$$\leq d(T^n x, T^{n+1} x) + d(T^{n+1} x, T^{n+2} x) + \cdots$$
$$+ d(T^{n+k-1} x, T^{n+k} x)$$
$$\leq (\lambda^n + \lambda^{n+1} + \cdots + \lambda^{n+k-1}) d(x, Tx)$$
$$= \lambda^n \frac{1 - \lambda^k}{1 - \lambda} d(x, Tx)$$
$$\leq \frac{\lambda^n}{1 - \lambda} d(x, Tx).$$

Since $\lambda \in [0, 1[$, clearly

$$\frac{\lambda^n}{1 - \lambda} d(x, Tx) \to 0 \quad \text{for} \quad n \to \infty.$$

Hence the above inequality for $d(T^n x, T^{n+k} x)$ shows that $(T^n x)$ is a Cauchy sequence in (M, d). Since (M, d) is a complete metric space, there exists therefore a point $x_0 \in M$ such that

$$T^n x \to x_0 \quad \text{for} \quad n \to \infty.$$

By the continuity of T it follows that

$$T^{n+1} x \to T x_0 \quad \text{for} \quad n \to \infty.$$

This shows that both x_0 and $T x_0$ are limit points for the sequence $(T^n x)$. In a metric space, a sequence can have at most one limit point and therefore $T x_0 = x_0$. Hence the point $x_0 \in M$ is indeed a fixed point for T, and thus the contraction T has at least one fixed point.

From the above it follows that the contraction T has exactly one fixed point $x_0 \in M$. Furthermore, it follows that $x_0 \in M$ can be determined as the limit point for the sequence $(T^n x)$ generated by an arbitrary point $x \in M$. This completes the proof of Theorem 3.5.10. $\square$

3.6 Continuous mappings of compact sets

As will soon be clear, the following theorem is a fundamental result in mathematical analysis.

Theorem 3.6.1. *Let $f : X \to Y$ be a continuous mapping between topological spaces X and Y, and let $K \subseteq X$ be a compact subset in X. Then the image $f(K) \subseteq Y$ of K by f is a compact subset in Y.*

Proof. Let $\{V_i \mid i \in I\}$ be an arbitrary open covering of $f(K)$ in Y. We need to prove that we can extract a finite subcovering from it. Since $f : X \to Y$ is continuous, $\{f^{-1}(V_i) \mid i \in I\}$ is an open covering of K in X. Since K is compact, it contains a finite subcovering, say $f^{-1}(V_{i_1}), \ldots, f^{-1}(V_{i_n})$, of K. Clearly, then, $V_{i_1}, \ldots, V_{i_k}$ covers $f(K)$. Thus we have shown that $\{V_i \mid i \in I\}$ contains a finite subcovering of $f(K)$ proving that $f(K)$ is compact. $\square$

If for $Y = \mathbb{R}$ we combine the above theorem with Theorem 3.2.4 (Heine-Borel), we get the following perfect generalization of the fundamental theorem from classical analysis stating that any continuous function on a closed and bounded interval is bounded; cf. Theorem 3.1.3.

Theorem 3.6.2. *Let $f : X \to \mathbb{R}$ be a continuous function on the topological space X. Then f is bounded in every compact subset K in X. In other words: For every compact subset K in X, there exists a constant $k \in \mathbb{R}^+$ such that $|f(x)| \leq k$ for all $x \in K$.*

Proof. According to Theorem 3.6.1, $f(K)$ is a compact subset in $\mathbb{R}$, and according to Theorem 3.2.4, $f(K)$ is therefore bounded. $\square$

In the proof of Theorem 3.6.2 we did not exploit the full power of Theorem 3.2.4. In fact, the set $f(K)$ is not only a bounded subset but also a closed subset in $\mathbb{R}$. Since $f(K)$ is bounded both the infimum $k_1 = \inf f(K)$ and the supremum $k_2 = \sup f(K)$ of $f(K)$ exist. Clearly, the numbers k_1 and k_2 are contact points for $f(K)$ in $\mathbb{R}$, and since $f(K)$ is a closed subset in $\mathbb{R}$, they have to belong to $f(K)$. Therefore there exist points $x_1, x_2 \in K$, such that $f(x_1) = k_1$ and $f(x_2) = k_2$. It is clear, that $f(x_1)$, respectively $f(x_2)$, is the minimum value, respectively the maximum value that f attains in K. This proves the following sharpening of Theorem 3.6.2.

Theorem 3.6.3. *Let $f : X \to \mathbb{R}$ be a continuous function on the topological space X. Then f attains both a maximum value and a minimum value in every compact subset K of X.*

Theorem 3.6.2 and Theorem 3.6.3 both have direct analogues in metric spaces substituting compactness with sequential compactness. The key to proving these results is the following easy theorem.

Theorem 3.6.4. *Let $f : X \to Y$ be a continuous mapping between metric spaces X and Y. If $K \subseteq X$ is a sequentially compact subset in X, then the image $f(K) \subseteq Y$ of K by f is a sequentially compact subset in Y.*

Proof. Let $(f(x_n))$ be an arbitrary sequence in the image $f(K) \subseteq Y$ of $K \subseteq X$ by $f : X \to Y$. Since K is a sequentially compact subset in X, we can extract a convergent subsequence (x_{n_k}) from the sequence (x_n) with limit point $x_0 \in K$. By Theorem 2.9.3, the subsequence $f(x_{n_k})$ of the sequence $(f(x_n))$ therefore has limit point $f(x_0) \in f(K)$. Since $(f(x_n))$ was an arbitrarily chosen sequence in $f(K)$ this proves that $f(K)$ is sequentially compact. $\square$

We finish this section by considering the concept of uniform continuity.

Definition 3.6.5. Let (X, d_X) and (Y, d_Y) be metric spaces. A mapping $f : X \to Y$ is said to be *uniformly continuous* in a subset K of X if

$$\forall \varepsilon > 0 \ \exists \delta > 0 \ \forall x, y \in K : \quad d_X(x, y) < \delta \quad \Rightarrow \quad d_Y(f(x), f(y)) < \varepsilon .$$

For a given $\varepsilon > 0$, continuity of f in K provides for every $x \in K$ a $\delta_x > 0$ that can be used in a neighbourhood of x only. Or, to be more precise, such that $d_Y(f(x), f(y)) < \varepsilon$ for $y \in K$ with $d_X(x, y) < \delta_x$. Uniform continuity on the other hand provides a $\delta > 0$ that can be used everywhere in K.

We have the following useful result on uniform continuity.

Theorem 3.6.6. *Let $f : X \to Y$ be a continuous mapping between metric spaces (X, d_X) and (Y, d_Y). Then f is uniformly continuous in every compact subset K of X.*

Proof. Let K be a compact subset of X. For a given $\varepsilon > 0$ we shall provide a $\delta > 0$, which can be used all over K as required by Definition 3.6.5.

So let $\varepsilon > 0$ be given. Since f is continuous, we can for every $x \in K$ choose a $\delta_x > 0$, such that $d_Y(f(x), f(y)) < \varepsilon/2$ for all $y \in K$ with $d_X(x, y) < \delta_x$. The system of open balls $\{B_{\delta_x/2}(x) \mid x \in K\}$ is an open covering of K. Since K is compact, this covering contains a finite subcovering, and hence there exist finitely many points $x_1, \ldots, x_n$ in K, such that $B_{\delta_{x_1}/2}(x_1), \ldots, B_{\delta_{x_n}/2}(x_n)$ cover K. Now, set

$$\delta = \min \{\delta_{x_1}/2, \ldots, \delta_{x_n}/2\} .$$

We claim that this $\delta > 0$ works. Indeed, consider any pair of points $x, y \in K$ with $d_X(x, y) < \delta$. For at least one of the finitely many points $x_1, \ldots, x_n$, say x_i, we have $d_X(x_i, x) < \delta_{x_i}/2$. The triangle inequality then gives

$$d_X(x_i, y) \leq d_X(x_i, x) + d_X(x, y) < \delta_{x_i}.$$

From the way δ_{x_i} was determined, we have

$$d_Y(f(x_i), f(x)) < \varepsilon/2 \quad \text{and} \quad d_Y(f(x_i), f(y)) < \varepsilon/2.$$

The triangle inequality now immediately implies

$$d_Y(f(x), f(y)) < \varepsilon,$$

which is all we had to show. This completes the proof. $\square$

3.7 Homeomorphisms

Recall that a mapping $f : X \to Y$ between sets X and Y is said to be *injective* if for any pair of points $x_1, x_2 \in X$,

$$f(x_1) = f(x_2) \quad \Rightarrow \quad x_1 = x_2,$$

or, equivalently,

$$x_1 \neq x_2 \quad \Rightarrow \quad f(x_1) \neq f(x_2).$$

The mapping f is said to be *surjective* if for every point $y \in Y$ there exists at least one point $x \in X$ such that $f(x) = y$.

If f is both injective and surjective, f is said to be *bijective*. A bijective mapping $f : X \to Y$ has an *inverse mapping*, denoted by $f^{-1} : Y \to X$. For $y \in Y$, we define $f^{-1}(y)$ to be the unique $x \in X$ for which $f(x) = y$.

Let $1_X : X \to X$, respectively $1_Y : Y \to Y$, denote the identity mapping (the identity) of X, respectively Y, defined by $1_X(x) = x$ for all $x \in X$, respectively $1_Y(y) = y$ for all $y \in Y$. Bijectivity can then be given the following alternative formulation.

Lemma 3.7.1. *A mapping $f : X \to Y$ between sets X and Y is bijective if and only if there exists a mapping $g : Y \to X$ such that*

$$g \circ f = 1_X \quad and \quad f \circ g = 1_Y.$$

If this is the case, then $g = f^{-1}$.

Proof. The condition $g \circ f = 1_X$ ensures that f is injective, and the condition $f \circ g = 1_Y$ ensures that f is surjective. $\square$

Now, let us return to topological spaces and continuous mappings.

Definition 3.7.2. Let X and Y be topological spaces. A *homeomorphism* is a bijective mapping $f : X \to Y$ for which both f and the inverse mapping f^{-1} are continuous.

Two topological spaces X and Y are said to be *homeomorphic*, or, *topologically equivalent*, if there exists a homeomorphism between them.

We note that a continuous mapping $f : X \to Y$ is a homeomorphism if and only if there exists a continuous mapping $g : Y \to X$ such that

$$g \circ f = 1_X \quad \text{and} \quad f \circ g = 1_Y.$$

We also note that if $f : X \to Y$ is a homeomorphism then it defines a bijective correspondence between the open sets in X and the open sets in Y. In particular, we would like to stress that the continuity of f^{-1} implies that if U is an open set in X then its image $f(U)$ is an open set in Y. Therefore the topological structures in two homeomorphic topological spaces X and Y are indistinguishable. In other words, homeomorphic topological spaces have the same topological properties. For instance, either both spaces satisfy the Hausdorff Axiom or neither of them do.

Often, it is difficult to determine the inverse mapping $f^{-1} : Y \to X$ of a bijective mapping $f : X \to Y$ explicitly, and hence it may be difficult to check whether it is continuous. But sometimes it follows automatically. In the theorem below, we show that if X is compact and Y is a Hausdorff space, then the inverse mapping $f^{-1} : Y \to X$ for a continuous, bijective mapping $f : X \to Y$ is always continuous. This result is very useful.

Yet, there are important situations, for example in connection with function spaces, in which a mapping is continuous and bijective but the inverse mapping fails to be continuous. One gets a trivial example of this phenomenon by considering $X = \mathbb{R}$ with the discrete metric and $Y = \mathbb{R}$ with the Euclidean metric. Then the identity map $1_\mathbb{R}$ on $\mathbb{R}$ is continuous, when considered as a mapping from X to Y, but it is not continuous when considered as a mapping from Y to X.

Theorem 3.7.3. *Let X and Y be topological spaces, where X is compact and Y is Hausdorff. If $f : X \to Y$ is a continuous, bijective mapping then the inverse mapping $f^{-1} : Y \to X$ is continuous. In this case, $f : X \to Y$ is therefore a homeomorphism.*

Proof. We shall prove that f^{-1} is continuous. According to Theorem 2.7.3, it suffices to show that $(f^{-1})^{-1}(A) = f(A)$ is a closed set in Y for any closed set A in X. Therefore, let A be a closed set in X. Since X is compact, it follows by Theorem 3.3.5, that A is a compact subset of X. Then Theorem 3.6.1 implies that $f(A)$ is a compact subset of Y since f is continuous. Finally, $f(A)$ is therefore closed in the Hausdorff space Y by Theorem 3.2.1. This completes the proof. $\square$

The next example demonstrates a typical way of using Theorem 3.7.3.

Example 3.7.4. Let $S^1 = \{(\cos\theta, \sin\theta) \mid \theta \in \mathbb{R}\}$ be the unit circle in the plane equipped with the induced topology. We shall give an alternative description of S^1.

In $\mathbb{R}$ we introduce the relation $\sim$ by setting $\theta \sim \theta'$, if $\theta - \theta'$ is an integral multiple of 2π; cf. Figure 3.2.

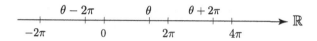

Fig. 3.2 An alternative description of the circle S^1 as a quotient space.

It is easy to show that $\sim$ is an equivalence relation in $\mathbb{R}$. This way the points of $\mathbb{R}$ are divided into equivalence classes of points, which are separated from each other by multiples of 2π. Note that every equivalence class has exactly one representative in the interval $[0, 2\pi[$. Let $\tilde{\mathbb{R}}$ denote the set of equivalence classes in $\mathbb{R}$ under $\sim$, and let $q : \mathbb{R} \to \tilde{\mathbb{R}}$ be the mapping which maps $\theta \in \mathbb{R}$ into its equivalence class $q(\theta) \in \tilde{\mathbb{R}}$. We equip $\tilde{\mathbb{R}}$ with the quotient topology induced from $\mathbb{R}$ by the mapping q. Our objective is to show that $\tilde{\mathbb{R}}$ and S^1 are homeomorphic topological spaces.

Define the mapping $p : \mathbb{R} \to S^1$ by $p(\theta) = (\cos\theta, \sin\theta)$ for $\theta \in \mathbb{R}$. Since Cosine and Sine are periodic functions of period 2π, it is easy to see that $p : \mathbb{R} \to S^1$ induces a mapping $\tilde{p} : \tilde{\mathbb{R}} \to S^1$ such that $p = \tilde{p} \circ q$. The situation is illustrated by the diagram:

Clearly $\tilde{p}$ is bijective. Since $\tilde{\mathbb{R}}$ is equipped with the quotient topology induced from $\mathbb{R}$ by the mapping q, and since p is continuous, it is easy to show that $\tilde{p}$ is also continuous. Since $\tilde{\mathbb{R}}$ is compact (because $\tilde{\mathbb{R}}$ is the image of the closed and bounded interval $[0, 2\pi]$ under the continuous mapping q), and since S^1 is a Hausdorff space (it is even metric), it follows by Theorem 3.7.3 that $\tilde{p}$ is a homeomorphism.

In other words, the topological spaces S^1 and $\tilde{\mathbb{R}}$ are homeomorphic. ◀

3.8   Connected sets

Informally speaking, a topological space is said to be connected if it does not consist of several disjoint pieces. However, we need to be more precise to make this into a rigorous definition. First a small preparation.

Lemma 3.8.1. *Let M be a topological space. Then the following statements are equivalent:*

(1) $\emptyset$ *and M are the only subsets of M which are both open and closed.*

(2) *If U_1 and U_2 are open sets in M which partition M into disjoint pieces, i.e. $M = U_1 \cup U_2$ and $U_1 \cap U_2 = \emptyset$, then either $U_1 = \emptyset$ or $U_2 = \emptyset$.*

(3) *If A_1 and A_2 are closed sets in M which partition M into disjoint pieces, i.e. $M = A_1 \cup A_2$ and $A_1 \cap A_2 = \emptyset$, then either $A_1 = \emptyset$ or $A_2 = \emptyset$.*

Proof. First we prove (1)$\Rightarrow$(2). Assume that U_1 and U_2 are open sets in M which partition M into disjoint pieces. Since $U_1 = M \setminus U_2$, we see that U_1 is also closed. From (1) it follows that either $U_1 = \emptyset$ or $U_1 = M$, but in the latter case $U_2 = M \setminus U_1 = \emptyset$.

Next we prove (2)$\Rightarrow$(3). Assume that A_1 and A_2 are closed sets in M which partition M into disjoint pieces. If we set $U_1 = M \setminus A_1$ and $U_2 = M \setminus A_2$ it is clear that U_1 and U_2 are open sets in M which partition M into disjoint pieces. According to (2) either $U_1 = \emptyset$, in which case $A_2 = \emptyset$, or $U_2 = \emptyset$, in which case $A_1 = \emptyset$.

Finally we prove (3)$\Rightarrow$(1). Assume that W is a subset of M which is both open and closed. If we set $A_1 = W$ and $A_2 = M \setminus W$, then it is clear that A_1 and A_2 are closed sets in M which partition M into disjoint pieces. Hence by (3), either $A_1 = \emptyset$ (thus $W = \emptyset$) or $A_2 = \emptyset$ (thus $W = M$).

Since all implications in the following cycle (1) $\Rightarrow$ (2) $\Rightarrow$ (3) $\Rightarrow$ (1) have now been proved, the proof is completed. $\square$

Definition 3.8.2. A topological space M, which satisfies one of the equivalent conditions (1), (2) or (3) in Lemma 3.8.1, and therefore all of them, is called *connected*.

A subset S in M is called connected if S is connected, when equipped with the topology induced from M, i.e. the open sets in S are obtained from the open sets in M by the intersection with S.

After this slightly abstract definition the following result is reassuring.

Lemma 3.8.3. *A closed and bounded interval* $[a, b]$ *in* $\mathbb{R}$ *is connected.*

Proof. First, note that since $[a, b]$ is a closed set in $\mathbb{R}$, any closed subset in $[a, b]$ equipped with the subspace topology, is also a closed set in $\mathbb{R}$.

The proof is by contradiction. Assume that $[a, b]$ is not connected. Then we can find closed sets A_1 and A_2 in $\mathbb{R}$, where $[a, b] = A_1 \cup A_2$, $A_1 \cap A_2 = \emptyset$, $A_1 \neq \emptyset$, and $A_2 \neq \emptyset$. Possibly after interchanging A_1 and A_2 we may assume that $b \in A_2$. Since A_1 is bounded from above, the number $c = \sup A_1$ exists. Now A_1 is closed, and hence $c \in A_1$. Since b is an upper bound for A_1, and $b \notin A_1$, we have $c < b$. But then $]c, b] \subseteq A_2$, since $c = \sup A_1$. Hence any open interval around c contains points from A_2. This shows that c is a contact point of A_2, and therefore $c \in A_2$, since A_2 is closed. Altogether $c \in A_1 \cap A_2$, which clearly contradicts that $A_1 \cap A_2 = \emptyset$. The assumption that $[a, b]$ is not connected must therefore be rejected, and the lemma follows. $\square$

Connectivity is - like compactness - a property of topological spaces that are preserved by continuous mappings. This is the content of the next theorem.

Theorem 3.8.4. *Let* $f : X \to Y$ *be a continuous mapping between topological spaces* X *and* Y, *and let* $S \subseteq X$ *be a connected subset in* X. *Then the image* $f(S) \subseteq Y$ *of* S *under* f *is a connected subset in* Y.

Proof. A subset in a topological space is connected, if it is connected when considered as a topological space equipped with the subspace topology. Hence it suffices to consider the topological spaces $X' = S$ and $Y' = f(S)$. The restriction of f to X' defines a continuous, surjective mapping $f' : X' \to Y'$ of X' onto Y'. Therefore it suffices to prove that if X' is connected then Y' is also connected.

Now the proof is by contradiction. Assume therefore that Y' is not connected. Then there exist open sets V_1' and V_2' in Y', such that $Y' = V_1' \cup V_2'$, $V_1' \cap V_2' = \emptyset$, $V_1' \neq \emptyset$, and $V_2' \neq \emptyset$. By the continuity and surjectivity of $f' : X' \to Y'$, the sets $U_1' = f'^{-1}(V_1')$ and $U_2' = f'^{-1}(V_2')$ constitutes a similar partitioning of X' into two disjoint, non-empty, open sets. This contradicts X' being connected. But then Y' must be connected, as was to be proved. $\square$

The following is perhaps a more intuitive concept of connectivity.

Definition 3.8.5. A topological space M is called *pathwise connected* if every pair of points $x, y \in M$ can be joined by a curve (path); i.e. if there is a continuous mapping $\varphi : [0, 1] \to M$, for which $\varphi(0) = x$ and $\varphi(1) = y$.

Theorem 3.8.6. *A pathwise connected topological space M is connected.*

Proof. Assume that M is pathwise connected. Let $W \subseteq M$ be a subset of M which is both open and closed, and assume that $W \neq \emptyset$. According to the first of the equivalent definitions of connectivity, we have to prove that $W = M$.

Choose a fixed point $x \in W$, and let $y \in M$ be an arbitrary point in M. Since M is pathwise connected, there exists a continuous mapping $\varphi : [0, 1] \to M$, such that $\varphi(0) = x$ and $\varphi(1) = y$. By the continuity of φ, the preimage $\varphi^{-1}(W)$ of W under φ is both open and closed in $[0, 1]$. Since $0 \in \varphi^{-1}(W)$, the set $\varphi^{-1}(W) \neq \emptyset$, and therefore $\varphi^{-1}(W) = [0, 1]$ since $[0, 1]$ is connected by Lemma 3.8.3. Hence $y = \varphi(1) \in W$. This shows that $W = M$, and therefore that M is connected. $\square$

As the following example shows, a connected topological space need not be pathwise connected.

Example 3.8.7. Consider in $\mathbb{R}^2$ the subset $S = L \cup E$, where

$$L = \left\{ (x_1, x_2) \in \mathbb{R}^2 \mid x_1 = 0, \ -1 \leq x_2 \leq 1 \right\},$$

and

$$E = \left\{ (x_1, \cos(1/x_1)) \in \mathbb{R}^2 \mid x_1 > 0 \right\};$$

cf. Figure 3.3.

Intuitively it is clear that $S = L \cup E$ is not pathwise connected, but S is actually connected.

That S is connected can be seen by noticing that S is exactly the closure of the pathwise connected, and therefore also connected, set E. Then the rest follows from a general result which states: *In an arbitrary topological space M, the closure $\overline{E}$ of a connected subset E in M is also connected.* We leave the proof of this statement to the reader. ◄

As an extension of the result stated in Lemma 3.8.3, we can now characterize the connected subsets of the real axis.

Theorem 3.8.8. *A subset S in $\mathbb{R}$ is connected if and only if it is an interval in the extended sense (i.e. half-lines and $\mathbb{R}$ itself are included).*

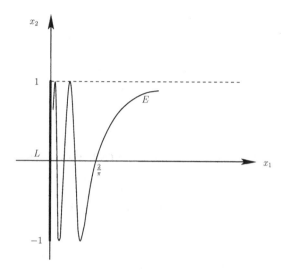

Fig. 3.3   An example of a set which is connected but not pathwise connected.

Proof. First, assume that S is connected. We shall prove that S is an interval in the extended sense. Assume that this is not the case. Then we can find points $x, y, z \in \mathbb{R}$, such that $x < y < z$, where $x, z \in S$ but $y \notin S$. Consider the sets

$$U_1 =]-\infty, y[\cap S \quad \text{and} \quad U_2 =]y, +\infty[\cap S.$$

Since S is equipped with the topology induced from $\mathbb{R}$, the sets U_1 and U_2 are open sets in S with $S = U_1 \cup U_2$ and $U_1 \cap U_2 = \emptyset$. From $x \in U_1$ and $z \in U_2$ it follows that we have a partitioning of S into disjoint, nonempty, open sets, which contradicts the connectivity of S. Therefore S is an interval in the extended sense.

Next, assume that S is an interval in the extended sense. Then S is clearly pathwise connected, and hence connected. $\square$

As a result of our efforts, we can now prove the following perfect generalization of the well-known and useful theorem from classical analysis stating that under a continuous function $f : [a, b] \to \mathbb{R}$, an interval $[a, b]$ is mapped onto an interval $[c, d]$.

Theorem 3.8.9. *Let $f : M \to \mathbb{R}$ be a continuous function defined on a compact, connected topological space M. Then the image $f(M)$ of M under f is a closed and bounded interval $[c, d]$ in $\mathbb{R}$.*

Proof. Using Theorem 3.8.4 it follows by the connectivity of M that $f(M)$ is a connected subset in $\mathbb{R}$, and hence according to Theorem 3.8.8 an interval in the extended sense.

Since M is compact, $f(M)$ is also compact according to Theorem 3.6.1, and thus by Theorem 3.2.4, $f(M)$ is a closed and bounded subset of $\mathbb{R}$.

Altogether, $f(M)$ is a closed and bounded interval $[c, d]$ as asserted. $\square$

Exercises and Further Results

Exercise 3.1. Let $f : \mathbb{R} \to \mathbb{R}$ be a real-valued function in the real variable $x \in \mathbb{R}$. Suppose that f is continuous and that $f(x)$ is convergent at infinity, i.e. for x going to $\pm\infty$. Prove that the function f is bounded.

Exercise 3.2. Let M be a topological space with the discrete topology. Prove that a subset K in M is compact if and only if K is a finite set.

Exercise 3.3. Construct an open covering of the open interval $]0, 1[$ that does not contain a finite subcovering.

Exercise 3.4. Consider the half-open interval $M =]0, 1]$. Show that (M, d) is not a complete metric space, when $d(x, y) = |x - y|$ is the Euclidean metric.

Exercise 3.5. Consider the half-open interval $M =]0, 1]$.
  Define the function $\widehat{d} : M \times M \to \mathbb{R}_0^+$ by

$$\widehat{d}(x, y) = |\log x - \log y|.$$

1) Show that $\widehat{d}$ is a metric on M.
2) Show that a subset $U \subseteq M$ is open with respect to d if and only if it is open with respect to $\widehat{d}$.
   Hint: The functions log and exp are both continuous.
3) Show that $(M, \widehat{d})$ is a complete metric space.

Exercise 3.6. Prove that if (X, d_X) and (Y, d_Y) are complete metric spaces, then the metric space $(X \times Y, d_{X \times Y})$ defined in Exercise 2.14, is also a complete metric space.

Exercise 3.7. Let $X =]0, 1]$ and $Y =]-\infty, 0]$. Show that by restriction to X, the function $\log : \mathbb{R}^+ \to \mathbb{R}$ defines a homeomorphism $f : X \to Y$.

Exercise 3.8. Let $f : X \to Y$ be a continuous injective mapping between two metric spaces. Prove that if $K \subseteq X$ is compact then $f|_K : K \to f(K)$ is a homeomorphism.

Exercise 3.9. Determine which of the following sets are pathwise connected:

1) $]0, 1[$.
2) $\mathbb{R} \setminus \mathbb{Z}$.
3) $\mathbb{Q}$.
4) $[0, 1]^2 \setminus \{(0, 0)\}$.
5) $\mathbb{R}^2 \setminus \mathbb{Q} \times \{0\}$.
6) $\{(x, \frac{1}{x}) \in \mathbb{R}^2 \mid x > 0\} \cup \{0\} \times \mathbb{R} \cup \mathbb{R} \times \{0\}$.
7) An open ball in n-dimensional Euclidean space.
8) The boundary of an open ball in n-dimensional Euclidean space.

Exercise 3.10. Let $A, B \subseteq M$ be pathwise connected subsets of a topological space M. Show that if $A \cap B \neq \emptyset$ then $A \cup B$ is pathwise connected.

Exercise 3.11. Let (M, d) be a metric space. For an arbitrary non-empty subset W in M, we define a function $\varphi : M \to \mathbb{R}$ by

$$\varphi(x) = \inf\{d(x, y) | y \in W\} \quad \text{for} \quad x \in M.$$

We call $\varphi(x)$ the *distance* from x to W, and write, in accordance with this, $\varphi(x) = d(x, W)$.

1) Let $x_1, x_2 \in M$ be arbitrary points in M.
First show that for an arbitrary point $y \in W$ it holds that

$$\varphi(x_1) \leq d(x_1, x_2) + d(x_2, y).$$

Next show that

$$|\varphi(x_1) - \varphi(x_2)| \leq d(x_1, x_2),$$

and conclude from this that φ is (uniformly) continuous on M.

2) Show that

$$d(x, W) = \varphi(x) = 0 \iff x \in \overline{W},$$

where $\overline{W}$ as usual denotes the closure of W.

3) Let A_1 and A_2 be disjoint, non-empty closed subsets in the metric space (M, d). Show that there exist disjoint, open sets U_1 and U_2 in M, such that $A_1 \subseteq U_1$ and $A_2 \subseteq U_2$.
<u>Hint:</u> You can make use of the distance functions $\varphi_1(x) = d(x, A_1)$ and $\varphi_2(x) = d(x, A_2)$.

Exercise 3.12. Let (M, d) be a metric space. For an arbitrary pair of non-empty subsets A and B in M, we define the *distance* from A to B, denoted $d(A, B)$, by

$$d(A, B) = \inf\{d(x, y)|x \in A, y \in B\}.$$

1) As in Exercise 3.11 we define for every $x \in M$ the distance to B by

$$d(x, B) = \inf\{d(x, y)|y \in B\}.$$

Argue that for arbitrary points $x \in A$ and $y \in B$ it holds that

$$d(A, B) \le d(x, B) \quad \text{and} \quad \inf\{d(x', B)|x' \in A\} \le d(x, y).$$

Utilize this to show that

$$d(A, B) = \inf\{d(x, B)|x \in A\}.$$

2) Show that if A is compact, then there exists a point $a_0 \in A$ such that $d(A, B) = d(a_0, B)$.

Hint: You can make use of the continuous function $\varphi : M \to \mathbb{R}$ defined by $\varphi(x) = d(x, B)$.

Next show that if B is also compact, then there exists a point $b_0 \in B$ such that $d(A, B) = d(a_0, b_0)$.

3) Let K be a compact subset in M contained in the open set U in M, i.e. $K \subseteq U \subseteq M$. Show that there exists an $r \in \mathbb{R}^+$, such that $B_r(x) \subseteq U$ for every $x \in K$.

Exercise 3.13. Let (M, d) be a metric space. Suppose that A and B are non-empty subsets of M. Consider the distance $d(A, B)$ defined in Exercise 3.12.

1) Prove that if A is compact and B is closed in M, and if $A \cap B = \emptyset$, then $d(A, B) > 0$.

2) Find an example of two closed subsets $A, B \subseteq \mathbb{R}^2$, where $A \cap B = \emptyset$, but $d(A, B) = 0$.

Exercise 3.14. Let $M = \{x \in \mathbb{R} \mid 0 \le x < 1\}$. Consider the family of subsets $\mathcal{T}$ in M consisting of the empty set $\emptyset$ and every subset $U \subseteq M$ of the form $U = \{x \in \mathbb{R} \mid 0 \le x < k\}$ for a number k with $0 < k \le 1$.

1) Show that $\mathcal{T}$ is a topology on M.

2) Show that in the topological space $(M, \mathcal{T})$, the sequence $(x_n = \frac{1}{n+1})$ will have every point in M as limit point.

3) Examine if the topology $\mathcal{T}$ stems from a metric on M.

Exercise 3.15. Let M be a topological space, and let (f_n), or in more detail $f_1, f_2, \ldots, f_n, \ldots$, be a sequence of continuous functions $f_n : M \to \mathbb{R}$, such that for all $x \in M$ it holds that

(i) $f_n(x) \geq 0$.

(ii) $f_1(x) \geq f_2(x) \geq \cdots \geq f_n(x) \geq \cdots$.

(iii) $\lim_{n \to \infty} f_n(x) = 0$.

In other words: The decreasing sequence of functions (f_n) converges pointwise to the 0-function.

For $\varepsilon > 0$ and $n \in \mathbb{N}$ we set

$$U_n(\varepsilon) = \{x \in M \mid 0 \leq f_n(x) < \varepsilon\} .$$

1) Show that $U_n(\varepsilon)$ is an open set in M.

2) Show that for fixed $\varepsilon > 0$, the collection of sets $\{U_n(\varepsilon) \mid n \in \mathbb{N}\}$ defines an open covering of M.

3) Now assume that M is compact. Show then that for every $\varepsilon > 0$ there is an $n_0 \in \mathbb{N}$, such that for all $n \geq n_0$ it holds that

$$0 \leq f_n(x) < \varepsilon \text{ for all } x \in M;$$

or written with quantifiers,

$$\forall \varepsilon > 0 \ \exists n_0 \in \mathbb{N} \ \forall n \in \mathbb{N} : n \geq n_0 \Rightarrow \forall x \in M : 0 \leq f_n(x) < \varepsilon.$$

We conclude that, under the given assumptions, the sequence of functions (f_n) converges uniformly to the 0-function.

This result is due to the Italian mathematician Ulisse Dini (1845–1918) and is known as *Dini's Theorem*.

4) Is it of importance that M is compact in 3)?

Exercise 3.16. Let $f, g : [a, b] \to \mathbb{R}$ be continuous functions defined in a closed and bounded interval $[a, b]$, for which $f(x) < g(x)$ for every $x \in [a, b]$. Show that

$$K = \{(x, y) \in \mathbb{R}^2 \mid a \leq x \leq b , \ f(x) \leq y \leq g(x)\}$$

is a compact subset in $\mathbb{R}^2$.

Exercise 3.17. Let $K_1 \supseteq K_2 \supseteq \cdots \supseteq K_n \supseteq \cdots$ be a descending sequence of non-empty, compact subsets in a Hausdorff space M. Show that the intersection of sets $\bigcap_{n=1}^{\infty} K_n$ is non-empty.

Exercise 3.18. Let (M, d) be a metric space. Show that if a subset A in M is a complete metric space in the induced metric from M, then A is a closed set in M.

Exercise 3.19. Let (M, d) be a metric space. A function $f : M \to \mathbb{R}$ is said to be *lower semicontinuous*, if the following condition is satisfied:

For every $x \in M$ and every $\varepsilon > 0$ there exists an open neighbourhood N of x in M, such that

$$f(x) - \varepsilon < f(y) \quad \text{for} \quad y \in N.$$

1) Show that a lower semicontinuous function $f : M \to \mathbb{R}$ is bounded from below on every sequentially compact subset K in M.

Hint: You can prove this indirectly, taking the proof of Theorem 3.1.3, page 53 (the first proof) as your source of inspiration.

2) Show that a lower semicontinuous function $f : M \to \mathbb{R}$ assumes a minimum value on every sequentially compact subset K in M.

Hint: Construct a sequence (x_n) in K for which

$$\lim_{n \to \infty} f(x_n) = \inf f(K) \,,$$

and use this to determine a point $x_0 \in K$, such that $f(x_0) = \inf f(K)$. (In a suitable setting this is the so-called *direct method* in the calculus of variations, and the sequence (x_n) is called a *minimizing sequence*.)

Exercise 3.20. A subset K in a metric space (M, d) is called *precompact* if for every $\epsilon > 0$ there exist finitely many points $x_1, \ldots, x_p \in K$ such that $K \subseteq B_\varepsilon(x_1) \cup \cdots \cup B_\varepsilon(x_p)$.

1) Show that a subset $K \subseteq \mathbb{R}^n$ in the space $\mathbb{R}^n$ (with Euclidean metric) is precompact if and only if it is bounded.

2) Let $f : X \to Y$ be a mapping between the metric spaces (X, d_X) and (Y, d_Y), and let $K \subseteq X$ be a precompact subset in X. Show that the image set $f(K) \subseteq Y$ is precompact in Y if f is uniformly continuous in K.

Exercise 3.21. Let S be a point set with more than one element equipped with the discrete topology.

1) Show that a topological space M is connected if and only if every continuous mapping $f : M \to S$ is constant.

2) Let $\{W_i | i \in I\}$ be a family of connected subsets in a topological space M, such that for every pair of sets W_i and W_j from the family it holds that $W_i \cap W_j \neq \emptyset$. Show that the union $\bigcup_{i \in I} W_i$ is a connected subset in M.

Exercise 3.22. Prove the following theorem: Let M be an arbitrary subset in the number space $\mathbb{R}^k$ with the usual topology, and let $\{U_i | i \in I\}$ be an arbitrary system of open sets in $\mathbb{R}^k$ that covers M. Then, either there exists a finite subsystem $\{U_{i_1}, \ldots, U_{i_n}\}$, or, there exists a countable subsystem $\{U_{i_1}, U_{i_2}, \ldots, U_{i_n}, \ldots\}$ that covers M.

The theorem is due to the Finnish mathematician Ernst Lindelöf (1870–1946) and is known as *Lindelöf's covering theorem*.

Hint: Utilize that the set of rational numbers is countable.

Exercise 3.23. Let E be a subset in the topological space M. Show that if E is connected, then the closure $\overline{E}$ of E is also connected.

Exercise 3.24. Let A be a bounded subset of $\mathbb{R}$, which contains infinitely many different points. Prove that there exists at least one point $x_0 \in \mathbb{R}$ for which all open intervals containing x_0 must contain infinitely many different points from A.

Exercise 3.25. Prove (e.g. by an indirect argument) that a continuous mapping $f : X \to Y$ between metric spaces (X, d_X) and (Y, d_Y) is uniformly continuous in every sequentially compact subset K of X.

Exercise 3.26. Let $f : [a, b] \to \mathbb{R}$ be a continuous function defined in a closed and bounded interval $[a, b]$ of $\mathbb{R}$.

1) Prove that the image $f([a, b])$ of $[a, b]$ under f is a closed and bounded interval $[c, d]$ of $\mathbb{R}$.

2) Prove that there exists a point $\xi \in [a, b]$ such that

$$\int_a^b f(x)dx = f(\xi)(b - a) .$$

This result is known as the *Mean Value Theorem for Integrals*.

Exercise 3.27. Let $f : X \to Y$ be a continuous map between metric spaces (X, d_X) and (Y, d_Y). Prove that if $S \subseteq X$ is a pathwise connected subset in X, then the image $f(S) \subseteq Y$ under f is a pathwise connected subset in Y.

Exercise 3.28. Consider the metric space (M, d) where $M = [1, \infty[$ and d is the usual distance in $\mathbb{R}$. Define the mapping $T : M \to M$ by

$$Tx = \frac{x}{2} + \frac{1}{x} \quad \text{for } x \in M .$$

Show that T is a contraction and determine the minimal contraction factor λ. Determine the fixed point for T.

Exercise 3.29. A mapping T of a metric space (M, d) into itself is called a *weak contraction* if

$$d(Tx, Ty) < d(x, y) \quad \text{for all } x, y \in M, x \neq y .$$

1) Show that T has at most one fixed point.
2) Show that T does not necessarily have a fixed point.

Hint: Consider the function $Tx = x + \frac{1}{x}$ for $x \geq 1$.

Exercise 3.30. Let T be a mapping from a complete metric space (M, d) into itself, and assume that there is a natural number m such that T^m is a contraction. Show that T has one and only one fixed point.

Exercise 3.31. Let $S \subseteq \mathbb{R}^n$ be an open set of n-dimensional Euclidean space $\mathbb{R}^n$. Prove that S is connected if and only if S is pathwise connected.

Exercise 3.32. Let $U \subseteq \mathbb{R}^n$ be an open set of n-dimensional Euclidean space $\mathbb{R}^n$, and let $f : U \to \mathbb{R}$ be a continuous function.

Make the identification $\mathbb{R}^{n+1} = \mathbb{R}^n \times \mathbb{R}$, and let $G(f)$ be the graph of the function f in $\mathbb{R}^{n+1}$, i.e.

$$G(f) = \{(x, f(x)) \in \mathbb{R}^n \times \mathbb{R} \,|\, x \in U\}.$$

1) Prove that $G(f)$ is homeomorphic to U.
2) Prove that the complement of $G(f)$ in $U \times \mathbb{R}$ is an open and dense subset in $U \times \mathbb{R}$, which has exactly two connectedness components if U is connected.

Chapter 4

Metric Structures in Vector Spaces

In this chapter we develop basic elements of the theory of metric structures in real vector spaces and continuity of linear mappings in this context. The metric structures defined by norms in vector spaces are of utmost importance in mathematical analysis and are known as normed vector spaces. Completeness of normed vector spaces is discussed and Banach spaces are defined. For later use we define the Riemann integral of continuous functions with values in a Banach space.

4.1 Normed vector spaces and their metric properties

We assume that the reader has a basic knowledge of the theory of vector spaces. In this book we shall be working over the scalar field of real numbers and accordingly all vector spaces are real vector spaces.

If the vector space V has dimension n, and $\{e_1, \ldots, e_n\}$ is a basis for V, then any vector $v \in V$ can in a unique way be written as a linear combination

$$v = \alpha_1 e_1 + \cdots + \alpha_n e_n, \quad \text{where} \quad \alpha_1, \ldots, \alpha_n \in \mathbb{R}.$$

The real numbers $\alpha_1, \ldots, \alpha_n$ are called the *coordinates* of the vector in the given basis for V.

The mapping that maps the vector $v \in V$ into its set of coordinates $(\alpha_1, \ldots, \alpha_n) \in \mathbb{R}^n$, is a bijective, linear mapping and defines an isomorphism of V and $\mathbb{R}^n$. From this we conclude: *Two finite dimensional real vector spaces of the same dimension n are isomorphic, and $\mathbb{R}^n$ is a model for the isomorphism class.*

We shall now introduce the concept of a norm in a vector space. To emphasize the connection to metric spaces, we note that the usual Euclidean metric d in $\mathbb{R}^n$ has the following additional properties:

(i) Relative distances are invariant under translation:
$$d(x + z, y + z) = d(x, y) , \quad \text{for all } x, y, z \in \mathbb{R}^n.$$

(ii) Multiplication by a real number enlarges or reduces distances uniformly:
$$d(\alpha x, \alpha y) = |\alpha| d(x, y) , \quad \text{for all } x, y \in \mathbb{R}^n, \alpha \in \mathbb{R}.$$

These properties are not consequences of the axioms for a metric, and they are not satisfied by all metrics in $\mathbb{R}^n$.

Now let V be a vector space equipped with a metric, which in addition to MET 1, MET 2, and MET 3 satisfies the properties corresponding to (i) and (ii). From property (i) it follows immediately that $d(x, y) = d(x - y, 0)$ for all $x, y \in V$. For an arbitrary vector $x \in V$, we therefore define the length of x as the number $||x|| = d(x, 0)$. With the given assumptions on the metric d, the quantity $||\cdot||$ has the following properties:

NORM 1 (positive definite)
$$||x|| \geq 0, \text{ for all } x \in V, \quad ||x|| = 0 \iff x = 0.$$

NORM 2 (uniform scaling)
$$||\alpha x|| = |\alpha| \, ||x|| \text{ for all } x \in V, \text{ and } \alpha \in \mathbb{R}.$$

NORM 3 (the triangle inequality)
$$||x + y|| \leq ||x|| + ||y|| \text{ for all } x, y \in V.$$

Proof. Since $||x|| = d(x, 0)$, NORM 1 follows directly from MET 1. Property NORM 2 is a consequence of the supplementary property (ii) of the metric d. Finally, property NORM 3 follows from MET 3 and the supplementary property (i) by the computations:

$$||x + y|| = d(x + y, 0) \leq d(x + y, y) + d(y, 0)$$
$$= d(x, 0) + d(y, 0)$$
$$= ||x|| + ||y||.$$

This completes the proof. $\square$

Inspired by the above considerations we go on to define a real-valued function $||\cdot|| : V \to \mathbb{R}$ by associating the real number $||x|| \in \mathbb{R}$ to the vector $x \in V$. Note that this function is determined by the metric d on V by the formula $||x|| = d(x, 0)$. Conversely, the metric d can on the other hand be recovered from $||\cdot||$ by the formula $d(x, y) = ||x - y||$.

Definition 4.1.1. Let V be a vector space. A function $||\cdot|| : V \to \mathbb{R}$ with the properties NORM 1, NORM 2 and NORM 3 is called a *norm* in V, and a vector space V together with a norm $||\cdot||$ is called a *normed vector space*.

Every norm in a vector space induces a metric in the vector space.

Theorem 4.1.2. *Let V be a normed vector space with norm $||\cdot||$. Then V has a canonical structure as a metric space with the metric d given by*

$$d(x,y) = ||x - y|| \quad for \ all \ x, y \in V.$$

Proof. According to NORM 1 we have $||x - y|| \geq 0$, and $||x - y|| = 0$ if and only if $x = y$. This shows that d satisfies MET 1. By using NORM 2, we get:

$$d(x,y) = ||x - y|| = ||(-1)(y - x)||$$
$$= |-1| \, ||y - x|| = ||y - x|| = d(y,x),$$

and hence d satisfies MET 2. Finally, MET 3 follows from NORM 3:

$$d(x,z) = ||x - z|| = ||(x - y) + (y - z)||$$
$$\leq ||x - y|| + ||y - z||$$
$$= d(x,y) + d(y,z).$$

This proves Theorem 4.1.2. $\square$

When a normed vector space is considered as a metric space, it is with the metric structure defined in Theorem 4.1.2. It is easy to show that the metric in Theorem 4.1.2 has additional properties (i) and (ii). Hence we are back to the point of departure for defining a norm.

Example 4.1.3. Let $\mathbb{R}^n$ be the space of real n-tuples with its usual vector space structure. For a vector $x = (x_1, \ldots, x_n) \in \mathbb{R}^n$ we define

$$||x||_1 = \sum_{i=1}^{n} |x_i| = |x_1| + \cdots + |x_n| \qquad \textit{the sum norm,}$$

$$||x||_2 = \sqrt{\sum_{i=1}^{n} x_i{}^2} = \sqrt{x_1{}^2 + \cdots + x_n{}^2} \qquad \textit{the Euclidean norm,}$$

$$||x||_\infty = \max\{|x_1|, \ldots, |x_n|\} \qquad \textit{the maximum norm.}$$

It is easy to show that $||\cdot||_1$ and $||\cdot||_\infty$ are norms in $\mathbb{R}^n$, and in Chapter 2, Example 2.2.3, we proved that also $||\cdot||_2$ is a norm in $\mathbb{R}^n$.

From each of the three norms we get a metric in $\mathbb{R}^n$, and each of these metrics induces a topology in $\mathbb{R}^n$. It is not difficult to show directly that these topologies are indeed the same topology in $\mathbb{R}^n$. This will, however, also be a consequence of a general result proved below in Theorem 4.1.6. ◄

Example 4.1.4. Let K be a compact topological space, e.g., a closed and bounded interval in $\mathbb{R}$. Denote by $C(K, \mathbb{R})$ the set of continuous real-valued functions $f : K \to \mathbb{R}$.

For $f, g \in C(K, \mathbb{R})$ and $\alpha \in \mathbb{R}$ we define the functions $f + g$ and $\alpha \cdot f$ on K by the pointwise definitions

$$(f + g)(x) = f(x) + g(x) \quad \text{and} \quad (\alpha \cdot f)(x) = \alpha \cdot f(x),$$

for $x \in K$. It is well known that $f + g$ and $\alpha \cdot f$ are continuous functions; hence they belong to $C(K, \mathbb{R})$. Often one simply says that the operations 'addition' and 'multiplication by a scalar' in $C(K, \mathbb{R})$ are *pointwise defined*. By these pointwise defined operations, $C(K, \mathbb{R})$ gets the structure of a real vector space. When K contains more than finitely many points, the vector space $C(K, \mathbb{R})$ usually has infinite dimension.

Since K is compact, any function $f \in C(K, \mathbb{R})$ is bounded, and thus we can define

$$\|f\| = \sup \left\{ |f(x)| \,\middle|\, x \in K \right\}.$$

Using the properties of supremum, it is not difficult to prove that $\|\cdot\|$ is a norm in $C(K, \mathbb{R})$. This norm is known as the *uniform norm*, or the *supremum norm*, in $C(K, \mathbb{R})$. ◄

As we have seen in Theorem 4.1.2 a norm defines a canonical metric, and hence a topology, in the vector space. Often it is only this topology which is of interest; hence the following definition.

Definition 4.1.5. We say that two norms $\|\cdot\|$ and $\|\cdot\|'$ in a vector space V are *equivalent* if the associated metrics define the same topology in V.

For finite dimensional vector spaces, the situation is very satisfying. By using Example 4.1.3, it is easy to construct a norm in an arbitrary vector space V of dimension n, simply by exploiting the fact that V is isomorphic to $\mathbb{R}^n$. Such a norm defines a topology in V. As the following theorem shows, this topology in V is independent of the choice of norm.

Theorem 4.1.6. *In a finite dimensional vector space V all norms are equivalent.*

Proof. By a choice of basis, V can be identified with one of the vector spaces $\mathbb{R}^n$ and so it is sufficient to show that an arbitrary norm $\|\cdot\|^*$ in $\mathbb{R}^n$ induces the same topology in $\mathbb{R}^n$ as the Euclidean norm $||\cdot||$ in $\mathbb{R}^n$. For this, it is enough to show that there exist constants $\alpha, \beta \in \mathbb{R}^+$ such that

$$\alpha||x|| \leq ||x||^* \leq \beta||x|| \quad \text{for all } x \in \mathbb{R}^n.$$

Indeed, if such constants $\alpha, \beta \in \mathbb{R}^+$ can be found, it follows easily that the metrics descending from $||\cdot||$ and $||\cdot||^*$ determine the same open sets in $\mathbb{R}^n$. The reason for this is that any open ball for one of the metrics contains an open ball (with the same centre) for the other metric, and vice-versa.

Our goal in the following is to prove the existence of such constants $\alpha, \beta \in \mathbb{R}^+$. First, however, the following assertion.

Assertion 4.1.7. *When $\mathbb{R}^n$ is equipped with the topology induced by the Euclidean norm $||\cdot||$ (the usual topology), then $||\cdot||^* : \mathbb{R}^n \to \mathbb{R}$ is a continuous function.*

Proof of Assertion 4.1.7. Let $\{e_1, \ldots, e_n\}$ be the canonical basis in $\mathbb{R}^n$. For an arbitrary vector $x = (x_1, \ldots, x_n)$ in $\mathbb{R}^n$ we have:

$$||x||^* = ||x_1 e_1 + \cdots + x_n e_n||^*$$
$$\leq |x_1| \, ||e_1||^* + \cdots + |x_n| \, ||e_n||^*$$
$$\leq (||e_1||^* + \cdots + ||e_n||^*)||x||.$$

Note that

$$k = ||e_1||^* + \cdots + ||e_n||^*$$

is then a fixed constant, for which

$$||x||^* \leq k||x|| \quad \text{for all } x \in \mathbb{R}^n.$$

For arbitrary vectors $x, y \in \mathbb{R}^n$ the triangle inequality yields:

$$||x||^* = ||(x - y) + y||^* \leq ||x - y||^* + ||y||^*,$$

or equivalently,

$$||x||^* - ||y||^* \leq ||x - y||^*.$$

By interchanging x and y we get the corresponding inequality

$$||y||^* - ||x||^* \leq ||y - x||^* = ||x - y||^*,$$

and consequently the inequality

$$\big| ||x||^* - ||y||^* \big| \leq ||x - y||^*.$$

If we combine the inequalities obtained, we get the inequality

$$\big| ||x||^* - ||y||^* \big| \leq k||x - y||,$$

which is valid for arbitrary vectors $x, y \in \mathbb{R}^n$.

This inequality shows directly that $||\cdot||^* : \mathbb{R}^n \to \mathbb{R}$ is a continuous function; in fact even uniformly continuous. This verifies the assertion. $\square$

We now start our search for constants $\alpha, \beta \in \mathbb{R}^+$ which relate the norms $||\cdot||$ and $||\cdot||^*$ in the manner described above. To this end, let

$$S^{n-1} = \left\{ x \in \mathbb{R}^n \mid ||x|| = 1 \right\}$$

denote the *unit sphere* in $\mathbb{R}^n$. Clearly, S^{n-1} is a closed and bounded subset in $\mathbb{R}^n$ equipped with the metric induced from the Euclidean norm in $\mathbb{R}^n$. Hence S^{n-1} is a compact subset in $\mathbb{R}^n$ by Theorem 3.2.4 (Heine-Borel).

Theorem 3.6.3 then shows that the function $||\cdot||^* : \mathbb{R}^n \to \mathbb{R}$, which is continuous according to the assertion just proved, attains a maximum value as well as a minimum value on the compact subset S^{n-1} in $\mathbb{R}^n$. Hence there exist constants $\alpha, \beta \in \mathbb{R}^+$, such that

$$\alpha \leq ||x||^* \leq \beta \quad \text{for} \quad ||x|| = 1.$$

Note in particular that $\alpha > 0$ since $||x||^* > 0$ for $x \neq 0$, thus especially when $||x|| = 1$.

An arbitrary vector $x \in \mathbb{R}^n$ can always be written in the form

$$x = ||x|| x_0 \quad \text{for a vector } x_0 \in \mathbb{R}^n \text{ with norm} \quad ||x_0|| = 1.$$

When x is expressed in this way, we get

$$||x||^* = ||x|| \, ||x_0||^*.$$

Multiplying the inequalities $\alpha \leq ||x_0||^* \leq \beta$ by $||x||$, we finally get the inequalities

$$\alpha||x|| \leq ||x||^* \leq \beta||x|| \quad \text{for all } x \in \mathbb{R}^n.$$

Thus we have proved the existence of constants $\alpha, \beta \in \mathbb{R}^+$ as requested. This completes the proof of Theorem 4.1.6. $\square$

4.2 Linearity and continuity

In the following, when we consider a normed vector space V as a topological space, it is implicitly understood that V is equipped with the topology defined by the metric on V, which by Theorem 4.1.2 is induced by the norm in V. Whenever V is finite dimensional, Theorem 4.1.6 shows that when considering topological questions we can use any norm in V. Note that a finite dimensional vector space can always be equipped with a norm, according to the remark preceding Theorem 4.1.6. Hence when we work with *finite dimensional vector spaces*, and we discuss *topological questions*, *we do not need to specify the norms*. We shall make use of this in the following theorem.

Theorem 4.2.1. *Let V and W be finite dimensional vector spaces. Then every linear mapping $T : V \to W$ is continuous.*

Proof. By choice of basis, any finite dimensional vector space can be identified with one of the vector spaces $\mathbb{R}^n$. Without loss of generality we can therefore assume that $T : \mathbb{R}^n \to \mathbb{R}^m$ is a linear mapping between the vector spaces $\mathbb{R}^n$ and $\mathbb{R}^m$, equipped with their respective maximum norms.

Now the proof is easy. Let $\{e_1, \ldots, e_n\}$ be the canonical basis in $\mathbb{R}^n$. For arbitrary vectors $x = (x_1, \ldots, x_n)$ and $y = (y_1, \ldots, y_n)$ in $\mathbb{R}^n$, the linearity of T implies that

$$T(x) - T(y) = T(x - y) = T\left(\sum_{i=1}^{n}(x_i - y_i)e_i\right)$$

$$= \sum_{i=1}^{n}(x_i - y_i)T(e_i) .$$

From which follows the inequality

$$\|T(x) - T(y)\| \le \sum_{i=1}^{n}|x_i - y_i|\|T(e_i)\|$$

$$\le \left(\sum_{i=1}^{n}\|T(e_i)\|\right)\|x - y\|$$

$$= k\|x - y\| ,$$

where we have set $k = \sum_{i=1}^{n}\|T(e_i)\|$.

Since k is a fixed constant, this inequality shows that T is continuous, in fact even uniformly continuous. $\square$

Remark 4.2.2. In Theorem 4.2.1 only the vector space V has to be finite dimensional. The vector space W can be an arbitrary normed vector space. ◁

In general, it is not true that a linear mapping between two normed vector spaces of infinite dimension is continuous.

Example 4.2.3. Let V be the vector space consisting of all infinite sequences of real numbers $a = (a_1, a_2, a_3, \ldots)$ with addition and multiplication by scalar defined position wise. Consider the subspace U in V consisting of the sequences $a = (a_1, a_2, a_3, \ldots)$ for which the series $\sum_{n=1}^{\infty}|a_n|$ is convergent. The comparison test for convergence of infinite series shows immediately that U is actually a subspace in V.

We can define a norm in U by

$$\|a\| = \sum_{n=1}^{\infty} \frac{1}{n!} |a_n|,$$

and a linear mapping $T : U \to U$ by

$$T(a_1, a_2, a_3, \dots) = (a_2, a_3, a_4, \dots).$$

It is easy to show that $\|\cdot\|$ really is a norm in U, and that T is a linear mapping. The linear mapping T is called a *shift operator*.

Although T is linear, it is not continuous at $0 = (0, 0, 0, \dots) \in U$. To see this, consider the sequence $(a^{(n)})$ in U, where $a^{(n)}$ is the infinite sequence of real numbers with $(n-1)!$ at the nth position and 0 at all the other positions. Then $\|a^{(n)}\| = \frac{1}{n}$, while $\|T(a^{(n)})\| = 1$. Hence the sequence $(a^{(n)})$ in U converges to the 0-sequence, without the image sequence $(T(a^{(n)}))$ converging to the 0-sequence. By Theorem 2.9.3, T fails therefore to be continuous at 0. ◄

In functional analysis linear mappings $T : V \to W$ are usually called *linear operators*. This terminology will often be adopted in the following. The basic facts about the continuity of linear operators are collected in the next theorem.

Theorem 4.2.4. *Let $T : V \to W$ be a linear operator between normed vector spaces $(V, \|\cdot\|_V)$ and $(W, \|\cdot\|_W)$. Then the following statements are equivalent:*

(1) *T is continuous.*

(2) *T is continuous at $0 \in V$.*

(3) *There is a constant k, such that $\|T(x)\|_W \le k$ for all unit vectors $x \in V$, i.e. for $\|x\|_V = 1$.*

(4) *There is a constant k, such that $\|T(x)\|_W \le k\|x\|_V$ for all $x \in V$.*

Proof. We prove the closed cycle of implications $(1) \Rightarrow (2) \Rightarrow (3) \Rightarrow (4) \Rightarrow (1)$.

$(1) \Rightarrow (2)$. Trivial.

$(2) \Rightarrow (3)$. Suppose that T is continuous at $0 \in V$. Corresponding to $\varepsilon = 1$, choose a $\delta > 0$, such that $\|T(x)\|_W \le 1$ for $\|x\|_V \le \delta$.

For any unit vector $x \in V$, i.e. $\|x\|_V = 1$, clearly $\|\delta x\|_V = \delta$, and hence $\|T(\delta x)\|_W \le 1$ by the choice of δ. Since T is linear we then get $\delta\|T(x)\|_W \le 1$, and consequently $\|T(x)\|_W \le k$ with $k = 1/\delta$.

(3)$\Rightarrow$(4). An arbitrary vector $x \in V$ can be written in the form $x = ||x||_V x_0$ with $||x_0||_V = 1$. For a constant k satisfying (3), we then have:

$$||T(x)||_W = ||T(||x||_V x_0)||_W = || \, ||x||_V T(x_0)||_W$$
$$= ||x||_V ||T(x_0)||_W \leq k||x||_V.$$

(4)$\Rightarrow$(1). If there exists a constant k satisfying (4), then for an arbitrary pair of vectors $x, y \in V$ we have:

$$||T(x) - T(y)||_W = ||T(x - y)||_W \leq k||x - y||_V.$$

This inequality shows that T is continuous. $\square$

The equivalent properties (3) and (4) of a continuous linear operator between normed vector spaces stated in Theorem 4.2.4 inspire the definition of the important notion of a bounded linear operator.

For any normed vector space $(V, ||\cdot||_V)$ denote by

$$C(V) = \{x \in V \mid ||x||_V \leq 1\}$$

the closed *unit ball* in V. Then the definition of boundedness of an operator can be formulated as follows.

Definition 4.2.5. A linear operator $T : V \to W$ between normed vector spaces $(V, ||\cdot||_V)$ and $(W, ||\cdot||_W)$ is called a *bounded* linear operator if it maps the closed unit ball $C(V)$ in V into a bounded set (a ball) in W.

The main result recorded in Theorem 4.2.4 can then be stated as follows.

Corollary 4.2.6. *A linear operator* $T : V \to W$ *between normed vector spaces* $(V, ||\cdot||_V)$ *and* $(W, ||\cdot||_W)$ *is continuous if and only if it is bounded.*

4.3 The operator norm for a bounded linear operator

The operator norm is an important number that can be assigned to every bounded linear operator $T : V \to W$ between normed vector spaces $(V, ||\cdot||_V)$ and $(W, ||\cdot||_W)$. It defines a norm in the vector space of bounded operators.

The norm of a bounded linear operator $T : V \to W$ measures, informally speaking, how much T distorts the unit sphere in V. The distortion of the unit sphere can be measured in several equivalent ways of which the most important are listed in the following lemma.

Lemma 4.3.1. *Let* $T : V \to W$ *be a bounded linear operator between normed vector spaces* $(V, ||\cdot||_V)$ *and* $(W, ||\cdot||_W)$. *Then each of the real numbers defined below exists:*

$$r = \inf \left\{ k \in \mathbb{R} \mid ||T(x)||_W \leq k \quad \text{for all vectors } x \in V \text{ with } ||x||_V = 1 \right\}$$
$$s = \inf \left\{ k \in \mathbb{R} \mid ||T(x)||_W \leq k||x||_V \quad \text{for all } x \in V \right\}$$
$$t = \sup \left\{ ||T(x)||_W \mid x \in V, \; ||x||_V = 1 \right\}$$
$$u = \sup \left\{ ||T(x)||_W \mid x \in V, \; ||x||_V \leq 1 \right\},$$

and $r = s = t = u$.

Proof. Since $T : V \to W$ is bounded it follows by Theorem 4.2.4 that there exists a constant $k \in \mathbb{R}$, such that $||T(x)||_W \leq k$ for all $x \in V$ with $||x||_V = 1$. Thus the set of real numbers

$$A = \left\{ k \in \mathbb{R} \mid ||T(x)||_W \leq k \quad \text{for all } x \in V \text{ with } ||x||_V = 1 \right\},$$

is non-empty. Furthermore, it is bounded below by 0. Hence the number $r = \inf A$ exists. The inequality $||T(x)||_W \leq k$ for all $x \in V$ with $||x||_V = 1$ immediately shows the existence of the number t.

An arbitrary vector $x \in V$ can be written in the form $x = ||x||_V x_0$ with $||x_0||_V = 1$. With x in this form, we get $||T(x)||_W = ||x||_V||T(x_0)||_W$. This shows that $||T(x_0)||_W \leq k$ for all $x_0 \in V$ with $||x_0||_V = 1$ if and only if $||T(x)||_W \leq k||x||_V$ for all $x \in V$. From this, it follows that

$$A = \left\{ k \in \mathbb{R} \mid ||T(x)||_W \leq k||x||_V \quad \text{for all } x \in V \right\}.$$

This shows that the number s exists, and furthermore, that $r = s$. The expression $||T(x)||_W = ||x||_V||T(x_0)||_W$ also shows immediately that, since the number t exists, the number u exists, and that $t = u$.

What is left to prove is that $r = t$. Indeed, since $||T(x)||_W \leq t$ for all $x \in E$ with $||x||_V = 1$, the number t belongs to the set A over which the infimum is taken when determining r. Hence we have $r \leq t$. At the same time $t \leq k$ for all $k \in A$, from which it follows that $t \leq r$. All in all $r = t$. This completes the proof of Lemma 4.3.1. $\square$

Based on Lemma 4.3.1 we introduce the following definition.

Definition 4.3.2. By *the operator norm*, or simply *the norm*, of a bounded linear operator $T : V \to W$ between normed vector spaces $(V, ||\cdot||_V)$ and $(W, ||\cdot||_W)$, we understand the real number $||T||$ determined in one of the following equivalent ways:

$$\|T\| = \inf\left\{k \in \mathbb{R} \mid \|T(x)\|_W \leq k \text{ for all } x \in V \text{ with } \|x\|_V = 1\right\}$$
$$= \inf\left\{k \in \mathbb{R} \mid \|T(x)\|_W \leq k\|x\|_V \text{ for all } x \in V\right\}$$
$$= \sup\left\{\|T(x)\|_W \mid x \in V, \|x\|_V = 1\right\}$$
$$= \sup\left\{\|T(x)\|_W \mid x \in V, \|x\|_V \leq 1\right\}.$$

The following estimate involving the operator norm of a bounded linear operator is crucial in many situations.

Lemma 4.3.3. *Let* $T : V \to W$ *be a bounded linear operator between normed vector spaces* $(V, \|\cdot\|_V)$ *and* $(W, \|\cdot\|_W)$. *Then*

$$\|T(x)\|_W \leq \|T\|\|x\|_V \quad \text{for all} \quad x \in V.$$

Proof. An arbitrary vector $x \in V$ can be written in the form $x = \|x\|_V x_0$ with $\|x_0\|_V = 1$. From the definition of the operator norm it follows immediately that $\|T(x_0)\|_W \leq \|T\|$. Then we get

$$\|T(x)\|_W = \|x\|_V\|T(x_0)\|_W \leq \|T\|\|x\|_V,$$

as claimed. $\square$

For an arbitrary pair of normed vector spaces V and W, we denote by

$$B(V, W) \quad \text{the space of bounded linear operators } T : V \to W.$$

For $T, T_1, T_2 \in B(V, W)$ and $\alpha \in \mathbb{R}$ we define $T_1 + T_2$ and αT by the usual pointwise defined operations

$$(T_1 + T_2)(x) = T_1(x) + T_2(x)$$
$$(\alpha T)(x) = \alpha T(x)$$

for $x \in V$.

The mappings $T_1 + T_2$ and αT are bounded linear operators and hence belong to $B(V, W)$, which thereby gets the structure of a real vector space. But $B(V, W)$ has a richer structure. The following theorem justifies in particular the name operator norm.

Theorem 4.3.4. *For arbitrary bounded linear operators* $T, T_1, T_2 : V \to W$ *and an arbitrary scalar* $\alpha \in \mathbb{R}$ *we have that*

NORM 1 $\|T\| \geq 0$, *and* $\|T\| = 0 \iff T = 0$.

NORM 2 $\|\alpha T\| = |\alpha|\|T\|$.

NORM 3 $||T_1 + T_2|| \leq ||T_1|| + ||T_2||$.

In other words: The operator norm is a norm in the vector space $B(V,W)$ and induces the structure of a normed vector space in this space.

Furthermore, if U, V and W are normed vector spaces, and $S : U \to V$ and $T : V \to W$ are bounded linear operators, it holds that

$$||T \circ S|| \leq ||T|| \, ||S||.$$

Proof. From Lemma 4.3.3 we get

$$0 \leq ||T(x)||_W \leq ||T|| \, ||x||_V$$

for all $x \in V$. From this NORM 1 follows immediately.

Since

$$||(\alpha T)(x)||_W = |\alpha| \, ||T(x)||_W$$

for all $||x||_V = 1$, in fact for all $x \in V$, NORM 2 is a direct consequence of the definition of the operator norm.

Since for $x \in V$ with $||x||_V = 1$ it is true that

$$
\begin{aligned}
||(T_1 + T_2)(x)||_W &= ||T_1(x) + T_2(x)||_W \\
&\leq ||T_1(x)||_W + ||T_2(x)||_W \\
&\leq ||T_1|| + ||T_2||,
\end{aligned}
$$

also NORM 3 follows directly from the definition of the operator norm.

If $S : U \to V$ and $T : V \to W$ are bounded linear operators between normed vector spaces, we have for $x \in U$ with $||x||_V = 1$:

$$||T \circ S(x)||_W = ||T(S(x))||_W \leq ||T|| \, ||S(x)||_V \leq ||T|| \, ||S||.$$

From this it follows immediately that $||T \circ S|| \leq ||T|| \, ||S||$. $\square$

4.4 Completeness of normed vector spaces

Let V be a normed vector space with norm $||\cdot||_V$. As usual the norm gives rise to a metric d on V, defined by $d(x,y) = ||x - y||_V$ for $x, y \in V$.

An important class of normed vector spaces was introduced by the Polish mathematician Stefan Banach (1892–1945).

Definition 4.4.1. A normed vector space V is called a *Banach space* if the associated metric space is complete.

By reformulation of the definitions we get:

(1) A sequence (x_n) in V is convergent with limit point $x_0 \in V$, if

$$\forall \varepsilon > 0 \; \exists n_0 \in \mathbb{N} \; \forall n \in \mathbb{N} \; : \quad n \geq n_0 \quad \Rightarrow \quad ||x_n - x_0||_V < \varepsilon.$$

(2) A sequence (x_n) in V is a Cauchy sequence, if

$$\forall \varepsilon > 0 \; \exists n_0 \in \mathbb{N} \; \forall n, m \in \mathbb{N} \; : \quad n, m \geq n_0 \quad \Rightarrow \quad ||x_n - x_m||_V < \varepsilon.$$

(3) A Banach space is a normed vector space V in which every Cauchy sequence is convergent.

In finite dimensions we have not introduced anything new, since we have the following theorem.

Theorem 4.4.2. *Every finite dimensional normed vector space is a Banach space.*

Proof. It suffices to consider the space $\mathbb{R}^n$ as the vector space.

Let $||\cdot||$ be the Euclidean norm in $\mathbb{R}^n$ and $||\cdot||^*$ an arbitrary norm in $\mathbb{R}^n$. In the proof of Theorem 4.1.6 we proved the existence of positive constants $\alpha, \beta \in \mathbb{R}^+$ such that

$$\alpha ||x|| \leq ||x||^* \leq \beta ||x|| \quad \text{for all} \quad x \in \mathbb{R}^n.$$

From this it is easily seen that a sequence (x_n) in $\mathbb{R}^n$ is convergent, respectively a Cauchy sequence, with respect to $||\cdot||^*$ if and only if it is convergent, respectively a Cauchy sequence, with respect to $||\cdot||$.

In the proof of Theorem 4.4.2 it is therefore sufficient to consider $\mathbb{R}^n$ with the Euclidean norm $||\cdot||$.

Now let (x_n) be an arbitrary Cauchy sequence in $\mathbb{R}^n$. Just as in the proof of Theorem 3.5.5 we can determine a constant $a \in \mathbb{R}^+$ such that $||x_n|| \leq a$ for all $n \in \mathbb{N}$. The ball $K = \{x \in \mathbb{R}^n \mid ||x|| \leq a\}$ in $\mathbb{R}^n$ is closed and bounded, and hence sequentially compact according to Theorem 3.4.3. Since (x_n) is a Cauchy sequence in the sequentially compact set K, it has a limit point in K according to Theorem 3.5.6, and thus, in particular, a limit point in $\mathbb{R}^n$. Hence any Cauchy sequence in $\mathbb{R}^n$ is convergent, and the proof is complete. $\square$

In a space of continuous linear mappings, the Banach property is transferred to the mapping space from the image space.

Theorem 4.4.3. *Let V be a normed vector spaces and let W be a Banach space. Then the vector space of bounded linear operators $B(V,W)$ is a Banach space with the operator norm as its norm.*

Proof. Consider $B(V,W)$ as a normed vector space with the operator norm as norm, and let (T_n) be an arbitrary Cauchy sequence in $B(V,W)$.

For all vectors $x \in V$, we have

$$||T_n(x) - T_m(x)||_W \leq ||T_n - T_m|| \, ||x||_V.$$

From this follows that the sequence $(T_n(x))$ is a Cauchy sequence in W for every fixed vector $x \in V$. Since W is a Banach space the sequence has a uniquely determined limit point in W which we denote by $T(x)$. Hence

$$T_n(x) \to T(x) \quad \text{for} \quad n \to \infty.$$

By this procedure we can define a mapping $T : V \to W$. It is easy to prove that T is linear.

Since

$$\big| ||T_n|| - ||T_m|| \big| \leq ||T_n - T_m||,$$

clearly $(||T_n||)$ is a Cauchy sequence in $\mathbb{R}$. Hence there exists a uniquely determined constant $c \in \mathbb{R}$, such that $||T_n|| \to c$ for $n \to \infty$.

For every vector $x \in V$, we have

$$||T_n(x)||_W \leq ||T_n|| \, ||x||_V.$$

By letting n go to infinity we get

$$||T(x)||_W \leq c ||x||_V.$$

This shows that T is bounded. Therefore T has an operator norm $||T||$, and the limiting process shows that $||T|| = c$.

We shall now prove that (T_n) converges to T. With this in mind consider an arbitrary vector $x \in V$, with $||x||_V = 1$. For all $n, k \in \mathbb{N}$, the triangle inequality gives the following estimate

$$||(T - T_n)(x)||_W \leq ||T(x) - T_{n+k}(x)||_W + ||T_{n+k}(x) - T_n(x)||_W$$
$$\leq ||T(x) - T_{n+k}(x)||_W + ||T_{n+k} - T_n||.$$

Given $\varepsilon > 0$, choose $n_0 \in \mathbb{N}$ so that

$$n \geq n_0 \quad \Rightarrow \quad ||T_{n+k} - T_n|| \leq \varepsilon.$$

Taking the limit for $k \to \infty$ it follows that

$$||(T - T_n)(x)||_W \leq \varepsilon \quad \text{for} \quad n \geq n_0.$$

From this we conclude that

$$n \geq n_0 \quad \Rightarrow \quad ||T - T_n|| \leq \varepsilon.$$

This shows that (T_n) converges to $T \in B(V,W)$ and hence completes the proof of Theorem 4.4.3. $\square$

If the norm in a Banach space is induced by an inner product in the underlying vector space, we get a particularly useful kind of normed vector space named after the German mathematician David Hilbert (1862–1943).

To introduce Hilbert spaces properly, we need to define the notion of an inner product in a vector space over the scalar field of real numbers $\mathbb{R}$.

In Example 2.2.3, page 20, we examined the natural vector space structure and the Euclidean metric in n-dimensional real number space $\mathbb{R}^n$. To that end we introduced an inner product and a norm (the Euclidean norm) in $\mathbb{R}^n$.

In many important real vector spaces V, but not all, we can introduce an inner product according to the following definition.

Definition 4.4.4. An *inner product* in a real vector space V is a mapping $\langle \cdot, \cdot \rangle : V \times V \to \mathbb{R}$ satisfying

IP 1 (symmetry)

$\langle x, y \rangle = \langle y, x \rangle$ all $x, y \in V$.

IP 2 (linearity)

$\langle \alpha x_1 + \beta x_2, y \rangle = \alpha \langle x_1, y \rangle + \beta \langle x_2, y \rangle$ all $x_1, x_2, y \in V, \alpha, \beta \in \mathbb{R}$.

IP 3 (positive definite)

$\langle x, x \rangle \geq 0$ and $\langle x, x \rangle = 0 \Leftrightarrow x = 0$ all $x \in V$.

Note that the symmetry assumption implies that an inner product is also linear in the second argument.

An inner product $\langle \cdot, \cdot \rangle$ in an arbitrary real vector space V induces a norm $||\cdot||$ in V, called the *induced norm*, by the definition

$$||x|| = \sqrt{\langle x, x \rangle} \quad \text{for all } x \in V .$$

The proof of this follows word by word the proofs in Example 2.2.3.

Then we can give the formal definition of a Hilbert space.

Definition 4.4.5. A Hilbert space is a vector space V equipped with an inner product such that the associated normed vector space is complete.

By Theorem 4.4.2, every finite dimensional inner product space is a Hilbert space. Hilbert spaces are studied in detail in Chapter 3 of my book [Functional Analysis-Entering Hilbert Space] listed in the bibliography. Here we limit ourself to describe an important example.

Example 4.4.6. The space $C([a, b], \mathbb{R})$ of real-valued, continuous functions $f : [a, b] \to \mathbb{R}$ defined in a closed interval $[a, b]$ admits an inner product defined by

$$\langle f, g \rangle = \int_a^b f(t)g(t)dt \ ,$$

for all $f, g \in C([a, b], \mathbb{R})$.

The space $C([a, b], \mathbb{R})$ is not complete in the induced norm, basically because the value of an integral cannot detect if the functions involved have discontinuities in sets of measure zero, e.g. countable sets. Hence $C([a, b], \mathbb{R})$ with the above inner product is not a Hilbert space.

There is, however, a general method for completing an inner product space to obtain a larger vector space V with an inner product, which is a Hilbert space. The original inner product space is dense in the completion. In the case of $C([a, b], \mathbb{R})$ with the above inner product, we have to add more functions to $C([a, b], \mathbb{R})$ to obtain the so-called quadratic integrable functions, which is the completion V of $C([a, b], \mathbb{R})$. ◄

4.5   Integration of functions with values in a Banach space

Let E be a Banach space with norm $|| \cdot ||_E$, and let $[a, b]$ be a closed and bounded interval in $\mathbb{R}$.

Our objective is to define the Riemann integral over $[a, b]$ of a continuous function $f : [a, b] \to E$. For $E = \mathbb{R}$, the integral was defined by the German mathematician Bernhard Riemann (1826–1866).

By a *partition* $\mathcal{D}$ of the interval $[a, b]$ we understand a finite collection of subintervals

$$[t_0, t_1], [t_1, t_2], \dots, [t_{n-1}, t_n]$$

of $[a, b]$ such that

$$\mathcal{D}: \qquad a = t_0 < t_1 < \cdots < t_{n-1} < t_n = b.$$

By the *norm* of the partition $\mathcal{D}$ we understand the number

$$\mu(\mathcal{D}) = \max \left\{ |t_i - t_{i-1}| \ \big| \ 1 \leq i \leq n \right\}.$$

For any selection of points $\tau_1, \tau_2, \dots, \tau_n$ in $[a, b]$ such that $\tau_i \in [t_{i-1}, t_i]$ for $1 \leq i \leq n$, we consider the following vector in E:

$$S(f, \mathcal{D}) = \sum_{i=1}^{n} f(\tau_i)(t_i - t_{i-1}) \ .$$

The vector $S(f, \mathcal{D})$ is called a *Riemann sum* of the continuous function $f : [a, b] \to E$, corresponding to the partition $\mathcal{D}$ of $[a, b]$. In the notation for the Riemann sum $S(f, \mathcal{D})$ we omit the points $\tau_1, \tau_2, \dots, \tau_n$ since they play a minor role only.

Lemma 4.5.1. *Let $f : [a, b] \to E$ be a continuous function. Then for every $\varepsilon > 0$, there exists a $\delta > 0$ such that*

$$\|S(f, \mathcal{D}') - S(f, \mathcal{D}'')\|_E < \varepsilon$$

for every pair of Riemann sums $S(f, \mathcal{D}')$ and $S(f, \mathcal{D}'')$ of f corresponding to partitions $\mathcal{D}'$ and $\mathcal{D}''$ of $[a, b]$ with norms $\mu(\mathcal{D}')$, $\mu(\mathcal{D}'') < \delta$.

Proof. Let $\varepsilon > 0$ be given. According to Theorem 3.6.6, $f : [a, b] \to E$ is uniformly continuous, and hence we can choose $\delta > 0$ such that

$$\|f(\tau') - f(\tau'')\|_E < \frac{\varepsilon}{b - a}$$

for all $\tau', \tau'' \in [a, b]$ with $|\tau' - \tau''| < 2\delta$.

Now let $\mathcal{D}'$ and $\mathcal{D}''$ be an arbitrary pair of partitions of $[a, b]$ with norms $\mu(\mathcal{D}')$, $\mu(\mathcal{D}'') < \delta$.

Together with $\mathcal{D}'$ and $\mathcal{D}''$ we consider the partition $\mathcal{D}$ of $[a, b]$, consisting of all division points from $\mathcal{D}'$ and $\mathcal{D}''$. We denote these partitions in the following way

$$\mathcal{D}' : \quad a = t_0' < t_1' < \cdots < t_{n'-1}' < t_{n'}' = b$$
$$\mathcal{D}'' : \quad a = t_0'' < t_1'' < \cdots < t_{n''-1}'' < t_{n''}'' = b$$
$$\mathcal{D} : \quad a = t_0 < t_1 < \cdots < t_{n-1} < t_n = b \ .$$

For every i with $1 \le i \le n$ there is exactly one i' and one i'' with $1 \le i' \le n'$ and $1 \le i'' \le n''$, such that

$$[t_{i-1}, t_i] = [t_{i'-1}', t_{i'}'] \cap [t_{i''-1}'', t_{i''}''] \ .$$

A small reflection reveals that

$$S(f, \mathcal{D}') - S(f, \mathcal{D}'') = \sum_{i=1}^{n} (f(\tau_{i'}') - f(\tau_{i''}''))(t_i - t_{i-1}) \ .$$

We note that $|\tau_{i'}' - \tau_{i''}''| < 2\delta$, since

$$\begin{cases} \tau_{i'}' \in [t_{i'-1}', t_{i'}'], \ \tau_{i''}'' \in [t_{i''-1}'', t_{i''}''] \\[2mm] [t_{i'-1}', t_{i'}'] \cap [t_{i''-1}'', t_{i''}''] \neq \emptyset \\[2mm] |t_{i'}' - t_{i'-1}'| < \delta, \ |t_{i''}'' - t_{i''-1}''| < \delta \ . \end{cases}$$

This gives us the estimate

$$\|S(f, \mathcal{D}') - S(f, \mathcal{D}'')\|_E \le \sum_{i=1}^{n} \|f(\tau_{i'}') - f(\tau_{i''}'')\|_E (t_i - t_{i-1})$$

$$\le \sum_{i=1}^{n} \frac{\varepsilon}{b - a}(t_i - t_{i-1})$$

$$= \varepsilon.$$

And hence we have proved Lemma 4.5.1 $\square$

Theorem 4.5.2. *Let $f : [a, b] \to E$ be a continuous function with values in a Banach space E. Then there exists exactly one vector $I(f) \in E$ which satisfies the following condition:*

For every $\varepsilon > 0$, there exists a $\delta > 0$ such that

$$\|I(f) - S(f, \mathcal{D})\|_E < \varepsilon$$

for every Riemann sum $S(f, \mathcal{D})$ of f corresponding to a partition $\mathcal{D}$ of $[a, b]$ with norm $\mu(\mathcal{D}) < \delta$.

Proof. Clearly there can be at most one vector $I(f) \in E$ satisfying the condition in the theorem. To prove the existence of such a vector, we proceed as follows.

Let $\mathcal{D}_1, \mathcal{D}_2, \ldots, \mathcal{D}_n, \ldots$ be an arbitrary sequence of partitions of $[a, b]$ such that the corresponding sequence of norms satisfies

$$\mu(\mathcal{D}_n) \to 0 \quad \text{for} \quad n \to \infty .$$

Let $(S(f, \mathcal{D}_n))$ be a sequence of corresponding Riemann sums of f.

Assertion 4.5.3. $(S(f, \mathcal{D}_n))$ *is a Cauchy sequence in E.*

Proof of Assertion 4.5.3. Let $\varepsilon > 0$ be given. Choose $\delta > 0$ according to Lemma 4.5.1. Then choose $n_0 \in \mathbb{N}$ such that $\mu(\mathcal{D}_n) < \delta$ for $n \geq n_0$. Now it is clear that

$$\|S(f, \mathcal{D}_n) - S(f, \mathcal{D}_m)\|_E < \varepsilon \quad \text{for} \quad n, m \geq n_0 ,$$

which proves the assertion. $\square$

Since E is a Banach space it follows that the Cauchy sequence $(S(f, \mathcal{D}_n))$ is convergent in E. Therefore there exists a vector $I(f) \in E$ such that

$$S(f, \mathcal{D}_n) \to I(f) \quad \text{for} \quad n \to \infty .$$

We shall now prove that the vector $I(f) \in E$ satisfies the condition in the theorem. To this end, let $\varepsilon > 0$ be given. Then, according to Lemma 4.5.1, we can choose a $\delta > 0$ such that

$$\|S(f, \mathcal{D}') - S(f, \mathcal{D}'')\|_E < \frac{\varepsilon}{2} \quad \text{when} \quad \mu(\mathcal{D}'), \mu(\mathcal{D}'') < \delta .$$

This $\delta > 0$ is sufficient. Indeed, choose $n_0 \in \mathbb{N}$ such that

$$\mu(\mathcal{D}_n) < \delta \quad \text{and} \quad \|I(f) - S(f, \mathcal{D}_n)\|_E < \frac{\varepsilon}{2} \quad \text{for} \quad n \geq n_0 .$$

If $S(f, \mathcal{D})$ is an arbitrary Riemann sum of f corresponding to a partition $\mathcal{D}$ of $[a, b]$ with norm $\mu(\mathcal{D}) < \delta$, then, for a fixed $n \geq n_0$, we get:

$$\|I(f) - S(f, \mathcal{D})\|_E \leq \|I(f) - S(f, \mathcal{D}_n)\|_E + \|S(f, \mathcal{D}_n) - S(f, \mathcal{D})\|_E$$

$$< \frac{\varepsilon}{2} + \frac{\varepsilon}{2} = \varepsilon ,$$

which completes the proof of Theorem 4.5.2. $\square$

There is a constructive element in the proof of Theorem 4.5.2 since we get the vector $I(f) \in E$ as the limit of a sequence of Riemann sums of f for which the norms of the underlying partitions of $[a,b]$ goes to 0. The advantage of the formulation chosen for Theorem 4.5.2 is that it shows immediately that the vector $I(f)$ is independent of the sequence of Riemann sums of f which goes into its construction.

Definition 4.5.4. Let $f : [a,b] \to E$ be a continuous function with values in a Banach space E. Then the unique vector $I(f)$ in E given by Theorem 4.5.2 is called the *integral* of f over $[a,b]$ and is denoted by

$$I(f) = \int_a^b f(t)dt.$$

From the proof of Theorem 4.5.2 we extract the following recipe for the evaluation of an integral:

If $(S(f, \mathcal{D}_n))$ is an arbitrary sequence of Riemann sums of f corresponding to the sequence of partitions $\mathcal{D}_1, \mathcal{D}_2, \ldots, \mathcal{D}_n, \ldots$ of $[a,b]$ such that the norms $\mu(\mathcal{D}_n) \to 0$ for $n \to \infty$, then

$$I(f) = \int_a^b f(t)dt = \lim_{n \to \infty} S(f, \mathcal{D}_n) .$$

Denote by $C([a,b], E)$ the space of continuous functions $f : [a,b] \to E$ equipped as a vector space with the pointwise defined operations; cf. Example 4.1.4. Then the integral assigns a vector $I(f) \in E$ to any continuous function $f : [a,b] \to E$ in $C([a,b], E)$, and hence defines a mapping

$$I : C([a,b], E) \to E.$$

We finish this section with some elementary results about the integral.

Theorem 4.5.5. *Let E be a Banach space with norm $|| \cdot ||_E$, and let $[a,b]$ be a closed and bounded interval in $\mathbb{R}$.*

(i) *For all continuous functions $f, g : [a,b] \to E$ and all $\alpha \in \mathbb{R}$:*

$$\int_a^b (f + g)(t)dt = \int_a^b f(t)dt + \int_a^b g(t)dt$$

and

$$\int_a^b (\alpha f)(t)dt = \alpha \int_a^b f(t)dt.$$

In other words: The integral $I : C([a,b], E) \to E$ is a linear mapping.

(ii) *For any continuous linear function $\varphi : E \to \mathbb{R}$ and any continuous function $f : [a, b] \to E$:*

$$\varphi\left(\int_a^b f(t)dt\right) = \int_a^b (\varphi \circ f)(t)dt.$$

(iii) *For any continuous function $f : [a, b] \to E$:*

$$\left\|\int_a^b f(t)dt\right\|_E \leq \int_a^b \|f(t)\|_E dt.$$

Proof. Let $\mathcal{D}_1, \mathcal{D}_2, \ldots, \mathcal{D}_n, \ldots$ be a sequence of partitions of $[a, b]$, for which the norms $\mu(\mathcal{D}_n) \to 0$ for $n \to \infty$.

Proof of (i). Let $(S(f, \mathcal{D}_n))$ and $(S(g, \mathcal{D}_n))$ be the corresponding sequences of Riemann sums of f and g, where we choose the same numbers τ_i in the subintervals of $[a, b]$ in the partitions $\mathcal{D}_n$ for the two functions. Clearly then,

$$S(f + g, \mathcal{D}_n) = S(f, \mathcal{D}_n) + S(g, \mathcal{D}_n)$$
$$S(\alpha f, \mathcal{D}_n) = \alpha\, S(f, \mathcal{D}_n)$$

are Riemann sums of $f + g$ and αf respectively, for all $n \in \mathbb{N}$. Elementary results on the convergence of sequences now show that the integral is a linear mapping.

Proof of (ii). Since $\varphi : E \to \mathbb{R}$ and $f : [a, b] \to E$ both are continuous functions, the composition $\varphi \circ f : [a, b] \to \mathbb{R}$ is a continuous function. Hence the integral $\int_a^b (\varphi \circ f)(t)dt$ exists.

Now let $(S(f, \mathcal{D}_n))$ be a sequence of Riemann sums of f corresponding to the sequence $\mathcal{D}_1, \mathcal{D}_2, \ldots, \mathcal{D}_n, \ldots$ of partitions of $[a, b]$.

From the linearity of φ it is immediate that

$$\varphi(S(f, \mathcal{D}_n)) = S(\varphi \circ f, \mathcal{D}_n)$$

is a Riemann sum of $\varphi \circ f$ for all $n \in \mathbb{N}$.

Since φ is continuous, by applying Theorem 2.9.3 we then get

$$\varphi\left(\int_a^b f(t)dt\right) = \varphi\left(\lim_{n \to \infty} S(f, \mathcal{D}_n)\right)$$

$$= \lim_{n \to \infty} \varphi\left(S(f, \mathcal{D}_n)\right)$$

$$= \lim_{n \to \infty} S(\varphi \circ f, \mathcal{D}_n)$$

$$= \int_a^b (\varphi \circ f)(t)dt\,,$$

which proves (ii).

Proof of (iii). For all $t, t_0 \in [a, b]$ we have

$$\big| \, \|f(t)\|_E - \|f(t_0)\|_E \, \big| \leq \|f(t) - f(t_0)\|_E.$$

This shows that the function $\|f\|_E : [a, b] \to \mathbb{R}$, which to $t \in [a, b]$ assigns $\|f(t)\|_E \in \mathbb{R}$, is continuous. Hence the integral $\int_a^b \|f(t)\|_E dt$ exists.

If

$$S(f, \mathcal{D}) = \sum_{i=1}^{n} f(\tau_i)(t_i - t_{i-1})$$

is a Riemann sum of f corresponding to the partition $\mathcal{D}$ of $[a, b]$, then the triangle inequality gives:

$$\|S(f, \mathcal{D})\|_E \leq \sum_{i=1}^{n} \|f(\tau_i)\|_E (t_i - t_{i-1})$$
$$= S\left(\|f\|_E, \mathcal{D}\right),$$

where $S(\|f\|_E, \mathcal{D})$ is a Riemann sum of $\|f\|_E$.

Now let $(S(f, \mathcal{D}_n))$ be a sequence of Riemann sums of f corresponding to the sequence $\mathcal{D}_1, \mathcal{D}_2, \ldots, \mathcal{D}_n, \ldots$ of partitions. Then

$$\|S(f, \mathcal{D}_n)\|_E \leq S\left(\|f\|_E, \mathcal{D}_n\right)$$

for all $n \in \mathbb{N}$. In the limit n going to ∞, it follows that

$$\left\| \int_a^b f(t) dt \right\|_E \leq \int_a^b \|f(t)\|_E dt,$$

which proves (iii).

This completes the proof of Theorem 4.5.4. $\square$

Exercises and Further Results

Exercise 4.1. Let K be a finite set with n elements equipped with the discrete topology. And let $C(K, \mathbb{R})$ denote the vector space of continuous functions $f : K \to \mathbb{R}$ with the pointwise defined vector space operations.

Show that one can construct a linear isomorphism between the vector spaces $C(K, \mathbb{R})$ and $\mathbb{R}^n$.

Exercise 4.2. Let K_1 and K_2 be compact topological spaces. For $i = 1, 2$, let $C(K_i, \mathbb{R})$ denote the vector space of continuous functions $f : K_i \to \mathbb{R}$ with the pointwise defined vector space operations. Turn $C(K_i, \mathbb{R})$ into a normed vector space with the uniform norm $\|f\| = \sup_{x \in K_i} |f(x)|$.

Suppose now that K_1 and K_2 are homeomorphic. Show that one can construct a norm preserving, linear isomorphism between the normed vector spaces $C(K_1, \mathbb{R})$ and $C(K_2, \mathbb{R})$.

Exercise 4.3. Let $E = C^\infty([0, 2\pi], \mathbb{R})$ be the vector space of differentiable functions $f : [0, 2\pi] \to \mathbb{R}$ of class C^∞.
   For $f \in E$ we set:

$$\|f\|_0 = \sup \left\{ |f(x)| \mid x \in [0, 2\pi] \right\}$$

$$\|f\|_1 = \sup\{ |f(x)| + |f'(x)| \mid x \in [0, 2\pi] \} .$$

1) Show that $\| \cdot \|_0$ and $\| \cdot \|_1$ are norms in E.

Define the linear mapping $D : E \to E$ by associating to $f \in E$ the derivative $f' \in E$ of f, i.e.

$$D(f) = f' \text{ for } f \in E.$$

2) Show that for every $n \in \mathbb{N}$ there exists a function $f_n \in E$ for which $\|f_n\|_0 = 1$ and $\|D(f_n)\|_0 = n$.
   Utilize this to show that $D : E \to E$ is not continuous, when E is equipped with the norm $\| \cdot \|_0$.

3) Show that $D : E_1 \to E_0$ is continuous, when E_1 is E equipped with the norm $\| \cdot \|_1$, and E_0 is E equipped with the norm $\| \cdot \|_0$.

Exercise 4.4. Let V be the space of continuous functions $f : \mathbb{R} \to \mathbb{R}$, such that $f(x) \to 0$ for $|x| \to \infty$.
   For a function $f \in V$ holds, in other words

$$\forall \varepsilon > 0 \; \exists a \in \mathbb{R}^+ \; \forall x \in \mathbb{R} : |x| > a \Rightarrow |f(x)| < \varepsilon .$$

Define the operations 'addition' and 'multiplication with scalars' in V by the obvious pointwise definitions.

1) Show that V is a vector space.

2) Show that every function $f \in V$ is bounded.

Making use of 2), we can define

$$\|f\| = \sup\{ |f(x)| \mid x \in \mathbb{R}\} \text{ for } f \in V .$$

3) Show that $\|\cdot\|$ is a norm in V.

Exercise 4.5. Let V be the space of polynomial functions $f : [0, 1] \to \mathbb{R}$ in the variable $t \in [0, 1]$, and define the operations 'addition' and 'multiplication with scalars' in V by the usual rules for polynomials.

1) Show that V is a normed vector space with the norm

$$||f|| = \int_0^1 |f(t)|\,dt \quad \text{for} \quad f \in V.$$

2) Define the mapping $B : V \times V \to \mathbb{R}$ by

$$B(f,g) = \int_0^1 f(t)g(t)\,dt \quad \text{for} \quad f,g \in V.$$

Prove that the mappings $B(f, g_0), B(f_0, g) : V \to \mathbb{R}$ are continuous linear mappings of $f \in V$, respectively $g \in V$, for fixed g_0, respectively f_0.

3) Show that the mapping $B : V \times V \to \mathbb{R}$ is not continuous, when $V \times V$ is given the product topology.

Exercise 4.6. Let V be the space of bounded sequences

$$x = (\alpha_1, \alpha_2, \ldots, \alpha_i, \ldots)$$

of real numbers $\alpha_i \in \mathbb{R}$.

Define the operations 'addition' and 'multiplication with scalars' in V by the obvious coordinate-wise definitions. Further, set $||x|| = \sup_{i \in \mathbb{N}} |\alpha_i|$.

1) Show that V is a vector space and that $||\cdot||$ is a norm in V.

2) Show that a sequence (x_n) of elements in V is convergent to the element $x_0 \in V$ if and only if the coordinates in x_n converges uniformly to the coordinates in x_0.

3) Show that V is a Banach space with the norm $||\cdot||$.

Exercise 4.7. Let V be the space of sequences

$$x = (\alpha_1, \alpha_2, \ldots, \alpha_i, \ldots)$$

of real numbers $\alpha_i \in \mathbb{R}$ in which at most finitely many $\alpha_i \neq 0$.

Define the operations 'addition' and 'multiplication with scalars' in V by the obvious coordinate-wise definitions. Further, set $||x|| = \sum_{i=1}^{\infty} |\alpha_i|$.

1) Show that V is a vector space and that $||\cdot||$ is a norm in V.

2) Consider an arbitrary infinite series of real numbers $\sum_{i=1}^{\infty} a_i$. Define the sequence (x_n) in V by $x_1 = (a_1, 0, 0, \ldots)$, $x_2 = (a_1, a_2, 0, \ldots)$, and, in general, $x_n = (a_1, a_2, \ldots, a_n, 0, \ldots)$.

Show that the series $\sum_{i=1}^{\infty} |a_i|$ is convergent, if and only if the sequence (x_n) is a Cauchy sequence in the normed vector space V, i.e.

$$\forall \varepsilon > 0 \; \exists n_0 \in \mathbb{N} \; \forall n, k \in \mathbb{N} : n \geq n_0 \Rightarrow ||x_{n+k} - x_n|| < \varepsilon .$$

3) Give an example of a Cauchy sequence in the normed vector space V that has no limit point in V.

Exercise 4.8. Let V be a finite dimensional, normed vector space with norm $||\cdot||$. Let $T : V \to V$ be an arbitrary linear mapping. Show that there exists a unit vector $x_0 \in V$, i.e. $||x_0|| = 1$, such that $||T(x_0)|| = ||T||$, where $||T||$ is the operator norm of T, i.e. $||T|| = \sup\{||T(x)|| \mid ||x|| = 1\}$.

Show by an example that this does not hold in general, when V has infinite dimension.

Exercise 4.9. Let $C([0, 1], \mathbb{R})$ be the vector space of continuous real-valued functions in the unit interval $[0, 1]$. For a continuous function $f : [0, 1] \to \mathbb{R}$ we set

$$||f||_1 = \int_0^1 |f(x)| \, dx \ .$$

1) Show that $||\cdot||_1$ is a norm in $C([0, 1], \mathbb{R})$.

We now equip $C([0, 1], \mathbb{R})$ as a normed vector space with the norm $||\cdot||_1$ and define the function

$$I : C([0, 1], \mathbb{R}) \to \mathbb{R} \quad \text{by} \quad I(f) = \int_0^1 f(x) dx \ .$$

2) Show that I is a continuous linear function.

3) Determine the operator norm of I.

4) In $C([0, 1], \mathbb{R})$ equipped with the norm $||\cdot||_1$, consider the sequences (f_n) and (g_n) defined by

$$f_n(x) = \begin{cases} 1 - nx & \text{for } 0 \le x \le \frac{1}{n} \\ 0 & \text{for } \frac{1}{n} \le x \le 1 \ , \end{cases}$$

$$g_n(x) = \begin{cases} n - n^2 x & \text{for } 0 \le x \le \frac{1}{n} \\ 0 & \text{for } \frac{1}{n} \le x \le 1 \ . \end{cases}$$

Examine the convergence of each of these sequences and, in the case of convergence, determine the limit function.

Exercise 4.10. Let $C([0, 1], \mathbb{R})$ be the vector space of continuous functions $f : [0, 1] \to \mathbb{R}$ equipped as a normed vector space with the norm

$$||f|| = \sup_{0 \le x \le 1} |f(x)| \ .$$

Let $\Phi = \Phi(x, y) : [0, 1] \times [0, 1] \to \mathbb{R}$ be a continuous function in two variables defined in the square $[0, 1] \times [0, 1]$ in $\mathbb{R}^2$. Assume that $\Phi(x, y) \ge 0$ for all $(x, y) \in [0, 1] \times [0, 1]$.

Define the function $\varphi = \varphi(x) : [0, 1] \to \mathbb{R}$ by

$$\varphi(x) = \sup_{0 \leq y \leq 1} \Phi(x, y) .$$

For $f \in C([0, 1], \mathbb{R})$ we define the function $f_\Phi = f_\Phi(y) : [0, 1] \to \mathbb{R}$ by

$$f_\Phi(y) = \int_0^1 \Phi(x, y) f(x) dx.$$

1) Show that for every $\varepsilon > 0$ there exists a $\delta > 0$ such that

a) $\quad \Phi(x_0, y) - \varepsilon \leq \Phi(x, y) \leq \Phi(x_0, y) + \varepsilon$

$\quad$ for $|x - x_0| \leq \delta$ and all $y \in [0, 1]$.

b) $\quad |\Phi(x, y) - \Phi(x, y_0)| \leq \varepsilon$

$\quad$ for $|y - y_0| \leq \delta$ and all $x \in [0, 1]$.

Make use of this to show that the functions $\varphi = \varphi(x)$ and $f_\Phi = f_\Phi(y)$ are continuous.

Since $f_\Phi \in C([0, 1], \mathbb{R})$, we can define the mapping

$$L : C([0, 1], \mathbb{R}) \to C([0, 1], \mathbb{R}) \quad \text{by} \quad L(f) = f_\Phi .$$

The mapping L is called an *integral operator* with *kernel* Φ.

2) Show that L is a continuous linear mapping.

3) Show that for all $f \in C([0, 1], \mathbb{R})$, we have

$$\|L(f)\| \leq \left(\int_0^1 \varphi(x) dx \right) \|f\| .$$

4) Show that the operator norm for L is given by

$$\|L\| = \sup_{0 \leq y \leq 1} \int_0^1 \Phi(x, y) dx .$$

Exercise 4.11. Denote by $C([-1, 1], \mathbb{R})$ the vector space of continuous real-valued functions defined in the closed interval $[-1, 1]$. For a continuous function $f : [-1, 1] \to \mathbb{R}$ we set $\|f\|_1 = \int_{-1}^1 |f(x)| dx$.

1) Show that $\|\cdot\|_1$ is a norm in $C([-1, 1], \mathbb{R})$.

2) Define the sequence (f_n) of continuous functions $f_n : [-1, 1] \to \mathbb{R}$ by

$$f_n(x) = \begin{cases} 1 & \text{for } -1 \leq x \leq 0 \\ 1 - nx & \text{for } 0 \leq x \leq \frac{1}{n} \\ 0 & \text{for } \frac{1}{n} \leq x \leq 1 . \end{cases}$$

Show that (f_n) is a Cauchy sequence in the normed vector space $(C([-1,1],\mathbb{R}), ||\cdot||_1)$.

3) Examine whether the normed vector space $(C([-1,1],\mathbb{R}), ||\cdot||_1)$ is a Banach space.

Exercise 4.12. Let l^1 denote the space of *absolute summable* sequences of real numbers, i.e. the space of real sequences (a_n) for which the infinite series $\sum_{n=1}^{\infty} |a_n|$ is convergent.

1) Prove that l^1 is a normed vector space with the definitions

$$(a_n) + (b_n) = (a_n + b_n) \quad \text{for} \quad (a_n), (b_n) \in l^1$$

$$\alpha(a_n) = (\alpha\, a_n) \quad \text{for} \quad \alpha \in \mathbb{R}, (a_n) \in l^1$$

$$||(a_n)||_1 = \sum_{n=1}^{\infty} |a_n| \quad \text{for} \quad (a_n) \in l^1 .$$

2) Prove that l^1 is a Banach space.

Exercise 4.13. Let l^∞ denote the space of bounded real sequences (a_n).

1) Prove that l^∞ is a normed vector space with the definitions of 'addition' and 'multiplication with scalars' as in Exercise 4.12 and

$$||(a_n)||_\infty = \sup_{n \in \mathbb{N}} |a_n| \quad \text{for} \quad (a_n) \in l^\infty .$$

2) Prove that l^∞ is a Banach space.

Exercise 4.14. Let l^2 denote the space of *square summable* sequences of real numbers, i.e. the space of real sequences (a_n) for which the infinite series $\sum_{n=1}^{\infty} a_n{}^2$ is convergent.

1) Prove that l^2 is a normed vector space with the definitions of 'addition' and 'multiplication with scalars' as in Exercise 4.12 and

$$||(a_n)||_2 = \sqrt{\sum_{n=1}^{\infty} a_n{}^2} \quad \text{for} \quad (a_n) \in l^2 .$$

2) Prove that l^2 is a Banach space.

3) Show that you can define an inner product in l^2 by which it becomes a Hilbert space.

Exercise 4.15. Let $||\cdot||$ and $||\cdot||^*$ be two norms in a vector space V. Prove that the two norms are equivalent (define the same topology in V) if and only if there exist constants $\alpha, \beta \in \mathbb{R}^+$ such that

$$\alpha||x|| \leq ||x||^* \leq \beta||x|| \quad \text{for all } x \in V.$$

Exercise 4.16. Let $V = C^\infty([a, b], \mathbb{R})$ be the vector space of differentiable functions $f : [a, b] \to \mathbb{R}$ of class C^∞.

For each integer $k \geq 0$, we define the C^k-norm $\| f \|_k$ of $f \in V$ as follows.

$$\| f \|_0 = \sup \{ | f(x) | \mid x \in [a, b] \}, \quad \text{for} \quad k = 0 \,,$$

$$\| f \|_k = \sup \{ | f(x) | + | f'(x) | + \cdots + | f^{(k)}(x) | \mid x \in [a, b] \}, \quad \text{for} \quad k \geq 1 \,.$$

Prove that none of these norms in V are equivalent.

Hint: You may wish to consider $[a, b] = [0, 2\pi]$ and suitable trigonometric functions.

Exercise 4.17. Let V be a Banach space with norm $\| \cdot \|$, and let U be a finite dimensional linear subspace of V.

1) Show that U is a Banach space when equipped with the norm $\| \cdot \|$ inherited from V.

2) Show that U is a closed subset in V in the topology induced by the norm $\| \cdot \|$.

Exercise 4.18. Let V be a Banach space with norm $\| \cdot \|$, and let $T : V \to V$ be a bounded linear operator in V.

Suppose there exists a constant $k \in \mathbb{R}$ such that

$$\| x \| \leq k \| T(x) \| \quad \text{for all} \quad x \in V.$$

Prove that the image space $T(V)$ of T is a closed subset in V.

Exercise 4.19. Let V be a Hilbert space with the inner product $\langle \cdot , \cdot \rangle$ and the induced norm $\| \cdot \|$.

Let $C(V)$ be the closed unit ball in V, i.e.

$$C(V) = \{ x \in V \mid \| x \| \leq 1 \}.$$

Prove that $C(V)$ is compact if and only if V is finite dimensional.

Hint: You may wish to use that if the vector space V is not spanned by a finite set of vectors, then you can find an infinite sequence (e_n) of pairwise orthogonal unit vectors in $C(V)$.

Exercise 4.20. Consider $\mathbb{R}^n$ with the canonical inner product defined by

$$\langle x, y \rangle = \sum_{i=1}^{n} x_i y_i \,,$$

for vectors $x = (x_1, \ldots, x_n)$ and $y = (y_1, \ldots, y_n)$ in $\mathbb{R}^n$.

Let $\{ e_1, \ldots, e_n \}$ be an arbitrary orthonormal basis in $\mathbb{R}^n$.

Show that with respect to the basis $\{e_1, \ldots, e_n\}$, the matrix representation $[T_{ij}]_{i,j=1,\ldots,n}$ of a linear mapping $T : \mathbb{R}^n \to \mathbb{R}^n$ is given by $T_{ij} = \langle T(e_j), e_i \rangle$.

Exercise 4.21. Let $L(\mathbb{R}^n, \mathbb{R}^m)$ be the vector space of linear mappings of $\mathbb{R}^n$ into $\mathbb{R}^m$.

1) Prove that $L(\mathbb{R}^n, \mathbb{R}^m)$ can be identified with the set of $m \times n$-matrices, and hence with $\mathbb{R}^{m \cdot n}$ by successively writing the m rows one after the other.

2) Let $[T_{ij}]_{\substack{i=1,\ldots,m \\ j=1,\ldots,n}}$ denote the matrix representation of the linear mapping $T : \mathbb{R}^n \to \mathbb{R}^m$ with respect to the canonical bases in $\mathbb{R}^n$ and $\mathbb{R}^m$.

Show that

$$\|T\|_1 = \sum_{i=1}^{m} \sum_{j=1}^{n} |T_{ij}|$$

defines a norm in $L(\mathbb{R}^n, \mathbb{R}^m)$.

3) Can you suggest other norms in $L(\mathbb{R}^n, \mathbb{R}^m)$?

Exercise 4.22. Let $C([0, 2\pi], \mathbb{R})$ be the space of continuous real-valued functions in the interval $[0, 2\pi]$, equipped with the inner product

$$\langle f, g \rangle = \int_0^{2\pi} f(x)g(x)dx \quad \text{for} \quad f, g \in C([0, 2\pi], \mathbb{R}),$$

and the associated norm

$$\|f\|_2 = \sqrt{\langle f, f \rangle} \quad \text{for} \quad f \in C([0, 2\pi], \mathbb{R}).$$

For each $n \in \mathbb{N}$ define the trigonometric function $f_n \in C([0, 2\pi], \mathbb{R})$ by $f_n(x) = \sin(nx)$.

1) Prove that $\|\cdot\|_2$ is a norm in $C([0, 2\pi], \mathbb{R})$.
   <u>Hint:</u> Use the Cauchy-Schwarz inequality $|\langle f, g \rangle| \leq \|f\|_2 \|g\|_2$.
2) Find for all $n, m \in \mathbb{N}$ the distance $\|f_n - f_m\|_2$.
3) Prove that the sequence (f_n) is not convergent.
4) Prove that the set $A = \{f_n \mid n \in \mathbb{N}\}$ is closed and bounded.
5) Prove that A is not compact.

Chapter 5

Differentiation in Normed Vector Spaces

It is convenient to develop the theory of differentiability in a sufficiently general setting in order that problems arising in infinite dimensional spaces can also be considered. Infinite dimensional problems arise naturally in the calculus of variations, since the underlying spaces on which relevant functions are to be minimized or maximized typically are function spaces, or more generally infinite dimensional manifolds. The configuration spaces for mechanical systems are often of infinite dimension. For instance, the configuration space of an oscillating string is an infinite dimensional linear space, while the configuration space of a liquid is a complicated infinite dimensional manifold.

We begin this chapter by analysing the classical definition of differentiability of a real-valued function of one real variable. By focusing on differentiability of a function as a property that allows the function to be locally approximated by linear functions, we arrive at a stage where we can naturally introduce differentiability in the setting of normed vector spaces, even of infinite dimension. It turns out that there are no extra difficulties in introducing the concept of differentiability in this more general setting. For the finite dimensional Euclidean spaces $\mathbb{R}^n$, we recover the case of functions of several variables.

In the theory of differentiable functions of several variables, two fundamental theorems stand out in particular, namely the *Chain Rule* on differentiability of a composition of functions, and the *Inverse Function Theorem*, which ensures the existence of a differentiable inverse function to a differentiable function in a neighbourhood of any point at which the differential has an inverse linear mapping.

The Chain Rule can be stated and proved for differentiable mappings between open sets in normed vector spaces in complete generality without any added difficulties.

For the statement and proof of an Inverse Function Theorem many new complications arise when we want to include also normed vector spaces of infinite dimension. The Inverse Function Theorem remains true, however, in complete generality for the class of Banach spaces. In this chapter we establish this for differentiability class C^1. As necessary preparations we prove a generalized Mean Value Theorem and introduce the notion of toplinear isomorphisms.

5.1 Differentiability in the classical setting

Let $U \subseteq \mathbb{R}$ be an open interval in $\mathbb{R}$, and let $f : U \to \mathbb{R}$ be a real-valued function. As should already be known, we say that f is differentiable at a point $x \in U$, if there exists a real number $a \in \mathbb{R}$, such that

$$\frac{f(x+h) - f(x)}{h} \to a \quad \text{as} \quad h \to 0.$$

The uniquely defined number $a \in \mathbb{R}$ is called the *derivative* of f at x, and is denoted by $f'(x)$ or $\dfrac{df}{dx}$. Strictly speaking, we should write $\dfrac{df}{dx}(x)$ in the latter notation.

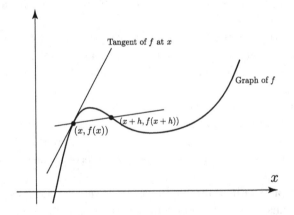

Fig. 5.1   Derivative and tangent. For differentiable functions of one real variable, the tangent line is the graph of the linear map approximating f near the point x. The value of the derivative gives the slope of the tangent.

Differentiability can be given a geometric interpretation by considering the graph of f; cf. Figure 5.1. Differentiability of $f : U \to \mathbb{R}$ at $x \in U$ is then equivalent to the secant through $(x, f(x))$ and $(x+h, f(x+h))$ having a limit position for h approaching 0, namely the line through $(x, f(x))$ with slope a. This line is called the *tangent* of the graph of f at the point $(x, f(x))$, or just the tangent of f at x. The tangent is a first order approximation of the graph in a neighbourhood of x.

The above definition is not immediately suitable for generalization, and therefore we seek a reformulation which more directly expresses that differentiability is concerned with the question of linear approximation. With this in mind we set

$$\varepsilon(h) = \begin{cases} \dfrac{f(x+h) - f(x)}{h} - a & \text{for } h \neq 0 \\ 0 & \text{for } h = 0. \end{cases}$$

Note here that ε is a function defined in $U_x = \{h \in \mathbb{R} \mid x + h \in U\}$, which is the open interval around $0 \in \mathbb{R}$ that we get by translating U by $-x$.

By introducing the linear function $\varphi : \mathbb{R} \to \mathbb{R}$ defined by $\varphi(h) = a \cdot h$, we get

$$f(x+h) - f(x) = \varphi(h) + \varepsilon(h)h \tag{5.1}$$

for all $h \in U_x$.

Clearly, $f : U \to \mathbb{R}$ is differentiable at $x \in U$ with derivative $a \in \mathbb{R}$ if and only if $\varepsilon = \varepsilon(h) : U_x \to \mathbb{R}$ is continuous at 0, or equivalently that

$$\varepsilon(h) \to 0 \quad \text{as} \quad h \to 0.$$

Since any linear function $\varphi : \mathbb{R} \to \mathbb{R}$ has the form $\varphi(h) = a \cdot h$ with $a = \varphi(1)$, the definition of differentiability can be given the following formulation:

The function $f : U \to \mathbb{R}$ is differentiable at $x \in U$ if there exists a linear function $\varphi : \mathbb{R} \to \mathbb{R}$ and a function $\varepsilon = \varepsilon(h) : U_x \to \mathbb{R}$, with $\varepsilon(0) = 0$ and $\varepsilon = \varepsilon(h)$ continuous at 0, such that equation (5.1) is satisfied for all $h \in U_x$.

Since $h/|h| = \pm 1$, for $h \neq 0$, we can by changing the sign of $\varepsilon(h)$ for $h < 0$, but keeping the name $\varepsilon(h)$, rewrite equation (5.1) in the form

$$f(x+h) - f(x) = \varphi(h) + \varepsilon(h)|h| \tag{5.2}$$

for all $h \in U_x$.

Note that the change of sign does not change the properties of continuity of $\varepsilon = \varepsilon(h)$ at 0, since $\varepsilon(0) = 0$.

Definition 5.1.1. For $U \subseteq \mathbb{R}$ and $x \in U$ we set $U_x = \{h \in \mathbb{R} \mid x + h \in U\}$. A function $\varepsilon = \varepsilon(h) : U_x \to \mathbb{R}$, which is continuous at 0 with value $\varepsilon(0) = 0$, is called an *ε-function*.

. It is now evident that the definition of differentiability can be formulated as follows.

Definition 5.1.2. A real-valued function $f : U \to \mathbb{R}$ defined in an open interval $U \subseteq \mathbb{R}$ is said to be *differentiable at a point* $x \in U$ if there exists a linear function $\varphi : \mathbb{R} \to \mathbb{R}$, and an ε-function $\varepsilon = \varepsilon(h) : U_x \to \mathbb{R}$ such that

$$f(x + h) - f(x) = \varphi(h) + \varepsilon(h)|h|$$

for all $h \in U_x$.

The linear function φ in Definition 5.1.2 is uniquely determined, and is called the *differential* of $f : U \to \mathbb{R}$ at $x \in U$. We denote the differential of f at x by $Df(x)$. Note that $Df(x) : \mathbb{R} \to \mathbb{R}$ is a linear function which depends on x, and that

$$\frac{df}{dx} = Df(x)(1).$$

Thus, as one would expect, there is a close relationship between the differential and the derivative of f at x.

The affine function

$$h \mapsto f(x) + Df(x)(h),$$

whose graph is precisely the tangent of f at x, is a first order approximation to the function

$$h \mapsto f(x + h).$$

After this reformulation, the concept of differentiability can easily be generalized to mappings between open sets in spaces which, like $\mathbb{R}$, have a linear structure and a topology. For things to work out, there has of course to be a connection between the linear structure and the topology. Maximal success is obtained when the topology is induced by a norm in a vector space, as in the case of $\mathbb{R}$ with absolute value $|\cdot|$. Hence we arrive at the conclusion that a good setting for the study of differentiability is mappings between open sets in vector spaces equipped with a norm (normed vector spaces). This setting comprises in particular the vector spaces $\mathbb{R}^n$ equipped with the Euclidean norm.

It does not impose extra difficulties at the basic level to allow the underlying normed vector spaces to be of infinite dimension; and in fact this generality is necessary in certain contexts.

Example 5.1.3 (sketchy). Consider a differentiable surface M in $\mathbb{R}^3$ (or more generally a Riemannian manifold). It is of interest to study whether there are closed geodesics on M. Informally speaking, a curve is called a *geodesic*, if it is the shortest connection on the surface between any two sufficiently close points on the curve. A *closed geodesic* is a geodesic which is periodic. A good example is a great circle on a sphere.

Now, consider a suitable class of mappings $H(S^1, M)$ of the circle S^1 into M. Specifically, $H(S^1, M)$ consists of the continuous mappings $\sigma : S^1 \to M$ that are almost everywhere differentiable and have a derivative σ' which is square integrable; cf. Figure 5.2.

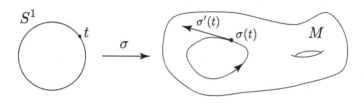

Fig. 5.2   An almost everywhere differentiable map σ of the circle S^1 into a space M. The space of all such mappings with a suitable topology is an example of a *Hilbert manifold*.

In the space of mappings $H(S^1, M)$, we can define an *energy function*

$$E : H(S^1, M) \to \mathbb{R}, \quad \text{by} \quad E(\sigma) = \int_{S^1} ||\sigma'(t)||^2 dt.$$

The mapping space $H(S^1, M)$ can be equipped with a topology in such a way that every point (mapping) $\sigma \in H(S^1, M)$ has an open neighbourhood which is homeomorphic to an open set in an infinite dimensional normed vector space of the particularly nice type called a *Hilbert space*; cf. Definition 4.4.5. We say that $H(S^1, M)$ locally looks like a Hilbert space. A topological space with this property is called a *Hilbert manifold*.

The notion of differentiability can be defined in a natural way in $H(S^1, M)$, and it turns out that the energy function $E : H(S^1, M) \to \mathbb{R}$ is indeed differentiable.

The points (mappings) $\sigma \in H(S^1, M)$, at which the differential of E is 0, are of particular interest. They are called *critical points* of E. It turns out that $\sigma \in H(S^1, M)$ is a critical point of E if and only if $\sigma : S^1 \to M$ represents a closed geodesic on M. This is a useful viewpoint since there exists a well-developed theory of critical points of differentiable functions. This theory, to which we shall give a short introduction in Section 9.6, page 267, was founded around 1930 by the American mathematician Marston Morse (1892–1977), and is known today as *Morse theory*.

Several other geometric variational problems can be handled effectively in such an infinite dimensional setting. ◄

Example 5.1.4 (sketchy). Let M be a differentiable surface in $\mathbb{R}^3$. Consider a liquid on the surface, and identify the points of the liquid at an initial position with the points of M. A new position of the liquid is then completely determined by a mapping $h : M \to M$ of M onto itself, that takes the point of the liquid, which initially is placed at $x \in M$, to the new position $h(x) \in M$. In fact, the mapping h is a homeomorphism, for which both h and the inverse mapping h^{-1} are differentiable. Such a mapping is called a *diffeomorphism* of M. The configuration space (i.e. the space of possible positions) for a liquid on M can in this way be identified with the space of diffeomorphisms $\mathcal{D}(M)$ of M. This space is a complicated infinite dimensional manifold, in the study of which it is of interest to use methods from mathematical analysis. ◄

Example 5.1.5. The configuration space for an oscillating string is an infinite dimensional space. If the string is identified with the graph of a function, its configuration space can be identified with the space V of continuous functions $f : [a, b] \to \mathbb{R}$, where $f(a) = f(b) = 0$. The space V can be equipped with the structure of an infinite dimensional normed vector space using the uniform norm; cf. Example 4.1.4, page 86. ◄

5.2 Differentiability in normed vector spaces

In the following, E and F denote normed vector spaces with norms $||\cdot||_E$ and $||\cdot||_F$, respectively.

For an open set $U \subseteq E$ in E, and a point $x \in U$, we denote by U_x the open set in E, which we get from U by translation by the vector $-x$, i.e.

$$U_x = \left\{ h \in E \mid x + h \in U \right\}.$$

Note that $0 \in U_x$.

We shall use the term *ε-mapping* for any mapping

$$\varepsilon = \varepsilon(h) : U_x \to F,$$

which is continuous at $0 \in U_x$, and has the value $\varepsilon(0) = 0$.

Definition 5.2.1. Let $U \subseteq E$ be an open set in E, and let $x \in U$ be a point in U. A mapping $f : U \to F$ is said to be *differentiable* at $x \in U$ if there exist a continuous linear mapping $L : E \to F$ and an ε-mapping $\varepsilon = \varepsilon(h) : U_x \to F$ such that

$$f(x + h) - f(x) = L(h) + \varepsilon(h)\|h\|_E,$$

for all $h \in U_x$.

Remark 5.2.2. If E is finite dimensional, then by Remark 4.2.2 a linear mapping $L : E \to F$ is automatically continuous, and therefore in this case we do not need to specify that L must be continuous in Definition 5.2.1. ◁

Remark 5.2.3. Since we require L to be continuous in Definition 5.2.1, clearly f is *continuous* at $x \in U$, if f is *differentiable* at $x \in U$. Conversely, we could also have required f to be continuous at $x \in U$, before we begin to talk about differentiability at $x \in U$. If we include continuity of f in the definition of differentiability, then if f is differentiable at $x \in U$, the linear map L is continuous at $0 \in E$, and hence continuous according to Theorem 4.2.4. So the latter definition is equivalent to the former. ◁

Lemma 5.2.4. *The linear mapping $L : E \to F$ in Definition 5.2.1 is uniquely determined.*

Proof. Assume that $L_1, L_2 : E \to F$ are linear mappings, with corresponding ε-mappings $\varepsilon_1, \varepsilon_2 : U_x \to F$, such that

$$f(x + h) - f(x) = L_1(h) + \varepsilon_1(h)\|h\|_E$$
$$= L_2(h) + \varepsilon_2(h)\|h\|_E$$

for all $h \in U_x$. This can be rewritten as

$$L_1(h) - L_2(h) = (\varepsilon_2(h) - \varepsilon_1(h))\|h\|_E$$

for all $h \in U_x$.

Let $h_0 \in E$ be an arbitrary vector in E. For $\alpha \in \mathbb{R}^+$ sufficiently small, the vector $h = \alpha h_0 \in U_x$, since U_x is an open set in E. For such an $\alpha \in \mathbb{R}^+$ we get

$$L_1(\alpha h_0) - L_2(\alpha h_0) = (\varepsilon_2(\alpha h_0) - \varepsilon_1(\alpha h_0))\alpha\|h_0\|_E,$$

and since L_1 and L_2 are linear, we can reduce by α and get

$$L_1(h_0) - L_2(h_0) = (\varepsilon_2(\alpha h_0) - \varepsilon_1(\alpha h_0))\|h_0\|_E.$$

Since

$$\varepsilon_2(\alpha h_0) - \varepsilon_1(\alpha h_0) \to 0 \in F \quad \text{for} \quad \alpha \to 0,$$

it follows that $L_1(h_0) = L_2(h_0)$.

The vector $h_0 \in E$ was arbitrarily chosen, and hence we conclude that $L_1 = L_2$. This shows that the linear mapping in the definition of differentiability is uniquely determined. $\square$

The uniquely determined continuous linear mapping $L : E \to F$ in Definition 5.2.1 is called the *differential* of $f : U \to F$ at $x \in U \subseteq E$. We denote the differential by $Df(x)$, or $df(x)$.

The differential $Df(x) : E \to F$ of f at $x \in U$ is a continuous linear mapping, acting on vectors $h \in E$. We write $Df(x)(h)$, $Df(x) \cdot h$, or more frequently just $Df(x)h$. The differential is uniquely determined by the equation

$$f(x + h) - f(x) = Df(x)h + \varepsilon(h)\|h\|_E$$

for all $h \in U_x$.

If $f : U \to F$ is differentiable at $x \in U \subseteq E$, then the affine mapping

$$h \mapsto f(x) + Df(x)h$$

is a first order approximation to the mapping

$$h \mapsto f(x + h).$$

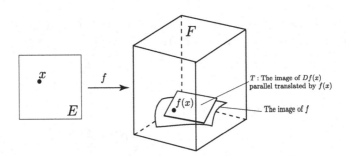

Fig. 5.3 Differentiability of a map f between normes vecor spaces. The image of the linear map $Df(x)$ parallel translated to the point $f(x)$ in F, approximates the image of f, and forms the tangent space T (an affine subspace of F) to the image of f.

Geometrically, we can think of the image of f as a distorted image of $U \subseteq E$ projected into F; see Figure 5.3. The image of E in F under the linear mapping $Df(x) : E \to F$, which is denoted by Im $Df(x)$, is a linear subspace of F. Let T be the affine subspace in F which is obtained by translation of Im $Df(x)$ in F by the vector $f(x)$, thus

$$T = \big\{ f(x) + Df(x)h \mid h \in E \big\}.$$

We see that T approximates (is 'tangential' to) the image of f in a neighbourhood of $f(x)$. This gives a geometric interpretation of differentiability, which remains valid even if E and F are infinite dimensional. In the special case $F = \mathbb{R}$, of course it gives a better insight to consider the graph of $f : U \to \mathbb{R}$, analogously to the situation known from calculus of a real-valued function of one real variable. The graph point of view can in fact be put into the general setting, since the graph of $f : U \to \mathbb{R}$ is precisely the image of $U \subseteq E$ in $E \times \mathbb{R}$ under the mapping $x \mapsto (x, f(x))$.

5.3   Interpretation of differentiability in special cases

In this section, we shall establish the connection to the classical formulation of the concept of differentiability of mappings between Euclidean spaces.

5.3.1   *Functions of one real variable*

First we consider the case $E = \mathbb{R}$ with the absolute value as the norm. The space F is an arbitrary normed vector space.

If $U \subseteq \mathbb{R}$ is an open interval in $\mathbb{R}$, a mapping $f : U \to F$ determines a *parameterized curve* in F, see Figure 5.4.

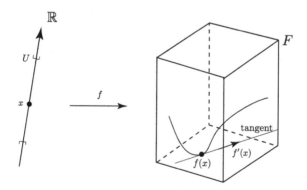

Fig. 5.4   Differentiability of a parametrized curve. The parametrized curve described by a differentiable mapping f from an open subset U of the real line into a normed vector space F has an approximating tangent line at a point $f(x)$ where $f'(x) \neq 0$.

If $f : U \to F$ is differentiable at $x \in U$, then the differential of f at x is a linear mapping $Df(x) : \mathbb{R} \to F$. Such a linear mapping is completely determined by the vector $Df(x)(1) \in F$, since

$$Df(x)(h) = Df(x)(h \cdot 1) = h \cdot Df(x)(1)$$

for all $h \in \mathbb{R}$.

When the equation for the definition of differentiability of f at $x \in U$ is rewritten as

$$\frac{f(x+h) - f(x)}{h} = Df(x)(1) + \varepsilon(h)\frac{|h|}{h} \quad \text{for} \quad h \neq 0,$$

we see that $Df(x)(1)$ can be interpreted as *the derivative* of $f : U \to F$ at $x \in U$. The derivative is denoted by $f'(x)$, or $\frac{df}{dx}(x)$. We note that $f'(x) \in F$, and stress that

$$f'(x) = \frac{df}{dx}(x) = Df(x)(1).$$

By translating $f'(x) \in F$ to $f(x)$, we get for $f'(x) \neq 0$ an oriented line segment generating the *tangent* to the parameterized curve at $f(x)$.

5.3.2 *Real-valued functions of several variables*

In this case $E = \mathbb{R}^n$ and $F = \mathbb{R}$. The norm in $\mathbb{R}^n$ is simply denoted by $||\cdot||$.

Let $U \subseteq \mathbb{R}^n$ be an open set in $\mathbb{R}^n$, and let $f : U \to \mathbb{R}$ be a function which is differentiable at $x \in U$. Then the differential of f at x is a linear function $Df(x) : \mathbb{R}^n \to \mathbb{R}$.

If $\{e_1, \ldots, e_n\}$ denotes the canonical basis in $\mathbb{R}^n$, then a vector $h \in \mathbb{R}^n$ can be written in a unique way as a linear combination

$$h = h_1 e_1 + \cdots + h_n e_n ,$$

or simply $h = (h_1, \ldots, h_n)$. Since the differential $Df(x)$ is linear, we get from this

$$Df(x)h = h_1 Df(x)e_1 + \cdots + h_n Df(x)e_n .$$

Now consider the defining equation for differentiability:

$$f(x+h) - f(x) = Df(x)h + \varepsilon(h)||h|| \quad \text{for} \quad h \in U_x .$$

If for fixed $i = 1, \ldots, n$ we consider f as a function of the ith variable, in other words if we consider only variations of the form $h = h_i e_i$, then by rewriting the defining equation for differentiability we get that

$$\frac{f(x + h_i e_i) - f(x)}{h_i} = Df(x)e_i + \varepsilon(h_i e_i)\frac{|h_i|}{h_i} \quad \text{for} \quad h_i \neq 0.$$

From this, evidently $Df(x)e_i$ can be interpreted as a derivative. We call this derivative the ith *partial derivative* of f at x, and denote it by $\dfrac{\partial f}{\partial x_i}(x)$, or by $D_i f(x)$. We stress that

$$\frac{\partial f}{\partial x_i}(x) = D_i f(x) = Df(x)e_i \ ,$$

and that

$$Df(x)h = \frac{\partial f}{\partial x_1}(x)h_1 + \cdots + \frac{\partial f}{\partial x_n}(x)h_n \ .$$

The vector

$$\nabla f(x) = \left(\frac{\partial f}{\partial x_1}(x), \ldots, \frac{\partial f}{\partial x_n}(x) \right)$$

is called the *gradient* of f at x.

If $\langle \cdot, \cdot \rangle$ denotes the canonical inner product in $\mathbb{R}^n$, see Example 2.2.3, then the differential can be written in the short form

$$Df(x)h = \langle \nabla f(x), h \rangle \ .$$

5.3.3 *Vector-valued functions of several variables*

Finally, we consider the general case $E = \mathbb{R}^n$ and $F = \mathbb{R}^m$.

Let $\{e_1, \ldots, e_n\}$ and $\{\bar{e}_1, \ldots, \bar{e}_m\}$ denote the canonical bases in $\mathbb{R}^n$ and $\mathbb{R}^m$, respectively, and let $U \subseteq \mathbb{R}^n$ be an open set in $\mathbb{R}^n$.

A mapping $f : U \to \mathbb{R}^m$ has m uniquely defined coordinate functions $f_1, \ldots, f_m : U \to \mathbb{R}$ defined by the equation

$$f(x) = f_1(x)\bar{e}_1 + \cdots + f_m(x)\bar{e}_m \quad \text{for} \quad x \in U.$$

To abbreviate the notation we simply write $f = (f_1, \ldots, f_m)$.

A linear mapping $L : \mathbb{R}^n \to \mathbb{R}^m$ has a corresponding coordinate-wise representation $L = (L_1, \ldots, L_m)$, where $L_1, \ldots, L_m : \mathbb{R}^n \to \mathbb{R}$ are linear functions.

It is easy to prove that a mapping $\varepsilon : U_x \to \mathbb{R}^m$ with coordinate representation $\varepsilon = (\varepsilon_1, \ldots, \varepsilon_m)$ is an ε-mapping if and only if all of its coordinate functions $\varepsilon_i : U_x \to \mathbb{R}$ are ε-functions.

If we write the equation in Definition 5.2.1 for the definition of differentiability of $f = (f_1, \ldots, f_m) : U \to \mathbb{R}^m$ at $x \in U$ out coordinate-wise, we get m equations

$$f_i(x + h) - f_i(x) = L_i(h) + \varepsilon_i(h)\|h\| \ , \quad h \in U_x \ ,$$

for $i = 1, \ldots, m$.

It follows immediately from this that $f : U \to \mathbb{R}^m$ is differentiable at $x \in U$ if and only if every coordinate function $f_i : U \to \mathbb{R}$ is differentiable at $x \in U$. At the same time we find that in the case of differentiability we have:

$$Df(x) = (Df_1(x), \dots, Df_m(x)).$$

From Subsection 5.3.2 we get for $h = (h_1, \dots, h_n) \in \mathbb{R}^n$ that

$$Df_i(x)h = \frac{\partial f_i}{\partial x_1}(x)h_1 + \dots + \frac{\partial f_i}{\partial x_n}(x)h_n,$$

for $i = 1, \dots, m$.

Thus the matrix representation of $Df(x) : \mathbb{R}^n \to \mathbb{R}^m$ with respect to the canonical bases in $\mathbb{R}^n$ and $\mathbb{R}^m$ is exactly the matrix of partial derivatives

$$\mathbf{D}f(x) = \begin{bmatrix} \frac{\partial f_1}{\partial x_1}(x) & \cdots & \frac{\partial f_1}{\partial x_n}(x) \\ \vdots & & \vdots \\ \frac{\partial f_m}{\partial x_1}(x) & \cdots & \frac{\partial f_m}{\partial x_n}(x) \end{bmatrix},$$

which is often abbreviated to

$$\frac{\partial(f_1, \dots, f_m)}{\partial(x_1, \dots, x_n)}.$$

The matrix $\mathbf{D}f(x)$ is called the *Jacobian matrix* for f at $x \in U$. Note that

$$\frac{\partial f_i}{\partial x_j}(x) = Df_i(x)e_j$$

for $i = 1, \dots, m;\ j = 1, \dots, n$.

In case $m = n$, the Jacobian matrix is a square matrix, and therefore it has a determinant. This determinant is called the *Jacobian* of f at $x \in U$, and is denoted by $Jf(x)$, or simply J when f and x are given by the context.

5.4 Some important rules from calculus

Let E and F be normed vector spaces with norms $||\cdot||_E$ and $||\cdot||_F$, respectively.

Let $U \subseteq E$ be an open set in E. For an arbitrary pair of mappings $f, g : U \to F$, we can, in the usual way, define the sum $f + g : U \to F$ and, for a scalar $\alpha \in \mathbb{R}$, the product $\alpha f : U \to F$.

Theorem 5.4.1. *Assume that the mappings $f, g : U \to F$ are both differentiable at the point $x \in U$. Then*

(1) *The sum, $f + g : U \to F$, is differentiable at $x \in U$ and*

$$D(f + g)(x) = Df(x) + Dg(x).$$

(2) *The product, $\alpha f : U \to F$, is differentiable at $x \in U$ for an arbitrary scalar $\alpha \in \mathbb{R}$ and*

$$D(\alpha f)(x) = \alpha Df(x).$$

Proof. Since $f, g : U \to F$ are differentiable at $x \in U$, there exist ε-mappings $\varepsilon_f, \varepsilon_g : U_x \to F$ such that

$$f(x + h) - f(x) = Df(x)h + \varepsilon_f(h)\|h\|_E$$
$$g(x + h) - g(x) = Dg(x)h + \varepsilon_g(h)\|h\|_E$$

for $h \in U_x$.

By adding these two equations we get:

$$(f + g)(x + h) - (f + g)(x) = (Df(x) + Dg(x))h + (\varepsilon_f + \varepsilon_g)(h)\|h\|_E$$

for $h \in U_x$. From this, (1) follows, since $\varepsilon = \varepsilon(h) = \varepsilon_f(h) + \varepsilon_g(h)$ is clearly an ε-mapping.

By multiplying the equation for f by the scalar $\alpha \in \mathbb{R}$ we get the equation:

$$(\alpha f)(x + h) - (\alpha f)(x) = (\alpha Df(x))h + (\alpha \varepsilon_f)(h)\|h\|_E$$

for $h \in U_x$. From this, (2) follows, since $\varepsilon = \varepsilon(h) = \alpha \varepsilon_f(h)$ is clearly an ε-mapping. $\square$

Theorem 5.4.2. *A continuous linear mapping $L : E \to F$ is differentiable at every point $x \in E$ and $DL(x) = L$.*

Proof. With $\varepsilon = \varepsilon(h) = 0 \in F$ for all $h \in E$, we get

$$L(x + h) - L(x) = L(h) + \varepsilon(h)\|h\|_E$$

for $h \in E$. Since $\varepsilon : E \to F$ is an ε-mapping, this equation shows that L is differentiable. $\square$

We shall now prove an extremely important result about the differentiability of a composite mapping. The result is known as the *Chain Rule*.

Let E, F and G be normed vector spaces with norms $\|\cdot\|_E$, $\|\cdot\|_F$ and $\|\cdot\|_G$, respectively.

Let $U \subseteq E$ and $V \subseteq F$ be open sets in E and F, respectively. Let $f : U \to F$ and $g : V \to G$ be mappings and assume that $f(U) \subseteq V$, so that the composite mapping $g \circ f : U \to G$ is defined.

Theorem 5.4.3 (The Chain Rule). *If $f : U \to F$ is differentiable at $x \in U$, and $g : V \to G$ is differentiable at $y = f(x) \in V$, then the composite mapping $g \circ f : U \to G$ is differentiable at $x \in U$, and*

$$D(g \circ f)(x) = Dg(f(x)) \circ Df(x).$$

Proof. There exist ε-mappings $\varepsilon_f : U_x \to F$ and $\varepsilon_g : V_{f(x)} \to G$ such that

$$f(x+h) - f(x) = Df(x)h + \varepsilon_f(h) \| h \|_E \quad \text{for} \quad h \in U_x$$
$$g(f(x)+k) - g(f(x)) = Dg(f(x))k + \varepsilon_g(k) \| k \|_F \quad \text{for} \quad k \in V_{f(x)} \;.$$

Since we assume that $f(U) \subseteq V$, the vector

$$k = f(x+h) - f(x) = Df(x)h + \varepsilon_f(h) \| h \|_E$$

will belong to $V_{f(x)}$ for $h \in U_x$. Using that $Dg(f(x))$ is linear, we get the following equation by inserting such a $k \in V_{f(x)}$ into the equation for g

$$g(f(x+h)) - g(f(x)) = Dg(f(x)) \circ Df(x)h + Dg(f(x))(\varepsilon_f(h) \| h \|_E)$$
$$+ \varepsilon_g(f(x+h) - f(x)) \| f(x+h) - f(x) \|_F.$$

With the aim of constructing a suitable ε-mapping, we set

$$\varepsilon(h) = \begin{cases} Dg(f(x)) \circ \varepsilon_f(h) + \varepsilon_g(f(x+h) - f(x)) \dfrac{\| f(x+h) - f(x) \|_F}{\| h \|_E}, & h \neq 0 \\ 0, & h = 0. \end{cases}$$

Note that $\varepsilon = \varepsilon(h) : U_x \to G$.
Since

$$\| f(x+h) - f(x) \|_F = \| Df(x)h + \varepsilon_f(h) \| h \|_E \|_F$$
$$\leq \| Df(x) \| \, \| h \|_E + \| \varepsilon_f(h) \|_F \| h \|_E$$
$$= (\| Df(x) \| + \| \varepsilon_f(h) \|_F) \, \| h \|_E \;,$$

where $\| Df(x) \|$ is the operator norm of the continuous linear mapping $Df(x)$, we get

$$\frac{\| f(x+h) - f(x) \|_F}{\| h \|_E} \leq \| Df(x) \| + \| \varepsilon_f(h) \|_F \quad \text{for} \quad h \neq 0.$$

This inequality shows that the quotient on the left-hand side is bounded for h in a neighbourhood of $0 \in U_x$, since $\varepsilon_f = \varepsilon_f(h)$ is continuous at $0 \in U_x$. But then it follows easily that $\varepsilon(h) \to 0$ as $h \to 0$, and hence $\varepsilon = \varepsilon(h) : U_x \to G$ is an ε-mapping.

The mapping $\varepsilon = \varepsilon(h) : U_x \to G$ is constructed such that

$$g(f(x+h)) - g(f(x)) = Dg(f(x)) \circ Df(x)h + \varepsilon(h)\|h\|_E$$

for all $h \in U_x$. Since $\varepsilon = \varepsilon(h)$ is an ε-mapping, the equation proves that $g \circ f$ is differentiable at $x \in U$ with the differential

$$D(g \circ f)(x) = Dg(f(x)) \circ Df(x).$$

This completes the proof of Theorem 5.4.3. $\square$

Example 5.4.4. Consider the Euclidean spaces $E = \mathbb{R}^n$, $F = \mathbb{R}^m$ and $G = \mathbb{R}^l$ with coordinates $x = (x_1, \ldots, x_n) \in \mathbb{R}^n$, $y = (y_1, \ldots, y_m) \in \mathbb{R}^m$ and $z = (z_1, \ldots, z_l) \in \mathbb{R}^l$, respectively.

Without further specification of the domains, we consider the mappings $y = f(x)$ and $z = g(y)$. In the case of differentiability, the differentials $Df(x)$, $Dg(f(x))$ and $D(g \circ f)(x)$ can be represented by their Jacobian matrices with respect to the canonical bases in $\mathbb{R}^n$, $\mathbb{R}^m$, and $\mathbb{R}^l$. If we furthermore identify the coordinate functions for the mappings f, g, and $g \circ f$ with the corresponding coordinates in the Euclidean spaces, then the Chain Rule takes the form:

$$\left[\frac{\partial z_i}{\partial x_j}(x)\right] = \left[\frac{\partial z_i}{\partial y_k}(f(x))\right]\left[\frac{\partial y_k}{\partial x_j}(x)\right],$$

or

$$\frac{\partial z_i}{\partial x_j}(x) = \frac{\partial z_i}{\partial y_1}(f(x))\frac{\partial y_1}{\partial x_j}(x) + \cdots + \frac{\partial z_i}{\partial y_m}(f(x))\frac{\partial y_m}{\partial x_j}(x),$$

for $i = 1, \ldots, l$; $j = 1, \ldots, n$. ◀

5.5 The first derivative of a differentiable mapping

Let E and F be normed vector spaces with norms $\|\cdot\|_E$ and $\|\cdot\|_F$, respectively. Denote by

$$L(E, F) \text{ the space of continuous linear mappings } L : E \to F.$$

Since a linear mapping is continuous if and only if it is bounded, it follows by Theorem 4.3.4 that $L(E, F)$ is a normed vector space when equipped with the operator norm.

Now let $U \subseteq E$ be an open set in E, and consider a mapping $f : U \to F$.

Definition 5.5.1. We say that f is differentiable in U if f is differentiable at every point $x \in U$.

When $f : U \to F$ is differentiable in U, we get a mapping

$$Df : U \to L(E, F),$$

defined by associating the differential $Df(x)$ of f at x to each $x \in U$. This mapping is called the *first derivative*, or simply the *differential*, of the differentiable mapping f.

Since a normed vector space $L(E, F)$ has a metric, the following definition makes sense.

Definition 5.5.2. The mapping $f : U \to F$ is said to be *differentiable of class C^1* if it is differentiable in U and its first derivative $Df : U \to L(E, F)$ is continuous.

Remark 5.5.3. The requirement that $Df : U \to L(E, F)$ is continuous should not be confused with the requirement that the linear mapping $Df(x) : E \to F$ is continuous for every $x \in U$. The latter requirement is part of the definition of differentiability of f in U, while the former requirement is yet another condition. ◁

As we shall see now, Definition 5.5.2 generalizes the corresponding well-known concept from classical analysis of functions of finitely many variables.

Let $E = \mathbb{R}^n$ and $F = \mathbb{R}^m$. Via the canonical bases in $\mathbb{R}^n$ and $\mathbb{R}^m$, a linear mapping $L \in L(\mathbb{R}^n, \mathbb{R}^m)$ can be identified with its corresponding $m \times n$-matrix $[a_{ij}]_{i=1,\ldots,m; j=1,\ldots,n}$. If we write the m rows of the matrix one after the other, we get a vector in $\mathbb{R}^{m \cdot n}$, and thereby $L(\mathbb{R}^n, \mathbb{R}^m)$ is identified with $\mathbb{R}^{m \cdot n}$ considered as vector spaces.

Now let $f : U \to \mathbb{R}^m$ be a differentiable mapping defined in an open set $U \subseteq \mathbb{R}^n$ in $\mathbb{R}^n$. By the identifications above, the differential $Df(x)$ of f at $x \in U$ is identified with the Jacobian matrix $\mathbf{D}f(x)$, and hence the first derivative $Df : U \to L(\mathbb{R}^n, \mathbb{R}^m)$ of $f = (f_1, \ldots, f_m)$ gets the partial derivatives $\dfrac{\partial f_i}{\partial x_j}$ as coordinate functions. We note this in the diagram:

$$x \in U \xrightarrow{\quad Df \quad} L(\mathbb{R}^n, \mathbb{R}^m) \ni Df(x)$$

$$\mathbf{D}f \searrow \qquad \| $$

$$\mathbb{R}^{m \cdot n} \ni \mathbf{D}f(x) = \left(\tfrac{\partial f_1}{\partial x_1}, \ldots, \tfrac{\partial f_1}{\partial x_n}, \ldots, \tfrac{\partial f_m}{\partial x_1}, \ldots, \tfrac{\partial f_m}{\partial x_n} \right).$$

Since all norms in a finite dimensional vector space are equivalent, we can use the maximum norm in $\mathbb{R}^{m \cdot n}$ instead of the operator norm in

$L(\mathbb{R}^n, \mathbb{R}^m)$ when considering topological questions. Then it is easy to show that the following statements are equivalent:

(1) $Df : U \to L(\mathbb{R}^n, \mathbb{R}^m)$ is continuous

(2) $\mathbf{D}f : U \to \mathbb{R}^{m \cdot n}$ is continuous

(3) $\dfrac{\partial f_i}{\partial x_j} : U \to \mathbb{R}$ is continuous for all $i = 1, \ldots, m;\ j = 1, \ldots, n$.

Statement (3) is the classical definition of differentiability of class C^1 for mappings between Euclidean spaces.

5.6 Mean value theorems

For the sake of completeness we first prove the classical mean value theorem.

Theorem 5.6.1 (The classical Mean Value Theorem). *If $f : [a, b] \to \mathbb{R}$ is a real-valued function which is continuous in the closed and bounded interval $[a, b]$ and differentiable in the open interval $]a, b[$, then there exists a $\zeta \in]a, b[$, such that $f(b) - f(a) = f'(\zeta)(b - a)$.*

Proof. Consider an arbitrary function $f : [a, b] \to \mathbb{R}$ satisfying the conditions given in the theorem, and define the function $g : [a, b] \to \mathbb{R}$ by

$$g(x) = f(x) - \left(f(a) + \frac{f(b) - f(a)}{b - a}(x - a) \right) \quad \text{for} \quad x \in [a, b].$$

Then $g(a) = g(b) = 0$ and we just have to prove that there exists a $\zeta \in]a, b[$, such that $g'(\zeta) = 0$. This special case of the mean value theorem is known as Rolle's theorem.

The proof of Rolle's theorem goes as follows. Since g is continuous in the closed and bounded interval $[a, b]$, it assumes a maximum value d as well as a minimum value c in $[a, b]$. If $c = d = 0$, clearly g is constant with value 0 and hence $g'(\zeta) = 0$ at any point $\zeta \in]a, b[$. If $c < 0$ or $d > 0$, there exists a point $\zeta \in]a, b[$ such that $g(\zeta) = c$ or $g(\zeta) = d$. By the definition of the derivative it follows immediately that $g'(\zeta) = 0$ at a point where a local minimum or maximum value of g is attained. This proves Rolle's theorem and thereby the mean value theorem. $\square$

From the classical mean value theorem we can derive the following, more general theorem.

Theorem 5.6.2. *Let U be an open set in the normed vector space E, and let $f : U \to \mathbb{R}$ be a differentiable function. If x and y are points in U so*

*that the line segment in E which connects x and y lies completely inside U,
then there exists a point z between x and y on this line segment such that*

$$f(y) - f(x) = Df(z)(y - x).$$

Proof. Let $\sigma : [0,1] \to U$ be the curve parameterized by $\sigma(t) = x + t(y-x)$
for $t \in [0,1]$, which describes the line segment from x to y; cf. Figure 5.5.
Then the composite function $g = f \circ \sigma : [0,1] \to \mathbb{R}$ is continuous in $[0,1]$
and differentiable in $]0,1[$.

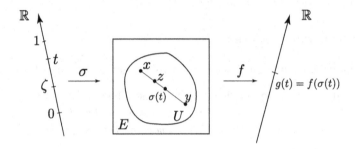

Fig. 5.5 The Mean Value Theorem in a general setting (Theorem 5.6.2) gives the existence of a point z between x and y such that $f(y) - f(x) = Df(z)(y - x)$.

By the classical mean value theorem there exists a $\zeta \in \,]0,1[$ such that

$$g(1) - g(0) = g'(\zeta).$$

The Chain Rule shows that

$$\begin{aligned} g'(\zeta) &= Dg(\zeta)(1) = Df(\sigma(\zeta)) \circ D\sigma(\zeta)(1) \\ &= Df(\sigma(\zeta))(\sigma'(\zeta)) = Df(\sigma(\zeta))(y - x). \end{aligned}$$

See also Subsection 5.3.1, page 119, and note that $\sigma'(\zeta) = y - x$.

The point $z = \sigma(\zeta)$ lies on the line segment from x to y. Since we
have $g'(\zeta) = Df(z)(y - x)$, and the values $g(0) = f(\sigma(0)) = f(x)$ and
$g(1) = f(\sigma(1)) = f(y)$, it follows that

$$f(y) - f(x) = Df(z)(y - x),$$

as asserted. $\square$

Definition 5.6.3. A subset C in a vector space E is said to be *convex*
if every line segment in E that connects two points in C is completely
contained in C.

Now assume in Theorem 5.6.2 also that there is a convex subset $C \subseteq U$ and a constant $k \in \mathbb{R}$ such that the differential Df of f is bounded in C by the constant k. We assume in other words that the operator norm of $Df(z)$ satisfies $||Df(z)|| \leq k$ for all $z \in C$. Then Theorem 5.6.2 immediately gives the inequality

$$|f(y) - f(x)| \leq k||y - x||_E \quad \text{for all} \quad x, y \in C.$$

We shall need such an inequality more generally, namely when the function f takes values in a normed vector space F. However, certain technical considerations involving the integration of continuous functions with values in F over an interval, require that F is a Banach space; cf. section 4.5.

We shall first discuss a general result which generalizes a well-known theorem from classical mathematical analysis.

Theorem 5.6.4 (The Fundamental Theorem of Calculus). *Suppose the function* $g : [a, b] \to F$ *is differentiable of class* C^1 *in the interval* $[a, b]$ *and has values in a Banach space* F. *Then*

$$g(b) - g(a) = \int_a^b g'(t)dt.$$

Proof. Consider the function $f : [a, b] \to F$ defined by

$$f(x) = g(a) + \int_a^x g'(t)dt.$$

We note that the integral exists since g' is continuous and F is a Banach space. As in the classical case with real-valued functions, it is easy to show that f is differentiable of class C^1 and that $f'(x) = g'(x)$ for all $x \in [a, b]$. Note also that $f(a) = g(a)$. We shall prove that $f(x) = g(x)$ for all $x \in [a, b]$.

With this in mind, consider the real-valued function $\varphi : [a, b] \to \mathbb{R}$ defined by $\varphi(x) = ||g(x) - f(x)||_F$. For an arbitrary $x \in [a, b]$ and an arbitrary increment $h \in \mathbb{R}$ such that $x + h \in [a, b]$ we have the inequality

$$\left| \frac{\varphi(x+h) - \varphi(x)}{h} \right| = \left| \frac{||g(x+h) - f(x+h)||_F - ||g(x) - f(x)||_F}{h} \right|$$

$$\leq \left| \left| \frac{(g(x+h) - f(x+h)) - (g(x) - f(x))}{h} \right| \right|_F$$

$$= \left| \left| \frac{g(x+h) - g(x)}{h} - \frac{f(x+h) - f(x)}{h} \right| \right|_F.$$

Since

$$\left|\left| \frac{g(x+h) - g(x)}{h} - \frac{f(x+h) - f(x)}{h} \right|\right|_F \to ||g'(x) - f'(x)||_F = 0$$

for $h \to 0$, we conclude that φ is differentiable with derivative $\varphi'(x) = 0$ for all $x \in [a, b]$. Then the classical mean value theorem shows that φ is constant in $[a, b]$, and since $\varphi(a) = 0$, therefore $\varphi(x) = ||g(x) - f(x)||_F = 0$ for all $x \in [a, b]$. From this follows that $g(x) = f(x)$ for all $x \in [a, b]$. In particular we have

$$g(b) = f(b) = g(a) + \int_a^b g'(t)dt,$$

as asserted. $\square$

Then we are ready to prove the estimate of the increment of a function that can replace the mean value theorem when the function under consideration takes values in a Banach space.

Theorem 5.6.5. *Let E be a normed vector space and let F be a Banach space. Let $f : U \to F$ be a differentiable mapping of class C^1 defined in an open set $U \subseteq E$. Assume that there is a convex subset $C \subseteq U$ and a constant $k \in \mathbb{R}$ such that the operator norm of the differential of f satisfies $||Df(z)|| \le k$ for all $z \in C$. Then we have the inequality*

$$||f(y) - f(x)||_F \le k||y - x||_E \quad \text{for all} \quad x, y \in C.$$

Proof. For fixed but arbitrarily chosen points $x, y \in C$ we consider the line segment joining x to y, parameterized by $\sigma : [0, 1] \to C$, defined by $\sigma(t) = x + t(y - x)$. Here we use the fact that C is convex. The composite function $g = f \circ \sigma : [0, 1] \to F$ is clearly of class C^1. First using Theorem 5.6.4 and the Chain Rule, we get

$$f(y) - f(x) = g(1) - g(0)$$
$$= \int_0^1 g'(t)dt$$
$$= \int_0^1 Df(\sigma(t))(y - x)dt.$$

Next using Theorem 4.5.4 (iii), we get:

$$\|f(y) - f(x)\|_F = \|\int_0^1 Df(\sigma(t))(y-x)dt\|_F$$

$$\leq \int_0^1 \|Df(\sigma(t))(y-x)\|_F dt$$

$$\leq \int_0^1 \left(\|Df(\sigma(t))\| \, \|y-x\|_E\right) dt$$

$$\leq \int_0^1 k\|y-x\|_E dt$$

$$= k\|y-x\|_E.$$

This is exactly the inequality asserted in the theorem. $\square$

5.7 Partial differentials

Let $E_1, \ldots, E_n$ and F be normed vector spaces equipped with norms $\|\cdot\|_{E_1}, \ldots, \|\cdot\|_{E_n}$ and $\|\cdot\|_F$, respectively.

The product space $\mathbf{E}^n = E_1 \times E_2 \times \cdots \times E_n$ containing all points $x = (x_1, x_2, \ldots, x_n)$, $x_i \in E_i$, can be turned into a normed vector space by the coordinate-wise defined vector space operations and with the norm

$$\|x\|_{\mathbf{E}^n} = \|(x_1, x_2, \ldots, x_n)\|_{\mathbf{E}^n} = \max\left\{\|x_1\|_{E_1}, \|x_2\|_{E_2}, \ldots, \|x_n\|_{E_n}\right\}.$$

This norm equips the product space $\mathbf{E}^n$ with a topology, which clearly coincides with the product topology defined in Example 2.4.10.

With $h = (h_1, h_2, \ldots, h_n) \in \mathbf{E}^n$, we define for each $i = 1, \ldots, n$, the inclusion mapping $\mathbf{e}_i : E_i \to \mathbf{E}^n$ by

$$\mathbf{e}_1(h_1) = (h_1, 0, \ldots, 0)$$

$$\mathbf{e}_2(h_2) = (0, h_2, \ldots, 0)$$

$$\vdots$$

$$\mathbf{e}_n(h_n) = (0, \ldots, 0, h_n).$$

All of the mappings $\mathbf{e}_i : E_i \to \mathbf{E}^n$ are clearly linear mappings. They are in fact, continuous (bounded) linear mappings, since they are norm preserving, i.e. $\|\mathbf{e}_i(h_i)\|_{\mathbf{E}^n} = \|h_i\|_{E_i}$ for $h_i \in E_i$.

For later use, note that by Theorem 5.4.2 the mapping $\mathbf{e}_i : E_i \to \mathbf{E}^n$ is differentiable at $0 \in E_i$ and that $D\mathbf{e}_i(0) = \mathbf{e}_i$ for all $i = 1, \ldots, n$.

Now let U be an open set in $\mathbf{E}^n$, and let $x = (x_1, x_2, \ldots, x_n) \in U$ be a point in U, which we keep fixed in the following definitions. Let also $f : U \to F$ be an arbitrary mapping.

Since U is open in the norm on $\mathbf{E}^n$, we can choose a radius $r > 0$ such that the product ball $\mathbf{B}_r^n(x)$ of the open balls $B_r(x_i) \subseteq E_i$, $i = 1, \ldots, n$, satisfies

$$\mathbf{B}_r^n(x) = B_r(x_1) \times B_r(x_2) \times \cdots \times B_r(x_n) \subseteq U.$$

The restrictions of the mapping $f : U \to F$ to the open balls $B_r(x_i)$ in the factors E_i in $\mathbf{E}^n$, $i = 1, \ldots, n$, i.e. the mappings

$$f_{E_i} : B_r(x_i) \to F \quad \text{defined by} \quad f_{E_i}(h_i) = f(x + \mathbf{e}_i(h_i)),$$

for $h = (h_1, h_2, \ldots, h_n) \in \mathbf{B}_r^n(0)$, are the key to define partial differentials.

In Section 5.3.2 we introduced partial derivatives of a function of several variables by keeping all but one of the variables fixed in turn. Correspondingly, we can now define partial differentials of a mapping $f : U \to F$, making use of the mappings f_{E_i}.

Definition 5.7.1. The mapping $f : U \to F$ is said to be *partial differentiable* at the point $x \in U$, or to have *partial differentials* at $x \in U$, if the mapping $f_{E_i} : B_r(x_i) \to F$, is differentiable at $0 \in E_i$, for all $i = 1, \ldots, n$.

If it exists, the differential of f_{E_i} at $0 \in E_i$, is called the ith *partial differential* of f at $x \in U$, and is denoted by $D_{E_i} f(x)$.

Differentiability always implies partial differentiability.

Theorem 5.7.2. *Let $\mathbf{E}^n = E_1 \times E_2 \times \cdots \times E_n$ be a product of normed vector spaces, and let F be a normed vector space. Let $f : U \to F$ be a mapping defined in the open set $U \subseteq \mathbf{E}^n$.*

Suppose that $f : U \to F$ is differentiable at the point $x \in U$. Then all partial differentials $D_{E_i} f(x)$, $i = 1, \ldots, n$, of f at $x \in U$ exist, and the differential of f at $x \in U$, is given by

$$Df(x)(h) = D_{E_1} f(x)(h_1) + D_{E_2} f(x)(h_2) + \cdots + D_{E_n} f(x)(h_n),$$

for $h = (h_1, h_2, \ldots, h_n) \in \mathbf{E}^n$.

Proof. Suppose the mapping $f : U \to F$ is differentiable at the point $x \in U$. Choose a radius $r > 0$, such that the product ball $\mathbf{B}_r^n(x)$ satisfies

$$\mathbf{B}_r^n(x) = B_r(x_1) \times B_r(x_2) \times \cdots \times B_r(x_n) \subseteq U.$$

For each $i = 1, \ldots, n$, the mapping

$$x + \mathbf{e}_i : E_i \to \mathbf{E}^n \quad \text{defined by} \quad (x + \mathbf{e}_i)(h_i) = x + \mathbf{e}_i(h_i),$$

for $h = (h_1, h_2, \ldots, h_n) \in \mathbf{B}_r^n(0)$, is differentiable at $0 \in E_i$ with differential

$$D(x + \mathbf{e}_i)(0) = D\mathbf{e}_i(0) = \mathbf{e}_i.$$

Since $f : U \to F$ is differentiable at $x \in U$, it follows by the chain rule (Theorem 5.4.3), that each of the mappings

$$f_{E_i} : B_r(x_i) \to F \quad \text{defined by} \quad f_{E_i}(h_i) = f(x + \mathbf{e}_i(h_i)),$$

for $h = (h_1, h_2, \ldots, h_n) \in \mathbf{B}_r^n(0)$, is differentiable at $0 \in E_i$ with differential

$$D_{E_i} f(x)(h_i) = \big(Df(x) \circ D(x + \mathbf{e}_i)(0)\big)(h_i) = Df(x)(\mathbf{e}_i(h_i)).$$

This proves that f is partial differentiable at $x \in U$.

With $h = (h_1, h_2, \ldots, h_n) \in \mathbf{E}^n$, the following computations reveal the asserted formula for the differential $Df(x) : \mathbf{E}^n \to F$ of f at $x \in U$.

$$Df(x)(h) = Df(x)(\mathbf{e}_1(h_1) + \mathbf{e}_2(h_2) + \cdots + \mathbf{e}_n(h_n))$$
$$= \sum_{i=1}^{n} Df(x)(\mathbf{e}_i(h_i))$$
$$= D_{E_1} f(x)(h_1) + D_{E_2} f(x)(h_2) + \cdots + D_{E_n} f(x)(h_n).$$

This completes the proof. $\square$

The opposite conclusion is more delicate. Making use of the Fundamental Theorem of Calculus in Banach spaces and estimates of operator norms, we can prove.

Theorem 5.7.3. *Let $\mathbf{E}^n = E_1 \times E_2 \times \cdots \times E_n$ be a product of normed vector spaces, and let F be a Banach space. Let $f : U \to F$ be a mapping defined in the open set $U \subseteq \mathbf{E}^n$.*

Suppose that all partial differentials of f exist and vary continuously in an open ball $\mathbf{B}_r^n(x)$ centred at the point $x \in U$. Then $f : U \to F$ is differentiable at $x \in U$, and the differential of f at $x \in U$, is given by

$$Df(x)(h) = D_{E_1} f(x)(h_1) + D_{E_2} f(x)(h_2) + \cdots + D_{E_n} f(x)(h_n),$$

for $h = (h_1, h_2, \ldots, h_n) \in \mathbf{E}^n$.

Proof. Suppose that the mapping $f : U \to F$ is partial differentiable at all points $y \in \mathbf{B}_r^n(x)$, where $\mathbf{B}_r^n(x)$ is an open ball in U centred at the point $x \in U$. Suppose in addition also, that the ith partial differential $D_{E_i} f : \mathbf{B}_r^n(x) \to L(E_i, F)$, which to $y \in \mathbf{B}_r^n(x)$ associates the partial differential $D_{E_i} f(y) : E_i \to F$, is continuous for all $i = 1, \ldots, n$.

With these conditions satisfied, we shall prove that $f : U \to F$ is differentiable at $x \in U$ with the differential

$$L(h) = D_{E_1} f(x)(h_1) + D_{E_2} f(x)(h_2) + \cdots + D_{E_n} f(x)(h_n),$$

for $h = (h_1, h_2, \ldots, h_n) \in \mathbf{E}^n$.

For simplicity in notation, we now restrict our considerations to $n = 2$.

For $h = (h_1, h_2) \in \mathbf{B}_r^2(0)$ we have the following computations using in turn Theorem 5.6.4 and Theorem 5.4.3.

$$f(x + h) - f(x) - L(h)$$
$$= [f(x + h) - f(x_1, x_2 + h_2)] + [f(x_1, x_2 + h_2) - f(x)] - L(h)$$
$$= \int_0^1 \frac{d}{dt} f(x_1 + th_1, x_2 + h_2) dt$$
$$+ \int_0^1 \frac{d}{ds} f(x_1, x_2 + sh_2) ds - [D_{E_1} f(x)(h_1) + D_{E_2} f(x)(h_2)]$$
$$= \int_0^1 \left(D_{E_1} f(x_1 + th_1, x_2 + h_2)(h_1) - D_{E_1} f(x)(h_1) \right) dt$$
$$+ \int_0^1 \left(D_{E_2} f(x_1, x_2 + sh_2)(h_2) - D_{E_2} f(x)(h_2) \right) ds$$
$$= \int_0^1 \left(D_{E_1} f(x_1 + th_1, x_2 + h_2) - D_{E_1} f(x) \right)(h_1) dt$$
$$+ \int_0^1 \left(D_{E_2} f(x_1, x_2 + sh_2) - D_{E_2} f(x) \right)(h_2) ds.$$

Using estimates of norms of integrals (Theorem 4.5.5), and estimates with operator norms (Lemma 4.3.3), we get from this

$$\| f(x + h) - f(x) - L(h) \|_F$$
$$\leq \left\| \int_0^1 \left(D_{E_1} f(x_1 + th_1, x_2 + h_2) - D_{E_1} f(x) \right)(h_1) dt \right\|_F$$
$$+ \left\| \int_0^1 \left(D_{E_2} f(x_1, x_2 + sh_2) - D_{E_2} f(x) \right)(h_2) ds \right\|_F$$
$$\leq \int_0^1 \left(\| D_{E_1} f(x_1 + th_1, x_2 + h_2) - D_{E_1} f(x) \| \, \| h_1 \|_{E_1} \right) dt$$
$$+ \int_0^1 \left(\| D_{E_2} f(x_1, x_2 + sh_2) - D_{E_2} f(x) \| \, \| h_2 \|_{E_2} \right) ds$$
$$\leq \left(\int_0^1 \| D_{E_1} f(x_1 + th_1, x_2 + h_2) - D_{E_1} f(x) \| dt \right) \| h \|_{\mathbf{E}^2}$$
$$+ \left(\int_0^1 \| D_{E_2} f(x_1, x_2 + sh_2) - D_{E_2} f(x) \| ds \right) \| h \|_{\mathbf{E}^2}.$$

There is a unique mapping $\varepsilon = \varepsilon(h) : \mathbf{B}_r^2(0) \to F$, such that $\varepsilon(0) = 0$ and for which the following equation is satisfied for all $h \in \mathbf{B}_r^2(0)$

$$f(x + h) - f(x) - L(h) = \varepsilon(h) \| h \|_{\mathbf{E}^2}.$$

For $h \in \mathbf{B}_r^2(0)$, we get the estimate

$$\|\varepsilon(h)\|_F \leq \int_0^1 \|D_{E_1}f(x_1 + th_1, x_2 + h_2) - D_{E_1}f(x)\|dt$$
$$+ \int_0^1 \|D_{E_2}f(x_1, x_2 + sh_2) - D_{E_2}f(x)\|ds.$$

Since the partial differentials are continuous in $\mathbf{B}_r^2(x)$, we can for any given $\varepsilon > 0$ choose a $\delta > 0$ such that for $\|h\|_{\mathbf{E}^2} < \delta < r$, we have

$$\|D_{E_1}f(x_1 + th_1, x_2 + h_2) - D_{E_1}f(x)\| < \varepsilon/2$$
$$\|D_{E_2}f(x_1, x_2 + sh_2) - D_{E_2}f(x)\| < \varepsilon/2.$$

Combining this with the estimate of $\|\varepsilon(h)\|_F$, we can conclude that

$$\forall \varepsilon > 0 \quad \exists \delta > 0 : \|h\|_{\mathbf{E}^2} < \delta \implies \|\varepsilon(h)\|_F < \varepsilon.$$

In other words, $\varepsilon = \varepsilon(h)$ is an ε-mapping.

Since the linear mapping $L = L(h) : \mathbf{E}^2 \to F$ is continuous, the equation

$$f(x + h) - f(x) - L(h) = \varepsilon(h)\|h\|_{\mathbf{E}^2}, \quad h \in \mathbf{B}_r^2(0),$$

shows that $f : U \to F$ is indeed differentiable at $x \in U$, and that the differential of f at $x \in U$, is given by

$$L(h) = D_{E_1}f(x)(h_1) + D_{E_2}f(x)(h_2).$$

This completes the proof for the case $n = 2$.

The general case can be proved in complete analogy, or, by a simple induction argument. $\square$

5.8   Toplinear isomorphisms

Let E and F be normed vector spaces.

Definition 5.8.1. A linear isomorphism $L : E \to F$, where both L and its inverse isomorphism L^{-1} are continuous, is called a *toplinear isomorphism*.

Remark 5.8.2. An immediate corollary to the famous 'Open Mapping Theorem', known as Banach's Theorem, states: *If E and F are Banach spaces and $L : E \to F$ is a continuous linear isomorphism, then the inverse isomorphism $L^{-1} : F \to E$ is also continuous.* For a proof see e.g. my book [Functional Analysis-Entering Hilbert Space, 2nd edition, Theorem 1.8.3]. ◁

In the following E is a Banach space with norm $||\cdot||_E$.

From Theorem 4.4.3, we know that the space of continuous linear mappings $L(E, E)$ of E into itself is a Banach space with the operator norm as its norm. In the Banach space $L(E, E)$ we consider the subset $Gl(E)$ of toplinear isomorphisms $L : E \to E$, i.e.

$$Gl(E) = \{L \in L(E, E) \mid L : E \to E \quad \text{toplinear isomorphism}\}.$$

In particular, $I \in Gl(E)$ denotes the identity isomorphism, defined by $I(x) = x$ for all $x \in E$. We mention that with the composition of toplinear isomorphisms as the group law, $Gl(E)$ is in fact a group, known as the *general linear group*.

It is our objective to prove the following theorem.

Theorem 5.8.3. *Let E be a Banach space. Then we have:*

(i) $Gl(E)$ *is an open set in* $L(E, E)$.

(ii) *The mapping* $\tau : Gl(E) \to Gl(E)$, *which assigns the inverse toplinear isomorphism* $\tau(L) = L^{-1} \in Gl(E)$ *to a toplinear isomorphism* $L \in Gl(E)$, *is differentiable of class* C^1. *The differential of* τ *is given by*

$$D\tau(L)H = -L^{-1} \circ H \circ L^{-1}$$

for $L \in Gl(E)$ and $H \in L(E, E)$.

For short the mapping τ is called the *inverse transformation*.

In the finite dimensional case, it is possible to give an elementary proof of the part of Theorem 5.8.3 needed in the proof of the Inverse Function Theorem.

Proof of Theorem 5.8.3 (E finite dimensional). If E is an n-dimensional vector space, the space of linear mappings $L(E, E)$ can be identified with the space of $n \times n$-matrices, and hence with $\mathbb{R}^{n^2}$, via matrix representation of the linear maps.

The set $Gl(E)$ is exactly the subset of $L(E, E)$ consisting of linear mappings with non-zero determinant, and therefore $Gl(E) = \det^{-1}(\mathbb{R} \setminus \{0\})$ is the preimage of the open set $(\mathbb{R} \setminus \{0\}) \subseteq \mathbb{R}$ under the continuous map $\det:L(E, E) \to \mathbb{R}$ defined by taking the determinant of matrices. Hence $Gl(E)$ is an open set in $L(E, E)$. That the inverse transformation $\tau : Gl(E) \to Gl(E)$ is differentiable of class C^1 follows immediately by inspection of the formula for finding the inverse matrix of an invertible matrix. $\square$

The infinite dimensional case of Theorem 5.8.3 requires considerably more work. The readers satisfied with a proof of the Inverse Function Theorem in finite dimensions can proceed directly to Section 5.9.

In the proof of Theorem 5.8.3 in the general case we shall need to consider series with terms in a Banach space E. Fortunately the theory is completely parallel to the theory of series with real numbers as terms.

By a *series* $\sum_{n=1}^{\infty} x_n$ in E, we understand a sequence $x_1, x_2, \ldots, x_n, \ldots$ of elements of E which we, to begin with, just formally sum. To the series there corresponds a sequence of *partial sums* $s_1, s_2, \ldots, s_n, \ldots$ defined by

$$s_n = \sum_{k=1}^{n} x_k = x_1 + \cdots + x_n.$$

We say that the series is *convergent* with *sum* $s \in E$ if the sequence of partial sums (s_n) is convergent with limit point s. In the case of convergence we write

$$s = \sum_{n=1}^{\infty} x_n.$$

The nth term x_n in a series $\sum_{n=1}^{\infty} x_n$ can be written as $x_n = s_n - s_{n-1}$, which gives the following necessary condition for convergence of a series.

Theorem 5.8.4. *In a convergent series $\sum_{n=1}^{\infty} x_n$ in the Banach space E, the nth term x_n converges to $0 \in E$ for n going to ∞.*

Since a Banach space considered as a metric space is complete, we also have the following important result.

Theorem 5.8.5 (The Cauchy condition for series). *A series $\sum_{n=1}^{\infty} x_n$ in a Banach space E is convergent if and only if*

$$\forall \varepsilon > 0 \ \exists n_0 \in \mathbb{N} \ \forall n, k \in \mathbb{N} : \quad n \geq n_0 \quad \Rightarrow \|s_{n+k} - s_n\|_E < \varepsilon,$$

or equivalently,

$$\forall \varepsilon > 0 \ \exists n_0 \in \mathbb{N} \ \forall n, k \in \mathbb{N} : \quad n \geq n_0 \quad \Rightarrow \| \sum_{i=n+1}^{n+k} x_i \|_E < \varepsilon.$$

From the Cauchy condition we deduce a very useful test for convergence.

Theorem 5.8.6 (The comparison test). *Let $\sum_{n=1}^{\infty} x_n$ be a series in a Banach space E. If there exists a convergent series of non-negative real numbers $\sum_{n=1}^{\infty} a_n$ such that $\|x_n\|_E \leq a_n$ for $n \geq n_0$ from a certain step n_0, then the series $\sum_{n=1}^{\infty} x_n$ is convergent.*

Proof. The triangle inequality gives the inequality

$$\Big|\Big| \sum_{i=n+1}^{n+k} x_i \Big|\Big|_E \le \sum_{i=n+1}^{n+k} a_i.$$

Since the Cauchy condition is satisfied in both E and $\mathbb{R}$, the result follows immediately. $\square$

We can now prove the following lemma.

Lemma 5.8.7. *Let $H : E \to E$ be a continuous linear mapping with operator norm $||H|| < 1$ in the Banach space E. Then $I + H$ is a toplinear isomorphism, or equivalently, $I + H \in Gl(E)$.*

Proof. It is obviously equivalent to show that $I - H \in Gl(E)$.

Define the iterates $H^n : E \to E$ of H inductively by $H^n = H \circ H^{n-1}$ for $n \ge 2$ and $H^1 = H$.

For a fixed $x \in E$ the comparison test shows that the series $\sum_{n=1}^{\infty} H^n(x)$ is convergent, since

$$||H^n(x)||_E \le ||H||^n ||x||_E \quad \text{for all} \quad n \ge 1,$$

and the geometric series $\sum_{n=1}^{\infty} ||x||_E ||H||^n$ with quotient $||H|| < 1$ is convergent.

Therefore, we can define the mapping $(I + \sum_{n=1}^{\infty} H^n) : E \to E$ by setting

$$\Big(I + \sum_{n=1}^{\infty} H^n\Big)(x) = x + \sum_{n=1}^{\infty} H^n(x) \quad \text{for} \quad x \in E.$$

It is easy to show that $I + \sum_{n=1}^{\infty} H^n$ is linear. By considering the corresponding bound on each of the partial sums and then taking limits it is also easy to show that

$$\Big|\Big|\Big(I + \sum_{n=1}^{\infty} H^n\Big)(x)\Big|\Big|_E \le ||x||_E + \sum_{n=1}^{\infty} ||H||^n ||x||_E$$

$$= \frac{1}{1 - ||H||} ||x||_E$$

for all $x \in E$. According to Theorem 4.2.4 it follows that $I + \sum_{n=1}^{\infty} H^n$ is continuous.

If we proceed by an informal calculation, we get

$$(I - H) \circ \Big(I + \sum_{n=1}^{\infty} H^n\Big) = (I - H) \circ (I + H + H^2 + \cdots) = I,$$

and

$$\left(I + \sum_{n=1}^{\infty} H^n\right) \circ (I - H) = (I + H + H^2 + \cdots) \circ (I - H) = I.$$

For each $x \in E$ we get more formally:

$$(I - H) \circ \left(I + \sum_{n=1}^{\infty} H^n\right)(x) = (I - H)\left(\lim_{k \to \infty}\left(x + \sum_{n=1}^{k} H^n(x)\right)\right)$$

$$= \lim_{k \to \infty} (I - H)\left(x + \sum_{n=1}^{k} H^n(x)\right)$$

$$= \lim_{k \to \infty} \left(x - H^{k+1}(x)\right) = x = I(x),$$

where we have used that $I - H$ is continuous, and that $\lim_{k \to \infty} H^{k+1}(x) = 0$ since $||H^{k+1}(x)||_E \leq ||H||^{k+1}||x||_E$ and $||H|| < 1$.

Similarly, the other equation can be formally verified.

The equations

$$(I - H) \circ \left(I + \sum_{n=1}^{\infty} H^n\right) = I$$

and

$$\left(I + \sum_{n=1}^{\infty} H^n\right) \circ (I - H) = I$$

show that $I - H$ is a continuous linear isomorphism with the continuous linear inverse

$$(I - H)^{-1} = I + \sum_{n=1}^{\infty} H^n.$$

Hence $I - H$ is a toplinear isomorphism. □

We are now ready to prove Theorem 5.8.3.

Proof of Theorem 5.8.3(i). Let $L \in Gl(E)$ be a toplinear isomorphism. Then an arbitrary continuous linear mapping $H \in L(E, E)$ can be written in the form

$$H = L + (H - L) = L \circ (I + L^{-1} \circ (H - L)).$$

Whenever $||L^{-1} \circ (H - L)|| < 1$, Lemma 5.8.7 shows that $H \in Gl(E)$.

According to Theorem 4.3.4 we have the inequality

$$||L^{-1} \circ (H - L)|| \leq ||L^{-1}|| \, ||H - L||.$$

It then follows immediately that

$$\left\{ H \in L(E,E) \;\Big|\; ||H - L|| < \frac{1}{||L^{-1}||} \right\} \subseteq Gl(E).$$

Hence there does indeed exist an open ball in $L(E,E)$ with centre $L \in Gl(E)$ completely contained in $Gl(E)$. This proves that $Gl(E)$ is an open set in $L(E,E)$. $\square$

Proof of Theorem 5.8.3(ii). First, we prove that $\tau : Gl(E) \to Gl(E)$ is differentiable at $I \in Gl(E)$.

In order to do this we must consider the increment in τ,

$$\tau(I + H) - \tau(I) = (I + H)^{-1} - I \;,$$

corresponding to the increment $H \in L(E,E)$ with $||H|| < 1$ (which ensures that $I + H$ is a toplinear isomorphism).

From the proof of Lemma 5.8.7 we know that

$$(I + H)^{-1} = I + \sum_{n=1}^{\infty} (-1)^n H^n,$$

and hence

$$\tau(I + H) - \tau(I) = -H + \sum_{n=2}^{\infty} (-1)^n H^n.$$

Define the mapping $\varepsilon = \varepsilon(H)$ by

$$\varepsilon(H) = \begin{cases} \sum_{n=2}^{\infty} (-1)^n \dfrac{H^n}{||H||} & \text{for} \quad H \neq 0 \\ 0 & \text{for} \quad H = 0. \end{cases}$$

For $0 < ||H|| < 1$ we have the inequality

$$||\varepsilon(H)|| \leq \sum_{n=2}^{\infty} \frac{||H||^n}{||H||} = \frac{||H||}{1 - ||H||} \;.$$

It follows that $\varepsilon = \varepsilon(H)$ is continuous at $0 \in L(E,E)$ and therefore $\varepsilon = \varepsilon(H)$ is an ε-mapping.

The equation

$$\tau(I + H) - \tau(I) = -H + \varepsilon(H)||H||$$

then shows that τ is differentiable at I, and that

$$D\tau(I)H = -H.$$

Next we will show that $\tau : Gl(E) \to Gl(E)$ is differentiable at an arbitrary toplinear isomorphism $L \in Gl(E)$.

For $H \in L(E, E)$ with $\|H\| < \frac{1}{\|L^{-1}\|}$, which ensures that

$$L + H = (I + H \circ L^{-1}) \circ L$$

is a toplinear isomorphism, the following calculation is valid:

$$
\begin{aligned}
\tau(L + H) - \tau(L) &= (L + H)^{-1} - L^{-1} \\
&= L^{-1} \circ [(I + H \circ L^{-1})^{-1} - I] \\
&= L^{-1} \circ [\tau(I + H \circ L^{-1}) - \tau(I)] \\
&= L^{-1} \circ [D\tau(I)(H \circ L^{-1}) + \varepsilon(H \circ L^{-1})\|H \circ L^{-1}\|].
\end{aligned}
$$

Now define the mapping $\varepsilon' = \varepsilon'(H)$ by

$$
\varepsilon'(H) = \begin{cases} L^{-1} \circ \varepsilon(H \circ L^{-1}) \dfrac{\|H \circ L^{-1}\|}{\|H\|} & \text{for} \quad H \neq 0 \\ 0 & \text{for} \quad H = 0. \end{cases}
$$

Then $\varepsilon' = \varepsilon'(H)$ is an ε-mapping, and we get the equation

$$\tau(L + H) - \tau(L) = -L^{-1} \circ H \circ L^{-1} + \varepsilon'(H)\|H\|,$$

which shows that τ is differentiable at $L \in Gl(E)$, and that

$$D\tau(L)H = -L^{-1} \circ H \circ L^{-1}.$$

It now only remains to prove that $D\tau : Gl(E) \to L(L(E, E), L(E, E))$ defined by

$$D\tau(L)H = -L^{-1} \circ H \circ L^{-1}, \text{ for } L \in Gl(E), \ H \in L(E, E),$$

is a continuous map.

To prove this, note that for fixed toplinear isomorphisms $L, L_0 \in Gl(E)$ we have the following estimates on operator norms in $L(E, E)$:

$$
\begin{aligned}
\|D\tau(L) - D\tau(L_0)\| &= \sup_{\|H\|=1} \|-L^{-1} \circ H \circ L^{-1} + L_0^{-1} \circ H \circ L_0^{-1}\| \\
&\leq \sup_{\|H\|=1} \|-L^{-1} \circ H \circ L^{-1} + L_0^{-1} \circ H \circ L^{-1}\| \\
&\quad + \sup_{\|H\|=1} \|-L_0^{-1} \circ H \circ L^{-1} + L_0^{-1} \circ H \circ L_0^{-1}\| \\
&\leq \|L^{-1} - L_0^{-1}\|\|L^{-1}\| + \|L_0^{-1}\|\|L^{-1} - L_0^{-1}\| \\
&= \|\tau(L) - \tau(L_0)\|\|\tau(L)\| + \|\tau(L_0)\|\|\tau(L) - \tau(L_0)\| \\
&\leq \|\tau(L) - \tau(L_0)\|(\|\tau(L) - \tau(L_0)\| + 2\|\tau(L_0)\|).
\end{aligned}
$$

Since $\tau : Gl(E) \to Gl(E)$ is differentiable, in particular continuous, it follows by these estimates that $D\tau : Gl(E) \to L(L(E, E), L(E, E))$ is indeed continuous.

Altogether this proves that $\tau : Gl(E) \to Gl(E)$ is differentiable of class C^1 and thereby completes the proof of Theorem 5.8.3. $\square$

5.9   The Inverse Function Theorem (C^1 mappings)

Let $U \subseteq E$ and $V \subseteq F$ be open sets in Banach spaces.

Definition 5.9.1. A bijective mapping $f : U \to V$ where both f and the inverse mapping $f^{-1} : V \to U$ are differentiable of class C^1, is called a *diffeomorphism of class C^1*. Provided such a diffeomorphism of class C^1 exists, we say that U and V are *diffeomorphic of class C^1*.

A diffeomorphism $f : U \to V$ of class C^1 can be characterized as a differentiable mapping of class C^1, for which there exists a differentiable mapping $g : V \to U$ of class C^1 such that $g \circ f = 1_U$ and $f \circ g = 1_V$, where 1_U and 1_V denote the identity mappings in U and V, respectively. Indeed, when such a mapping g exists, Lemma 3.7.1 shows that f is bijective with inverse $f^{-1} = g$.

If $f : U \to V$ is a diffeomorphism with inverse $g = f^{-1} : V \to U$, and $x_0 \in U$ is a point in U with image $y_0 = f(x_0) \in V$, we get by the Chain Rule

$$Dg(y_0) \circ Df(x_0) = 1_E \quad \text{and} \quad Df(x_0) \circ Dg(y_0) = 1_F,$$

where 1_E and 1_F are the identity isomorphisms in E and F, respectively. This shows that $Df(x_0) : E \to F$ is a toplinear isomorphism with the inverse toplinear isomorphism $Dg(y_0) : F \to E$, and hence

$$Df(x_0)^{-1} = Dg(y_0) = Df^{-1}(f(x_0)).$$

Therefore, if the differentiable mapping $f : U \to V$ has a differentiable inverse $f^{-1} : V \to U$, then the differential $Df(x_0) : E \to F$ of f at any point $x_0 \in U$ is a toplinear isomorphism, with the inverse toplinear isomorphism $Df^{-1}(f(x_0)) : F \to E$. In short form, this can be expressed as follows: If the differentiable mapping f has a differentiable inverse, then the differential $Df(x_0)$ has an inverse. The Inverse Function Theorem, the proof of which is the object of this section, states that in the setting of Banach spaces, locally the converse statement is also true. To make this more precise it is appropriate to define the concept of a local diffeomorphism.

Definition 5.9.2. A differentiable mapping (of class C^1) $f : U \to V$ is said to be a *local diffeomorphism (of class C^1)* at a point $x_0 \in U$, if there exist open neighbourhoods $U(x_0) \subseteq U$ of x_0 and $V(f(x_0)) \subseteq V$ of $f(x_0)$ such that the restriction $f|_{U(x_0)}$ of f to $U(x_0)$ defines a diffeomorphism (of class C^1) $f|_{U(x_0)} : U(x_0) \to V(f(x_0))$.

We are now ready for the main theorem of this chapter.

Theorem 5.9.3 (The Inverse Function Theorem). *Let E and F be Banach spaces, and let $f : U \to F$ be a differentiable mapping of class C^1 defined in an open set $U \subseteq E$. Assume that $x_0 \in U$ is a point in U at which the differential $Df(x_0) : E \to F$ of f is a toplinear isomorphism. Then f is a local diffeomorphism of class C^1 at $x_0 \in U$.*

In other words: There exist open neighbourhoods $U(x_0) \subseteq U$ of x_0 and $V(f(x_0)) \subseteq V$ of $f(x_0)$ such that f maps $U(x_0)$ diffeomorphically of class C^1 onto $V(f(x_0))$, i.e. the restriction $f|_{U(x_0)}$ of f to $U(x_0)$ defines a diffeomorphism $f|_{U(x_0)} : U(x_0) \to V(f(x_0))$ of class C^1; cf. Figure 5.6.

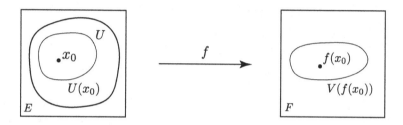

Fig. 5.6   The inverse function theorem (Theorem 5.9.3) gives a condition for a mapping f between two normed vector spaces E and F to be a local diffeomorphism at a point x_0.

Proof. By composing f with appropriate translations which, for their part, clearly are diffeomorphisms of class C^1, we can, without loss of generality, assume that $x_0 = 0 \in E$ and $f(x_0) = 0 \in F$. Next, by replacing f with $Df(x_0)^{-1} \circ f$, we can also assume that $F = E$ and that $Df(x_0) = 1_E$ is the identity isomorphism in E. In the following we put these extra assumptions on f. Let $||\cdot||$ denote the norm in E.

Set $T(x) = x - f(x)$. Then $T(0) = 0$ and $DT(0) = 0$. By hypothesis, f is differentiable of class C^1, and so DT is continuous. Hence we can choose $\rho \in \mathbb{R}^+$ such that the open ball $B_\rho(0) \subseteq U$ and

$$||DT(x)|| \leq \frac{1}{2} \quad \text{for} \quad ||x|| \leq \rho.$$

The condition on $||DT(x)||$ implies that $Df(x) = I - DT(x)$ is a toplinear isomorphism for $||x|| \leq \rho$; cf. Lemma 5.8.7.

Theorem 5.6.5 gives the inequality

$$||T(x)|| \leq \frac{1}{2}||x|| \quad \text{for} \quad ||x|| \leq \rho.$$

For an arbitrary $\sigma \in \mathbb{R}^+$, let

$$K_\sigma(0) = \{x \in E \mid \|x\| \leq \sigma\}$$

denote the closed ball in E with centre $0 \in E$ and radius σ. Since $K_\sigma(0)$ is a closed set in the Banach space E, Theorem 3.5.7 shows that $K_\sigma(0)$ is a complete metric space.

Assertion 5.9.4. *For every $y \in K_{\rho/2}(0)$ there exists a uniquely determined $x \in K_\rho(0)$, such that $f(x) = y$.*

Proof of Assertion 5.9.4. Let $y \in K_{\rho/2}(0)$ be given. Then $\|y\| \leq \frac{\rho}{2}$. Consider the mapping

$$T_y(x) = y + T(x) = y + x - f(x),$$

and note that $y = f(x)$ if and only if x is a fixed point for T_y.

For $\|x\| \leq \rho$ we have the inequality

$$\|T_y(x)\| = \|y + T(x)\| \leq \|y\| + \|T(x)\| \leq \frac{\rho}{2} + \frac{\rho}{2} = \rho \, .$$

From this follows that T_y can be considered as a continuous mapping of the complete metric space $K_\rho(0)$ into itself. If we make use of the bound on the differential DT of T, Theorem 5.6.5 gives the inequality

$$\|T_y(x_1) - T_y(x_2)\| = \|T(x_1) - T(x_2)\| \leq \frac{1}{2}\|x_1 - x_2\|$$

for all $x_1, x_2 \in K_\rho(0)$. Therefore the mapping $T_y : K_\rho(0) \to K_\rho(0)$ is a contraction in a complete metric space, and hence it has a uniquely determined fixed point according to Theorem 3.5.10. This fixed point is the point $x \in K_\rho(0)$ such that $f(x) = y$ and Assertion 5.9.4 is proved. $\square$

Now let

$$U_\rho = B_\rho(0) \cap f^{-1}(B_{\rho/2}(0)).$$

Then U_ρ is an open neighbourhood of $0 \in E$. From Assertion 5.9.4 follows

Assertion 5.9.5. *After possibly first slightly decreasing ρ, the mapping f defines a bijective mapping of U_ρ onto $B_{\rho/2}(0)$.*

Proof of Assertion 5.9.5. The reason that it might be necessary first to decrease ρ slightly is to avoid the situation where a point x, with $\|x\| = \rho$, is mapped to a point $f(x)$ with $\|f(x)\| < \frac{\rho}{2}$.

Let ρ' be a number such that $0 < \rho' < \rho$. Then it is clear that Assertion 5.9.4 is also satisfied for ρ'. We shall prove that the above mentioned problem does not exist for ρ'.

The proof is by contradiction. Assume there is an x with $||x|| = \rho'$ and $||f(x)|| < \frac{\rho'}{2}$. Since f is continuous, there exists an x_1 with $\rho' < ||x_1|| < \rho$, such that $||f(x_1)|| < \frac{\rho'}{2}$. Since Assertion 5.9.4 is valid for ρ', there also exists an x_2 with $||x_2|| \leq \rho'$ such that $f(x_2) = f(x_1)$. This contradicts that Assertion 5.9.4 is valid for ρ and so proves Assertion 5.9.5. $\square$

Now assume that $\rho \in \mathbb{R}^+$ is chosen so that f defines a bijective mapping of U_ρ onto $B_{\rho/2}(0)$. The restriction of f to U_ρ, $f|_{U_\rho} : U_\rho \to B_{\rho/2}(0)$, therefore has an inverse

$$g = f^{-1} : B_{\rho/2}(0) \to U_\rho.$$

We write $g = f^{-1}$ since g is a local inverse for f.

Assertion 5.9.6. $g = f^{-1} : B_{\rho/2}(0) \to U_\rho$ *is continuous.*

Proof of Assertion 5.9.6. Any $x \in K_\rho(0)$ can be written in the form

$$x = x - f(x) + f(x) = T(x) + f(x).$$

For all $x_1, x_2 \in K_\rho(0)$ we therefore get the inequality

$$||x_1 - x_2|| \leq ||T(x_1) - T(x_2)|| + ||f(x_1) - f(x_2)||$$
$$\leq \frac{1}{2}||x_1 - x_2|| + ||f(x_1) - f(x_2)||,$$

which can be reduced to

$$||x_1 - x_2|| \leq 2||f(x_1) - f(x_2)||.$$

This immediately gives us the inequality

$$||g(y_1) - g(y_2)|| \leq 2||y_1 - y_2||,$$

which is valid for all $y_1, y_2 \in B_{\rho/2}(0)$. This shows that g is continuous and hence proves Assertion 5.9.6. $\square$

Assertion 5.9.7. $g = f^{-1} : B_{\rho/2}(0) \to U_\rho$ *is differentiable.*

Proof of Assertion 5.9.7. We shall prove that g is differentiable at any point $y \in B_{\rho/2}(0)$.

We already have a candidate for the differential of g at y, namely the isomorphism $L = Df(g(y))^{-1}$. We need to show that there is an ε-mapping $\varepsilon = \varepsilon(h)$ such that

$$g(y + h) - g(y) = L(h) + \varepsilon(h)||h||$$

for an arbitrary increment $h \in E$ with $y + h \in B_{\rho/2}(0)$.

Set $k = g(y + h) - g(y)$. Since f is differentiable at $g(y)$ there exists an ε-mapping $\varepsilon' = \varepsilon'(k)$ such that

$$f(g(y) + k) - f(g(y)) = Df(g(y))k + \varepsilon'(k)\|k\|.$$

By using the identities $f(g(y)+k) = f(g(y+h)) = y+h$ and $f(g(y)) = y$, we get

$$h = Df(g(y))k + \varepsilon'(k)\|k\|,$$

and next, by applying $L = Df(g(y))^{-1}$ to both sides of the equation,

$$L(h) = k + L(\varepsilon'(k))\|k\|.$$

Since $k = g(y + h) - g(y)$ it is clear that we have to set

$$\varepsilon(h) = \begin{cases} -L(\varepsilon'(k))\frac{\|k\|}{\|h\|} & \text{for} \quad h \neq 0 \\ 0 & \text{for} \quad h = 0. \end{cases}$$

From the proof of Assertion 5.9.6 we have the inequality

$$\|k\| = \|g(y + h) - g(y)\| \leq 2\|y + h - y\| = 2\|h\|.$$

Hence by the continuity of g, $\varepsilon' = \varepsilon'(k)$ and L, clearly $\varepsilon = \varepsilon(h)$ is an ε-mapping. Furthermore, $\varepsilon = \varepsilon(h)$ is constructed so that it satisfies the equation

$$g(y + h) - g(y) = L(h) + \varepsilon(h)\|h\|,$$

which shows that g is differentiable at $y \in B_{\rho/2}(0)$ with the differential

$$Dg(y) = L = Df(g(y))^{-1}.$$

This proves Assertion 5.9.7. $\square$

So far we have proved that $f|_{U_\rho} : U_\rho \to B_{\rho/2}(0)$ has a differentiable inverse $g = f^{-1} : B_{\rho/2}(0) \to U_\rho$, and that the differential of g at $y \in B_{\rho/2}(0)$ is given by

$$Dg(y) = Df(g(y))^{-1}.$$

Thus, the first derivative of g,

$$Dg : B_{\rho/2}(0) \to L(E, E),$$

is exactly the composite mapping

$$Dg = \tau \circ Df \circ g,$$

where $\tau : Gl(E) \to Gl(E)$ is the mapping which assigns the inverse toplinear isomorphism to a toplinear isomorphism of E. Recall that τ is differentiable of class C^1, according to Theorem 5.8.3.

Since f is differentiable of class C^1, the differential Df is continuous. Hence Dg is also continuous, being the composition of continuous mappings, and therefore g is differentiable of class C^1. This completes the proof of Theorem 5.9.3. $\square$

Exercises and Further Results

Exercise 5.1. Let $f : \mathbb{R} \to \mathbb{R}$ be a differentiable function with bounded differential quotient. Show that f is uniformly continuous.

Hint: You can use the classical Mean Value Theorem; cf. p. 127.

Exercise 5.2. Show that

$$\left| \frac{a}{\sqrt{1+a^2}} - \frac{b}{\sqrt{1+b^2}} \right| \leq |a-b|$$

for all real numbers $a, b \in \mathbb{R}$.

Exercise 5.3. Let $(E, ||\cdot||)$ be a normed vector space, and let $U \subseteq E$ be an open set in E.

For an arbitrary pair of real-valued functions $f, g : U \to \mathbb{R}$, we define the product function $f \cdot g : U \to \mathbb{R}$ by the obvious pointwise defined operation.

Assume that the functions $f, g : U \to \mathbb{R}$ are both differentiable at the point $x \in U$. Prove that the product function $f \cdot g : U \to \mathbb{R}$ is differentiable at $x \in U$ with the differential $D(f \cdot g)(x) : E \to \mathbb{R}$ given by the formula

$$D(f \cdot g)(x) = f(x) \cdot Dg(x) + g(x) \cdot Df(x) .$$

Exercise 5.4. Let $F_1, \ldots, F_m$ be normed vector spaces with norms $||\cdot||_{F_1}, \ldots, ||\cdot||_{F_m}$, respectively.

Consider the product space $\mathbf{F}^m = F_1 \times F_2 \times \cdots \times F_m$ as a normed vector space with the obvious vector space operations and with the norm

$$||y||_{\mathbf{F}^m} = ||(y_1, y_2, \ldots, y_m)||_{\mathbf{F}^m} = \max\{||y_1||_{F_1}, ||y_2||_{F_2}, \ldots, ||y_m||_{F_m}\} .$$

Let $U \subseteq E$ be an open set in the normed vector space E, and consider a mapping

$$f = (f_1, \ldots, f_m) : U \to \mathbf{F}^m,$$

defined by mappings $f_i : U \to F_i$, $i = 1, \ldots, m$, into the factors of $\mathbf{F}^m$.

Prove that $f : U \to \mathbf{F}^m$ is differentiable at a point $x \in U$ if and only if all of the mappings $f_i : U \to F_i$, $i = 1, \ldots, m$, are differentiable at $x \in U$.

In the case of differentiability find an expression for the differential of f.

Exercise 5.5. Let X, Y and F be normed vector spaces. Turn $X \times Y$ into a normed vector space with norm $||(x, y)|| = \max\{||x||_X, ||y||_Y\}$.

Let $f : U \to F$ and $g : V \to F$ be mappings defined in open sets $U \subseteq X$ and $V \subseteq Y$, respectively. Suppose that $f : U \to F$ is differentiable at the point $x_0 \in U$, and that $g : V \to F$ is differentiable at the point $y_0 \in V$.

Define the mapping $h : U \times V \to F$ by

$$h(x, y) = f(x) + g(y) \quad \text{for} \quad (x, y) \in U \times V.$$

Show that the mapping $h : U \times V \to F$ is differentiable at the point $(x_0, y_0) \in U \times V$, and determine the differential.

Exercise 5.6. Let E and F be normed vector spaces, and let $f : U \to F$ be a continuous mapping defined in an open neighbourhood U of a point $x_0 \in E$.

Suppose that f is differentiable in $U \setminus \{x_0\}$ and that the differential $Df(x)$ for $x \in U \setminus \{x_0\}$ approaches a continuous linear mapping $L \in L(E, F)$ when x approaches x_0.

Prove that f is differentiable at x_0 with differential $Df(x_0) = L$.

Exercise 5.7. Let $U \subseteq E$ be an open set in the normed vector space E. Let $f : U \to \mathbb{R}$ be a real-valued function, which is differentiable at a point $x \in U$, and let $g : \mathbb{R} \to \mathbb{R}$ be a real-valued function, which is differentiable at the point $f(x) \in \mathbb{R}$.

Show that the composite function $g \circ f : U \to \mathbb{R}$ is differentiable at $x \in U$, and that the differential is given by

$$D(g \circ f)(x) = g'(f(x))Df(x).$$

Exercise 5.8. Let $f : E \to F$ be a mapping between normed vector spaces E and F which is differentiable at $0 \in E$ and has the property $f(\alpha h) = \alpha f(h)$ for all $\alpha \in \mathbb{R}$ and all $h \in E$. Show that f is linear.

Exercise 5.9. Let $\mathbb{R}^n$ denote an n-dimensional Euclidean space equipped with the usual inner product

$$\langle x, y \rangle = \sum_{i=1}^{n} x_i y_i \,,$$

and the associated norm

$$\|x\| = \sqrt{\langle x, x \rangle} = \sqrt{\sum_{i=1}^{n} x_i x_i} \,.$$

Denote by $E = C^1([a, b], \mathbb{R}^n)$ the space of differentiable real-valued functions $f : [a, b] \to \mathbb{R}^n$ of class C^1 defined on the interval $[a, b]$ in $\mathbb{R}$. We can equip E with the structure of a normed vector space with norm

$$\|f\|_1 = \sup_{a \leq t \leq b} (\|f(t)\| + \|f'(t)\|) \,.$$

Define the (kinetic) energy function $K : E \to \mathbb{R}$ by

$$K(f) = \frac{1}{2} \int_a^b ||f'(t)||^2 dt = \frac{1}{2} \int_a^b \langle f'(t), f'(t) \rangle dt \quad \text{for} \ f \in E.$$

1) Prove that K is differentiable at every $f \in E$ with differential $DK(f)$: $E \to \mathbb{R}$ given by

$$DK(f)(h) = \int_a^b \langle f'(t), h'(t) \rangle dt \quad \text{for all} \ h \in E.$$

2) Prove that the differential of K at $f \in E$ is zero, i.e. $DK(f) = 0$, if and only if f is a constant function. (<u>Hint</u>: Try to set $h = f$.)

3) Provide a physical interpretation of the result in 2).

Exercise 5.10. Let $\mathcal{H}$ denote a vector space with inner product $\langle \cdot, \cdot \rangle$ and associated norm $||\cdot||$ defined by $||x|| = \sqrt{\langle x, x \rangle}$ for $x \in \mathcal{H}$. (Example: $\mathcal{H} = \mathbb{R}^n$ equipped with the Euclidean inner product $\langle x, y \rangle = \sum_{i=1}^n x_i y_i$.)

Let E denote a finite dimensional proper subspace of $\mathcal{H}$ and let $u \in \mathcal{H}$ be a fixed point in $\mathcal{H}$ outside E.

Define the function $f : E \to \mathbb{R}$ by

$$f(x) = ||x - u||^2 = \langle x - u, x - u \rangle \quad \text{for} \ x \in E .$$

1) Prove that f is differentiable at every point $x \in E$ with differential $Df(x) : E \to \mathbb{R}$ given by

$$Df(x)(h) = 2\langle x - u, h \rangle \quad \text{for all} \ h \in E.$$

2) Prove that the differential of f is zero at exactly one point $x_0 \in E$.

Hint: The differential of f is zero at $x_0 \in E$, i.e. $Df(x_0) = 0$, if and only if the vector $x_0 - u$ is orthogonal to E.

Exercise 5.11. Let E and F be Banach spaces, and let $f : U \to F$ be a differentiable mapping defined in a connected, open set $U \subseteq E$.

Prove that if the differential $Df(x) = 0 \in L(E, F)$ for all $x \in U$, then $f : U \to F$ is a constant mapping.

Exercise 5.12. Let $L(E, E)$ be the normed vector space of continuous linear mappings in a normed vector space E.

Let $f : L(E, E) \to L(E, E)$ be the map defined by squaring of linear mappings, i.e.

$$f(L) = L^2 = L \circ L \quad \text{for} \ L \in L(E, E).$$

Prove that f is differentiable at any point $X \in L(E,E)$ with differential

$$Df(X)(H) = X \circ H + H \circ X \quad \text{for} \quad H \in L(E,E).$$

Exercise 5.13. Let $f : \mathbb{R}^2 \to \mathbb{R}$ be the real-valued function in two real variables $(u,v) \in \mathbb{R}^2$ defined by $f(u,v) = |u| + |v|$.

Show that f is differentiable at $(u,v) \in \mathbb{R}^2$ except if $u = 0$ or $v = 0$.

Exercise 5.14. Let $f : \mathbb{R}^2 \to \mathbb{R}$ be the real-valued function in two real variables $(u,v) \in \mathbb{R}^2$ defined by

$$f(u,v) = \frac{uv}{u^2 + v^2}, \quad \text{for} \quad (u,v) \neq (0,0),$$

and $f(0,0) = 0$.

1) Show that both partial derivatives $D_u f$ and $D_v f$ exist at every point $(u,v) \in \mathbb{R}^2$, and evaluate them explicitly in terms of u and v.

2) Explain why the function f is not differentiable at $(0,0)$.

Exercise 5.15. Let $f : \mathbb{R}^2 \to \mathbb{R}$ be the real-valued function in two real variables $(u,v) \in \mathbb{R}^2$ defined by

$$f(u,v) = \frac{u^2 v^2}{u^4 + v^4}, \quad \text{for} \quad (u,v) \neq (0,0),$$

and $f(0,0) = 0$.

1) Show that both partial derivatives $D_u f$ and $D_v f$ exist at every point $(u,v) \in \mathbb{R}^2$, and evaluate them explicitly in terms of u and v.

2) Explain why the function f is not differentiable at $(0,0)$.

Exercise 5.16. Let $f : \mathbb{R}^2 \to \mathbb{R}$ be the real-valued function in two real variables $(u,v) \in \mathbb{R}^2$ defined by

$$f(u,v) = \frac{uv}{\sqrt{u^2 + v^2}} \; \sin(\frac{1}{\sqrt{u^2 + v^2}}), \quad \text{for} \quad (u,v) \neq (0,0),$$

and $f(0,0) = 0$.

1) Show that both partial derivatives $D_u f$ and $D_v f$ exist at every point $(u,v) \in \mathbb{R}^2$, and that they vary continuously in both variables separately.

2) Explain why the function f is not differentiable at $(0,0)$.

Exercise 5.17. Let $f : \mathbb{R}^2 \to \mathbb{R}$ be a differentiable function in two real variables, and let $g_1, g_2 : \mathbb{R}^3 \to \mathbb{R}$ be two functions in three real variables $x = (x_1, x_2, x_3) \in \mathbb{R}^3$ defined by the expressions

$$g_1(x_1, x_2, x_3) = x_1^2 + x_2^2 + x_3^2 \quad \text{and} \quad g_2(x_1, x_2, x_3) = x_1 + x_2 + x_3.$$

Consider the vector-valued mapping $g = (g_1, g_2) : \mathbb{R}^3 \to \mathbb{R}^2$, and the composite function $h = f \circ g : \mathbb{R}^3 \to \mathbb{R}$.

Prove that h is a differentiable function and that the gradient vector

$$\nabla h(x) = ((D_1 h)(x), (D_2 h)(x), (D_3 h)(x))$$

satisfies the formula

$$||\nabla h(x)||^2 =$$
$$4(D_1 f(g(x)))^2 g_1(x) + 4(D_1 f(g(x)))(D_2 f(g(x))) g_2(x) + 3(D_2 f(g(x)))^2.$$

Exercise 5.18. Let E be a finite dimensional normed vector space, and let U be an open subset in E with compact closure.

Suppose $f : E \to \mathbb{R}$ is a real-valued, continuous function, which is differentiable in U, and vanishes in $E \setminus U$. Prove that there exists a point $x_0 \in U$ where the differential $Df(x_0) = 0$.

Exercise 5.19. Let l^1 be the Banach space defined in Exercise 4.12, page 108. It consists of all sequences of real numbers (a_n) for which the infinite series $\sum_{n=1}^{\infty} |a_n|$ is convergent.

Show that the norm function $||\cdot||_1 : l^1 \to \mathbb{R}$, defined by

$$||(a_n)||_1 = \sum_{n=1}^{\infty} |a_n| \quad \text{for} \quad (a_n) \in l^1 ,$$

is not differentiable at any point of l^1.

Exercise 5.20. Let E and F be Banach spaces. Let $U \subseteq E$ be an open set in E, and let $[a, b] \subseteq \mathbb{R}$ be a closed interval in $\mathbb{R}$.

For any continuous mapping $f : U \times [a, b] \to F$ we can define the mapping $g : U \to F$ by

$$g(x) = \int_a^b f(x, t)dt \quad \text{for} \quad x \in U.$$

1) Prove that $g : U \to F$ is a continuous mapping.

2) Now assume further that the partial differential $D_E f(x, t) : E \to F$ of f exists in all points $(x, t) \in U \times [a, b]$ and varies continuously with (x, t).

Prove then that the mapping $g : U \to F$ is differentiable of class C^1 in U, and that the first derivative $Dg : U \to L(E, F)$ of g is given by

$$Dg(x) = \int_a^b D_E f(x, t)dt \quad \text{for} \quad x \in U.$$

Exercise 5.21. Let X be a compact topological space, and let M be a complete metric space with metric d.

By $C(X, M)$ we denote the space of continuous mappings $f : X \to M$. For $f, g \in C(X, M)$ we put

$$D(f, g) = \sup_{x \in X} d(f(x), g(x)) .$$

Then D is a metric in $C(X, M)$.

1) Show that (f_n) is a Cauchy sequence in the metric D on $C(X, M)$ if and only if $(f_n : X \to M)$ is a uniform Cauchy sequence, i.e.

$$\forall \varepsilon > 0 \ \exists n_0 \in \mathbb{N} \ \forall n, m \in \mathbb{N} : n, m \geq n_0 \implies \forall x \in X : d(f_n(x), f_m(x)) < \varepsilon .$$

Now let (f_n) be a Cauchy sequence in $(C(X, M), D)$.

2) Show that for every $x \in X$, there exists a uniquely determined $y \in M$, such that $f_n(x) \to y$ for $n \to \infty$.

Define a mapping $f : X \to M$ by setting $f(x) = y$ for all $x \in X$, where $y \in M$ is determined as in 2). In other words, the mapping f is defined by

$$f(x) = \lim_{n \to \infty} f_n(x) \quad \text{for} \quad x \in X .$$

3) First show that (f_n) converges uniformly to f for n going to ∞, i.e.

$$\forall \varepsilon > 0 \ \exists n_0 \in \mathbb{N} \ \forall n \in \mathbb{N} : n \geq n_0 \implies \forall x \in X : d(f_n(x), f(x)) < \varepsilon .$$

Next show that $f : X \to M$ is continuous.

Hint:

$$d(f(x), f(x_0)) \leq d(f(x), f_n(x)) + d(f_n(x), f_n(x_0)) + d(f_n(x_0), f(x_0)) .$$

4) Show that $(C(X, M), D)$ is a complete metric space.

Exercise 5.22. (Local Existence and Uniqueness Theorem for Autonomous Ordinary Differential Equations.)

Let E be a Banach space, and let $U \subseteq E$ be an open set in E. The norm in E is denoted by $||\cdot||$.

A mapping $f : U \to E$ is said to be *Lipschitz continuous* in U, if there exists a constant k, such that

$$||f(x_1) - f(x_2)|| \leq k||x_1 - x_2|| \quad \text{for all} \quad x_1, x_2 \in U.$$

Now assume that $f : U \to E$ is a Lipschitz continuous mapping in U. (One can think of f as a vector field in U by placing the vector $f(x) \in E$ at every point $x \in U$.)

Consider the Initial Value Problem posed below consisting of the differential equation (i) together with the initial value condition (ii):

$$\text{(i)} \quad \frac{dx}{dt} = f(x) \qquad \text{(ii)} \quad x(0) = x_0 \in U \ .$$

By a *solution*, or, an *integral curve*, to the Initial Value Problem (i) and (ii) we understand a differentiable curve $\varphi : J \to U$ defined in an interval J around $0 \in \mathbb{R}$, such that

$$\frac{d\varphi}{dt} = f(\varphi(t)) \quad \text{for all} \quad t \in J,$$

and such that $\varphi(0) = x_0$.

1) Show that $\varphi : J \to U$ solves (i) and (ii) if and only if φ satisfies the integral equation

$$\varphi(t) = x_0 + \int_0^t f(\varphi(\tau))d\tau \ .$$

For $a > 0$, let J_a denote the interval $J_a = [-a, a]$, and for $b > 0$, let $S_b = \{x \in E \,|\, \|x - x_0\| \le b\}$ denote the closed ball in E with centre $x_0 \in U$ and radius b. For $b > 0$ sufficiently small, we have $S_b \subseteq U$, and we shall only consider such b.

Let $C(J_a, S_b)$ denote the space of continuous mappings $\varphi : J_a \to S_b$ equipped with the metric D as in Exercise 5.21.

To $\varphi \in C(J_a, S_b)$ we associate $\psi : J_a \to E$ defined by

$$\psi(t) = x_0 + \int_0^t f(\varphi(\tau))d\tau \quad \text{for} \quad t \in J_a \ .$$

2) Show that for sufficiently small $a > 0$, the mapping $\psi \in C(J_a, S_b)$.

3) Show that for sufficiently small $a > 0$, the mapping

$$T : C(J_a, S_b) \to C(J_a, S_b) \ ,$$

which assigns $\psi = T(\varphi) \in C(J_a, S_b)$ to $\varphi \in C(J_a, S_b)$, is a contraction.

4) Show that with $a > 0$ as in 3), there exists a unique solution $\varphi \in C(J_a, S_b)$ to the differential equation

$$\frac{dx}{dt} = f(x) \ ,$$

such that $\varphi(0) = x_0$.

5) Show that if $\varphi_1 : J_1 \to U$ and $\varphi_2 : J_2 \to U$ are two solutions to the differential equation

$$\frac{dx}{dt} = f(x) \ ,$$

defined in overlapping open intervals J_1 and J_2, such that $\varphi_1(t_0) = \varphi_2(t_0)$ at a point $t_0 \in J_1 \cap J_2$, then $\varphi_1(t) = \varphi_2(t)$ at all points $t \in J_1 \cap J_2$.

6) Show that there exists a unique maximal solution to the Initial Value Problem (i) and (ii). (A maximal solution is a solution with an open interval of definition that cannot be extended.)

Exercise 5.23. Let $U \subseteq E$ be an open set in the normed vector space E.

We say that a real-valued function $f : U \to \mathbb{R}$ is *homogeneous* of degree $p \geq 1$ over U, if
$$f(\lambda x) = \lambda^p f(x),$$
for every $\lambda \in \mathbb{R}$ and for every $x \in U$ for which $\lambda x \in U$.

Suppose now that $f : U \to \mathbb{R}$ is differentiable in the open set $U \subseteq E$.

Prove that $f : U \to \mathbb{R}$ is homogeneous of degree $p \geq 1$ over U, if and only if the differential of f satisfies the equation
$$Df(x)(x) = pf(x) \quad \text{for all} \quad x \in U.$$
This result is known as *Euler's theorem for homogeneous functions.*

Hint: For fixed $x \in U$, introduce the function $g(\lambda) = f(\lambda x) - \lambda^p f(x)$. You can then exploit that with the condition on the differential of f satisfied, the function g satisfies the following first order linear differential equation
$$g'(\lambda) - \frac{p}{\lambda} g(\lambda) = 0,$$
and the initial condition $g(1) = 0$.

Exercise 5.24. Let E and F be Banach spaces with norms $\|\cdot\|_E$ and $\|\cdot\|_F$, respectively, and let $f : B \to F$ be a differentiable mapping defined in an open ball $B \subseteq E$. Suppose there exists a continuous, linear mapping $L \in L(E, F)$ and a constant $k > 0$, such that the operator norm
$$\|Df(x) - L\| \leq k \quad \text{for all} \quad x \in B.$$

Prove that
$$\|f(y) - f(x) - L(y - x)\|_F \leq k \|y - x\|_E \quad \text{for all} \quad x, y \in B.$$

Exercise 5.25. Let E and F be Banach spaces.

Suppose that $f : U \to F$ is a differentiable mapping of class C^1 defined in an open set $U \subseteq E$, and suppose that the differential $Df(x)$ is a toplinear isomorphism for all $x \in U$.

Let $M = f^{-1}(0)$ be the preimage of $0 \in F$.

1) Prove that M contains only isolated points.

2) Prove that the interior of M ($\operatorname{int} M$) is the empty set.

3) Prove that if $K \subseteq U$ is a compact set, then $M \cap K$ is a finite set.

Chapter 6

Higher Order Derivatives

In Chapter 5, we investigated two fundamental theorems in the theory of differentiability in Banach spaces, namely the *Chain Rule* on differentiability of a composition of mappings, and the *Inverse Function Theorem,* which locally ensures the existence of a differentiable inverse to a differentiable mapping at a point where the differential is invertible. We proved these theorems for differentiability class C^1.

In this chapter, we introduce higher order derivatives. This requires that we first develop the theory of spaces of continuous multilinear mappings and their operator norms. In the class of Banach spaces, we construct Banach spaces of multilinear mappings. Then we can define higher order derivatives of differentiable mappings by iteration of derivatives of first order.

The basic results for higher order derivatives of mappings in Banach spaces are presented. In particular we prove that higher order derivatives of differentiable mappings of class C^r are symmetric r-linear mappings if $r \geq 2$. We also prove Taylor's formula for functions with values in a Banach space, both the case with integral remainder and the case with asymptotic remainder. We finish the chapter with a proof of the Inverse Function Theorem for differentiable mappings of class C^r, for every $r \geq 1$.

6.1 Multilinear mappings

In connection with the introduction of higher order derivatives of a mapping we need some notions from the theory of multilinear mappings.

Let $E_1, \ldots, E_r$ and F be normed vector spaces with norms $||\cdot||_{E_1}, \ldots, ||\cdot||_{E_r}$ and $||\cdot||_F$, respectively.

A mapping

$$L : E_1 \times \cdots \times E_r \to F,$$

is called *multilinear* if it is linear in each variable, i.e.

$$L(x_1,\ldots,\alpha x_i'+\beta x_i'',\ldots,x_r)=\alpha L(x_1,\ldots,x_i',\ldots,x_r)+\beta L(x_1,\ldots,x_i'',\ldots,x_r)$$

for every $i = 1,\ldots,r$, and for all $x_j \in E_j$, $x_i', x_i'' \in E_i$, $\alpha, \beta \in \mathbb{R}$.

As noted in Section 5.7, the product space $E_1 \times \cdots \times E_r$ is itself a normed vector space by the obvious coordinate-wise defined vector space operations and with the norm

$$||(x_1,\ldots,x_r)|| = \max\{||x_1||_{E_1},\ldots,||x_r||_{E_r}\}.$$

This norm equips $E_1 \times \cdots \times E_r$ with a topology, which clearly coincides with the product topology defined in Example 2.4.10.

Corresponding to Theorem 4.2.4 for linear mappings, the following theorem can be proved.

Theorem 6.1.1. *For a multilinear mapping* $L : E_1 \times \cdots \times E_r \to F$, *the four statements below are equivalent:*

(1) L *is continuous.*

(2) L *is continuous at* $(0,\ldots,0) \in E_1 \times \cdots \times E_r$.

(3) *There exists a constant* k, *such that* $||L(x_1,\ldots,x_r)||_F \le k$ *for all* $x_i \in E_i$ *with* $||x_1||_{E_1} = \cdots = ||x_r||_{E_r} = 1$.

(4) *There exists a constant* k *such that*

$$||L(x_1,\ldots,x_r)||_F \le k||x_1||_{E_1} \cdot \ldots \cdot ||x_r||_{E_r}$$

for all $x_i \in E_i$, $i = 1,\ldots,r$.

For a continuous multilinear mapping

$$L : E_1 \times \cdots \times E_r \to F,$$

it can be shown, corresponding to Lemma 4.3.1, that the four numbers below exist and coincide:

$$\begin{aligned}||L|| &= \inf\left\{k \mid ||L(x_1,\ldots,x_r)||_F \le k,\ ||x_1||_{E_1} = \cdots = ||x_r||_{E_r} = 1\right\}\\ &= \inf\left\{k \mid ||L(x_1,\ldots,x_r)||_F \le k||x_1||_{E_1} \cdot \ldots \cdot ||x_r||_{E_r},\ \text{all}\ x_i \in E_i\right\}\\ &= \sup\left\{||L(x_1\ldots,x_r)||_F \mid ||x_1||_{E_1} = \cdots = ||x_r||_{E_r} = 1\right\}\\ &= \sup\left\{||L(x_1\ldots,x_r)||_F \mid ||x_i||_{E_i} \le 1;\ i = 1,\ldots,r\right\}.\end{aligned}$$

We call the common value of these numbers the *operator norm*, or simply the *norm*, of the continuous multilinear mapping L, and we use the notation $||L||$ as in the linear case.

For a continuous multilinear mapping L and all $x_i \in E_i$, $i = 1, \ldots, r$, we have the following useful inequality corresponding to Lemma 4.3.3,

$$\| L(x_1, \ldots, x_r) \|_F \le \| L \| \, \| x_1 \|_{E_1} \cdot \ldots \cdot \| x_r \|_{E_r}.$$

For any collection of normed vector spaces $E_1, \ldots, E_r$ and F, denote by

$$L(E_1, \ldots, E_r; F) \qquad \text{the space of continuous multilinear}$$
$$\text{mappings} \ \ L : E_1 \times \cdots \times E_r \to F.$$

By introducing the usual pointwise defined operations, $L(E_1, \ldots, E_r; F)$ becomes a vector space, and corresponding to Theorem 4.3.4 we have

Theorem 6.1.2. *The operator norm is a norm in $L(E_1, \ldots, E_r; F)$, which thereby is equipped with the structure of a normed vector space.*

We now consider more closely the space of continuous bilinear mappings $L(E_1, E_2; F)$, which we shall compare with the space of continuous linear mappings $L(E_1, L(E_2, F))$. The latter space makes sense, since $L(E_2, F)$ is a normed vector space when viewed with the operator norm, and then we can state the following theorem.

Theorem 6.1.3. *The spaces $L(E_1, E_2; F)$ and $L(E_1, L(E_2, F))$ are related by the following results:*

(1) A continuous bilinear mapping

$$L : E_1 \times E_2 \to F$$

induces a continuous linear mapping

$$L_1 : E_1 \to L(E_2, F)$$

defined by $L_1(x_1)(x_2) = L(x_1, x_2)$ for $x_1 \in E_1$, $x_2 \in E_2$.

(2) A continuous linear mapping

$$L_1 : E_1 \to L(E_2, F)$$

induces a continuous bilinear mapping

$$L : E_1 \times E_2 \to F$$

defined by $L(x_1, x_2) = L_1(x_1)(x_2)$ for $x_1 \in E_1$, $x_2 \in E_2$.

(3) The mappings

$$L(E_1, E_2; F) \quad \longleftrightarrow \quad L(E_1, L(E_2, F))$$
$$L \qquad\qquad\qquad\qquad L_1$$

by which $L_1 \in L(E_1, L(E_2, F))$ and $L \in L(E_1, E_2; F)$ correspond to one another, are (linear) isomorphisms. When $L(E_1, E_2; F)$ and $L(E_1, L(E_2, F))$ are equipped with their respective operator norms, these isomorphisms preserve the norms, i.e. $||L_1|| = ||L||$.

The spaces $L(E_1, E_2; F)$ and $L(E_1, L(E_2, F))$ can in other words be identified as normed vector spaces in a canonical way.

Proof. Proof of (1). Initially assume that $L : E_1 \times E_2 \to F$ is a continuous bilinear mapping. If $||L||$ denotes the operator norm for L in $L(E_1, E_2; F)$ then we have

$$||L(x_1, x_2)||_F \leq ||L|| \, ||x_1||_{E_1} ||x_2||_{E_2} \qquad \text{for all} \qquad x_1 \in E_1, \ x_2 \in E_2.$$

Therefore, for a fixed $x_1 \in E_1$ we have

$$||L_1(x_1)(x_2)||_F = ||L(x_1, x_2)||_F \leq (||L|| \, ||x_1||_{E_1}) ||x_2||_{E_2}$$

for all $x_2 \in E_2$.

This inequality shows that $L_1(x_1) : E_2 \to F$ is a continuous linear mapping for fixed $x_1 \in E_1$, and hence $L_1(x_1) \in L(E_2, F)$ for all $x_1 \in E_1$. Thus we have proved that L induces a mapping

$$L_1 : E_1 \to L(E_2, F).$$

It is easy to show that L_1 is linear. We shall now see that L_1 is continuous when using the operator norm in $L(E_2, F)$.

If we denote the operator norm in $L(E_2, F)$ by $|| \cdot ||_{(2)}$, then we have the following inequality for all $x_1 \in E_1$,

$$
\begin{aligned}
||L_1(x_1)||_{(2)} &= \sup \left\{ ||L_1(x_1)(x_2)||_F \mid ||x_2||_{E_2} = 1 \right\} \\
&= \sup \left\{ ||L(x_1, x_2)||_F \mid ||x_2||_{E_2} = 1 \right\} \\
&\leq ||L|| \, ||x_1||_{E_1} .
\end{aligned}
$$

In particular $||L_1(x_1)||_{(2)} \leq ||L||$ for all $x_1 \in E_1$ with $||x_1||_{E_1} = 1$. This shows that the linear mapping $L_1 : E_1 \to L(E_2, F)$ is continuous, and that its operator norm satisfies $||L_1|| \leq ||L||$.

Proof of (2). Assume now that $L_1 : E_1 \to L(E_2, F)$ is a continuous linear mapping. Define $L : E_1 \times E_2 \to F$ by $L(x_1, x_2) = L_1(x_1)(x_2)$. It is clear that L is bilinear.

Since L_1 is continuous, we have

$$||L_1(x_1)||_{(2)} \leq ||L_1|| \, ||x_1||_{E_1} \qquad \text{for all} \qquad x_1 \in E_1.$$

Since $L_1(x_1) : E_2 \to F$ is continuous for any $x_1 \in E_1$, we get

$$\begin{aligned}
||L(x_1, x_2)||_F &= ||L_1(x_1)(x_2)||_F \\
&\leq ||L_1(x_1)||_{(2)} ||x_2||_{E_2} \\
&\leq ||L_1|| \, ||x_1||_{E_1} ||x_2||_{E_2} \, ,
\end{aligned}$$

for all $x_1 \in E_1$, $x_2 \in E_2$.

This shows that $L : E_1 \times E_2 \to F$ is a continuous bilinear mapping, and that $||L|| \leq ||L_1||$.

Proof of (3). Based on the results proved in (1) and (2), we can define the mappings

$$\begin{array}{ccc}
L(E_1, E_2; F) & \longleftrightarrow & L(E_1, L(E_2, F)). \\
L & & L_1
\end{array}$$

These mappings are clearly linear and mutually inverse mappings. Since we have already proved that $||L_1|| \leq ||L||$ and $||L|| \leq ||L_1||$, we conclude that $||L|| = ||L_1||$, proving that the mappings are norm preserving isomorphisms. This proves (3), and completes the proof of Theorem 6.1.3. $\square$

Similarly, we can define norm-preserving isomorphisms:

$$\begin{aligned}
L(E_1, E_2, \ldots, E_r; F) &\cong L(E_1, L(E_2, \ldots, E_r; F)) \\
&\cong L(E_1, L(E_2, L(E_3, \ldots, E_r; F)))
\end{aligned}$$

etc.

If all the spaces $E_1 = E_2 = \cdots = E_r = E$ are the same normed vector space E, we use the notation

$$L^r(E, F) = L(\underbrace{E, E, \ldots, E}_{r}; F) \, ,$$

and call $L^r(E, F)$ *the space of r-linear continuous mappings of E into F.*

As we have seen above, there are canonical norm preserving isomorphisms between the spaces

$$\begin{aligned}
L^r(E, F) &\cong L(E, L^{r-1}(E, F)) \\
&\cong L(E, L(E, L^{r-2}(E, F)))
\end{aligned}$$

etc.

## 6.2	Banach spaces of multilinear mappings

By the construction of spaces of continuous multilinear mappings, the Banach property is inherited by the mapping space from the image space. This is the content of the following theorem corresponding to Theorem 4.4.3.

Theorem 6.2.1. *Let $E_1, \ldots, E_r$ be normed vector spaces ($r \geq 1$), and let F be a Banach space. Then the vector space of continuous multilinear mappings $L(E_1, \ldots, E_r; F)$ is a Banach space with the operator norm as its norm.*

Proof. Consider $L(E_1, \ldots, E_r; F)$ as a normed vector space with the operator norm as norm, and let (L_n) be an arbitrary Cauchy sequence in $L(E_1, \ldots, E_r; F)$.

For all r-tuples of vectors $x_i \in E_i$, $i = 1, \ldots, r$, we have

$$||L_n(x_1, \ldots, x_r) - L_m(x_1, \ldots, x_r)||_F \leq ||L_n - L_m|| \, ||x_1||_{E_1} \cdots ||x_r||_{E_r}.$$

From this, it follows that for fixed vectors $x_i \in E_i$, the sequence $(L_n(x_1, \ldots, x_r))$ is a Cauchy sequence in F. Since F is a Banach space the sequence has a uniquely determined limit point in F which we denote by $L(x_1, \ldots, x_r)$. Hence

$$L_n(x_1, \ldots, x_r) \to L(x_1, \ldots, x_r) \quad \text{for} \quad n \to \infty.$$

This procedure defines a mapping

$$L : E_1 \times \cdots \times E_r \to F.$$

It is easy to prove that L is multilinear.

Since

$$\big| ||L_n|| - ||L_m|| \big| \leq ||L_n - L_m||,$$

clearly $(||L_n||)$ is a Cauchy sequence in $\mathbb{R}$. Hence there exists a uniquely determined constant $k \in \mathbb{R}$, such that $||L_n|| \to k$ for $n \to \infty$.

For an arbitrary collection of vectors $x_i \in E_i$, $i = 1, \ldots, r$, we have

$$||L_n(x_1, \ldots, x_r)||_F \leq ||L_n|| \, ||x_1||_{E_1} \cdots ||x_r||_{E_r}.$$

If we let n go to infinity, it follows that

$$||L(x_1, \ldots, x_r)||_F \leq k ||x_1||_{E_1} \cdots ||x_r||_{E_r}.$$

This shows that L is continuous. Therefore L has an operator norm $||L||$, and the limiting process shows that $||L|| = k$.

We shall now prove that (L_n) converges to L. With this in mind consider an arbitrary collection of vectors $x_i \in E_i$, $i = 1, \ldots, r$, with $\|x_i\|_{E_i} = 1$. For all $n, k \in \mathbb{N}$, the triangle inequality gives the following estimate

$$
\begin{aligned}
\|(L - L_n)(x_1, \ldots, x_r)\|_F &\leq \|L(x_1, \ldots, x_r) - L_{n+k}(x_1, \ldots, x_r)\|_F \\
&\quad + \|L_{n+k}(x_1, \ldots, x_r) - L_n(x_1, \ldots, x_r)\|_F \\
&\leq \|L(x_1, \ldots, x_r) - L_{n+k}(x_1, \ldots, x_r)\|_F \\
&\quad + \|L_{n+k} - L_n\|.
\end{aligned}
$$

Given $\varepsilon > 0$, choose $n_0 \in \mathbb{N}$ so that

$$
n \geq n_0 \quad \Rightarrow \quad \|L_{n+k} - L_n\| \leq \varepsilon.
$$

Taking the limit for $k \to \infty$ it follows that

$$
\|(L - L_n)(x_1, \ldots, x_r)\|_F \leq \varepsilon \quad \text{for} \quad n \geq n_0.
$$

From this we deduce

$$
n \geq n_0 \quad \Rightarrow \quad \|L - L_n\| \leq \varepsilon.
$$

This shows that (L_n) converges to $L \in L(E_1, \ldots, E_r; F)$ and hence completes the proof of Theorem 6.2.1. $\square$

6.3 Higher order derivatives

Let E and F be normed vector spaces with norms $\|\cdot\|_E$ and $\|\cdot\|_F$, respectively.

Let $U \subseteq E$ be an open set in E, and let $f : U \to F$ be a mapping.

Whenever $f : U \to F$ is differentiable in U, we can consider the first derivative of f, i.e. the mapping

$$
Df : U \to L(E, F),
$$

which to $x \in U$ associates the differential $Df(x)$ of f at x.

Since $L(E, F)$ is a normed vector space with the operator norm as the norm, it is logical to ask whether Df is differentiable in U. If this is the case, we say that f is two times differentiable in U, and we get a mapping

$$
D(Df) : U \to L(E, L(E, F)).
$$

We call this mapping the *second derivative* of f. The normed vector space $L(E, L(E, F))$ can be identified with the space of continuous bilinear mappings $L^2(E, F)$ of E into F. If we set $D^2 f = D(Df)$, the second derivative of f is by this identification a mapping

$$
D^2 f : U \to L^2(E, F).
$$

Since $L^2(E, F)$ is a normed vector space with the operator norm as the norm, it is logical to ask whether $D^2 f$ is differentiable in U. If this is the case, we say that f is three times differentiable in U, and we get a mapping

$$D(D^2 f) : U \to L(E, L^2(E, F)),$$

called the *third derivative* of f. The normed vector space $L(E, L^2(E, F))$ can be identified with the normed vector space $L^3(E, F)$, and if we set $D^3 f = D(D^2 f)$, the third derivative of f is by this identification a mapping

$$D^3 f : U \to L^3(E, F).$$

By the same procedure, we can define that f is r *times differentiable* in U, and the rth *derivative* of f is a mapping

$$D^r f : U \to L^r(E, F).$$

Definition 6.3.1. The mapping $f : U \to F$ is said to be *differentiable of class C^r* for $r \geq 1$, if it is r times differentiable in U and if the rth derivative $D^r f : U \to L^r(E, F)$ is continuous. If f is differentiable of class C^r for every $r \geq 1$, we say that f is *differentiable of class C^∞*.

We have the following result on differentiability of composite mappings extending Theorem 5.4.3 to any differentiability class C^r, for r in the range $1 \leq r \leq \infty$ (including ∞). This result is also referred to as the *Chain Rule*.

Theorem 6.3.2 (The Chain Rule). *Let E, F and G be normed vector spaces, and let $U \subseteq E$ and $V \subseteq F$ be open sets in E and F, respectively. Let $f : U \to F$ and $g : V \to G$ be mappings for which $f(U) \subseteq V$, such that the composite mapping $g \circ f : U \to G$ is defined. Then, if f and g are differentiable of class C^r for any $1 \leq r \leq \infty$, also $g \circ f$ is differentiable of class C^r.*

The first derivative is given by

$$D(g \circ f)(x) = (Dg \circ f)(x) \circ Df(x), \quad for \ x \in U.$$

Proof. The Chain Rule for differentiability class C^r for all $1 \leq r \leq \infty$, follows by induction making use of the Chain Rule for class C^1 proved in Theorem 5.4.3.

The proof goes as follows. Suppose the Chain Rule has been proved for differentiability class C^{r-1} for $r \geq 2$, and suppose the mappings f and g are differentiable of class C^r. Then the mappings Df and $Dg \circ f$ are differentiable of class C^{r-1}, and hence the mapping $D(g \circ f)(x) = (Dg \circ f)(x) \circ Df(x)$ is differentiable of class C^{r-1} by assumption. This proves that $g \circ f$ is differentiable of class C^r, and thereby the Chain Rule for differentiability class C^r. $\square$

6.4 Higher order derivatives in finite dimensions

In this section, we clarify how the above concepts generalize the corresponding concepts from classical analysis of functions in finitely many variables.

Let $E = \mathbb{R}^n$ and $F = \mathbb{R}^m$. By using the canonical bases in $\mathbb{R}^n$ and $\mathbb{R}^m$, $L(\mathbb{R}^n, \mathbb{R}^m)$ can be identified with the vector space of $m \times n$-matrices, and furthermore with $\mathbb{R}^{m \cdot n}$ by successively writing the m rows one after the other. In the same way, the vector space

$$L^2(\mathbb{R}^n, \mathbb{R}^m) \cong L(\mathbb{R}^n, L(\mathbb{R}^n, \mathbb{R}^m))$$

can be identified with $\mathbb{R}^{(m \cdot n) \cdot n} = \mathbb{R}^{m \cdot n^2}$. In general, $L^r(\mathbb{R}^n, \mathbb{R}^m)$ can be identified with $\mathbb{R}^{m \cdot n^r}$.

Let $U \subseteq \mathbb{R}^n$ be an open set in $\mathbb{R}^n$, and let $f : U \to \mathbb{R}^m$ be a mapping with coordinate description $f = (f_1, \ldots, f_m)$.

If $f : U \to \mathbb{R}^m$ is differentiable in U, then we get the first derivative $Df : U \to L(\mathbb{R}^n, \mathbb{R}^m)$. By the identification $L(\mathbb{R}^n, \mathbb{R}^m) \cong \mathbb{R}^{m \cdot n}$, as we saw in Section 5.3, Df has the coordinate description

$$Df = \left(\frac{\partial f_1}{\partial x_1}, \ldots, \frac{\partial f_1}{\partial x_n}, \ldots, \frac{\partial f_m}{\partial x_1}, \ldots, \frac{\partial f_m}{\partial x_n} \right),$$

or with alternative notation

$$Df = (D_1 f_1, \ldots, D_n f_1, \ldots, D_1 f_m, \ldots, D_n f_m).$$

When $f : U \to \mathbb{R}^m$ is two times differentiable, we get the second derivative $D^2 f : U \to L^2(\mathbb{R}^n, \mathbb{R}^m)$. By the identification $L^2(\mathbb{R}^n, \mathbb{R}^m) \cong \mathbb{R}^{m \cdot n^2}$, $D^2 f$ has the coordinate description

$$D^2 f = \left(\left[\frac{\partial^2 f_1}{\partial x_i \partial x_j} \right]_{i=1,\ldots,n; j=1,\ldots,n}, \ldots, \left[\frac{\partial^2 f_m}{\partial x_i \partial x_j} \right]_{i=1,\ldots,n; j=1,\ldots,n} \right),$$

or with alternative notation

$$D^2 f = \left([D_{ij} f_1]_{i=1,\ldots,n; j=1,\ldots,n}, \ldots, [D_{ij} f_m]_{i=1,\ldots,n; j=1,\ldots,n} \right).$$

In both cases the individual matrices are to be written up row after row. Hence the coordinate functions of $D^2 f$ are exactly the *second order partial derivatives* of the coordinate functions of f.

In the alternative notation, note that D_{ij} indicates differentiation of a function first with respect to its ith variable and next with respect to its jth variable. D_{ji} indicates the opposite order of differentiations.

Provided that $f : U \to \mathbb{R}^m$ is r times differentiable, we get the rth derivative $D^r f : U \to L^r(\mathbb{R}^n, \mathbb{R}^m)$. Correspondingly, by the identification

$L^r(\mathbb{R}^n, \mathbb{R}^m) \cong \mathbb{R}^{m \cdot n^r}$, the coordinate functions of $D^r f$ are exactly the rth order partial derivatives of the coordinate functions of f.

By using the maximum norm in $L^r(\mathbb{R}^n, \mathbb{R}^m) \cong \mathbb{R}^{m \cdot n^r}$, it is not difficult to prove the following theorem from calculus of functions in several variables.

Theorem 6.4.1. *The mapping $f : U \to \mathbb{R}^m$ is differentiable of class C^r if and only if all partial derivatives of order $\leq r$ of the coordinate functions of f exist and are continuous in $U \subseteq \mathbb{R}^n$.*

6.5   Higher order derivatives of multilinear mappings

For continuous linear mappings we have the following result.

Theorem 6.5.1. *A continuous linear mapping $L : E \to F$ between normed vector spaces is differentiable of class C^∞.*
The first derivative is given by $DL = L$.
The rth derivative $D^r L = 0$ for $r \geq 2$.

Proof. The first derivative was determined in Theorem 5.4.2. Since the first derivative is a constant mapping, all higher derivatives of L exist with values $D^r L = 0$ for $r \geq 2$, proving that L is differentiable of class C^∞. □

In connection with finding higher order derivatives of composite mappings involving continuous linear mappings, the following result is useful knowledge.

Theorem 6.5.2. *Let E, F and G be normed vector spaces, and let $U \subseteq E$ be an open subset of E. Let $f : U \to F$ be a differentiable mapping of class C^r, $r \geq 1$, and let $\varphi : F \to G$ be a continuous linear mapping.*
Then the composite mapping $\varphi \circ f : U \to G$ is differentiable of class C^r, and the derivative of order r is given by

$$D^r(\varphi \circ f) = \varphi \circ D^r f.$$

Proof. By Theorem 6.5.1, the linear mapping φ is differentiable of class C^∞. By Theorem 6.3.2, the composite mapping $\varphi \circ f : U \to G$ is therefore differentiable of class C^r, and the first derivative is given by $D(\varphi \circ f) = \phi \circ Df$. By induction we get then $D^r(\phi \circ f) = \phi \circ D^r f$ for all $r \geq 1$. □

Now let $E_1, \ldots, E_n$ and F be normed vector spaces. Then the product space $\mathbf{E}^n = E_1 \times \cdots \times E_n$ is a normed vector space when equipped with the obvious coordinate-wise defined vector space operations and with the norm of vectors $x = (x_1, \ldots, x_n) \in \mathbf{E}^n = E_1 \times \cdots \times E_n$ defined by

$$\|x\| = \max\{\|x_1\|_{E_1}, \ldots, \|x_n\|_{E_n}\}.$$

For continuous multilinear mappings we have the following result.

Theorem 6.5.3. *For every $n \geq 2$, a continuous multilinear mapping of normed vector spaces $L : E_1 \times \cdots \times E_n \to F$, is differentiable of class C^∞. The first derivative of L is given by the mapping*

$$DL : E_1 \times \cdots \times E_n \to L(\mathbf{E}^n, F)$$

defined by

$$DL(x)(h)$$
$$= L(h_1, x_2, \cdots, x_n) + L(x_1, h_2, x_3, \cdots, x_n) + \cdots + L(x_1, x_2, \cdots, x_{n-1}, h_n),$$

for $x = (x_1, \ldots, x_n)$, $h = (h_1, \ldots, h_n) \in \mathbf{E}^n = E_1 \times \cdots \times E_n$. The nth-derivative $D^n L$ is a constant mapping. The rth-derivative $D^r L = 0$ for $r > n$.

Proof. We prove that L is differentiable by manipulating the equation of definition of differentiability along the lines in Section 5.3.2.

First we consider the case $n = 2$.

Using that L is multilinear, we get the following equations

$$L(x + h) - L(x) = L(x_1 + h_1, x_2 + h_2) - L(x_1, x_2)$$
$$= L(x_1, x_2) + L(x_1, h_2) + L(h_1, x_2) + L(h_1, h_2) - L(x_1, x_2)$$
$$= \big(L(h_1, x_2) + L(x_1, h_2)\big) + L(h_1, h_2).$$

Now define the mapping $\varepsilon = \varepsilon(h)$ by

$$\varepsilon(h) = \begin{cases} \frac{1}{\|h\|} L(h_1, h_2) & h \neq 0 \\ 0 & h = 0. \end{cases}$$

Since $\|L(h_1, h_2)\|_F \leq \|L\| \|h_1\|_{E_1} \|h_2\|_{E_2}$, where $\|L\|$ is the operator norm of the continuous multilinear mapping L developed in Section 6.1, it is easy to prove that $\varepsilon(h)$ is an ε-mapping, i.e. $\varepsilon(h) \to 0$ for $h \to 0$.

Altogether we get

$$L(x + h) - L(x) = \big(L(h_1, x_2) + L(x_1, h_2)\big) + \varepsilon(h)\|h\|$$
$$= DL(x)(h) + \varepsilon(h)\|h\|,$$

where $DL(x) : \mathbf{E}^2 \to F$ is the continuous linear mapping defined by

$$DL(x)(h) = L(h_1, x_2) + L(x_1, h_2),$$

for $x = (x_1, x_2)$, $h = (h_1, h_2) \in \mathbf{E}^2 = E_1 \times E_2$.

Hence L is differentiable with first derivative as stated for $n = 2$.

Next we consider the case $n = 3$.

By multilinearity of L, we get in this case the following equations

$$L(x + h) - L(x) = L(x_1 + h_1, x_2 + h_2, x_3 + h_3) - L(x_1, x_2, x_3)$$
$$= \big(L(h_1, x_2, x_3) + L(x_1, h_2, x_3) + L(x_1, x_2, h_3)\big)$$
$$+ \big(L(x_1, h_2, h_3) + L(h_1, x_2, h_3) + L(h_1, h_2, x_3) + L(h_1, h_2, h_3)\big).$$

Now define the mapping $\varepsilon = \varepsilon(h)$ by $\varepsilon(0) = 0$ and for $h \neq 0$ by

$$\varepsilon(h)\|h\| = \big(L(x_1, h_2, h_3) + L(h_1, x_2, h_3) + L(h_1, h_2, x_3) + L(h_1, h_2, h_3)\big).$$

For $h \neq 0$, we get the estimates

$$\|\varepsilon(h)\|_F \leq \frac{\|L\|}{\|h\|} \big(\|x_1\|_{E_1}\|h\|^2 + \|x_2\|_{E_2}\|h\|^2 + \|x_3\|_{E_3}\|h\|^2 + \|h\|^3\big)$$
$$= \|L\| \big(\|x_1\|_{E_1}\|h\| + \|x_2\|_{E_2}\|h\| + \|x_3\|_{E_3}\|h\| + \|h\|^2\big),$$

from which follows that $\varepsilon(h) \to 0$ for $h \to 0$. In other words an ε-mapping.

Altogether we get

$$L(x + h) - L(x) = \big(L(h_1, x_2, x_3) + L(x_1, h_2, x_3) + L(x_1, x_2, h_3)\big) + \varepsilon(h)\|h\|$$
$$= DL(x)(h) + \varepsilon(h)\|h\|,$$

where $DL(x) : \mathbf{E}^3 \to F$ is the continuous linear mapping defined by

$$DL(x)(h) = L(h_1, x_2, x_3) + L(x_1, h_2, x_3) + L(x_1, x_2, h_3),$$

for $x = (x_1, x_2, x_3)$, $h = (h_1, h_2, h_3) \in \mathbf{E}^3 = E_1 \times E_2 \times E_3$.

Hence L is differentiable with first derivative as stated for $n = 3$.

Finally notice that for an arbitrary $n \geq 2$, we can by the same procedure prove that L is differentiable with first derivative as stated in the theorem.

By applying the same procedure to the n parts of $DL(x)(h)$, we can prove that the first derivative of L is differentiable with a derivative given by exchanging in each part of $DL(x)(h)$ a vector $x_i \in E_i$ with the corresponding vector $k_i \in E_i$, where $k = (k_1, \ldots, k_n) \in \mathbf{E}^n = E_1 \times \cdots \times E_n$. Thereby we get the second derivative $D^2 L(x)(h, k)$ of L.

Continuing this way step by step exchanging vectors $x_i \in E_i$ with vectors $k_i \in E_i$, we prove in the nth step that L is differentiable of class C^n and that the nth-derivative $D^n L : \mathbf{E}^n \to L^n(\mathbf{E}^n, F)$ is a constant mapping with a multiple of L as value.

Since the nth derivative is given by a constant mapping, the multilinear mapping L is indeed differentiable of class C^∞, and the rth-derivative $D^r L = 0$ for $r > n$. $\square$

Theorem 6.5.3 has the following useful corollary.

Corollary 6.5.4. *For E, F and G normed vector spaces define the map*

$$f : L(E, F) \times L(F, G) \to L(E, G),$$

by the composition of mappings, i.e.

$$f(L, M) = M \circ L, \quad for \quad L \in L(E, F), \; M \in L(F, G).$$

Then f is a bilinear mapping of normed vector spaces, and hence it is differentiable of class C^∞.

6.6 Symmetry of higher order derivatives

In this section we shall prove the following remarkable and deep result for the higher order derivatives of a differentiable map $f : U \to F$ defined in an open set $U \subseteq E$, where E and F are Banach spaces with norms $||\cdot||_E$ and $||\cdot||_F$, respectively.

Theorem 6.6.1. *If $f : U \to F$ is differentiable of class C^r for $r \geq 2$, then $D^r f$ is a symmetric, r-linear mapping for all $x \in U$, i.e.*

$$D^r f(x)(h_{\sigma(1)}, \ldots, h_{\sigma(r)}) = D^r f(x)(h_1, \ldots, h_r)$$

for an arbitrary permutation σ of the numbers $1, \ldots, r$ and arbitrary vectors $h_1, \ldots, h_r \in E$.

In finite dimensions, Theorem 6.6.1 corresponds to the following classical theorem about mixed partial derivatives.

Theorem 6.6.2 (Equality of mixed partial derivatives). *Let $U \subseteq \mathbb{R}^n$ be an open set in $\mathbb{R}^n$, and let $f : U \to \mathbb{R}$ be a function in finitely many variables, which is differentiable of class C^r for $r \geq 2$. Then all mixed partial derivatives of order $\leq r$ of $f : U \to \mathbb{R}$, which involve the same differentiations, but in permuted succession, are equal.*

We give a proof of Theorem 6.6.2 since we shall make use of this special case in the proof of Theorem 6.6.1.

Proof. (Theorem 6.6.2) An arbitrary permutation of a sequence of r partial differentiations is composed of finitely many interchanges of two partial differentiations. Hence it suffices to prove that for a differentiable function $f : U \to \mathbb{R}$ of class C^2 in two real variables $x = (u, v) \in \mathbb{R}^2$ defined in an open set $U \subseteq \mathbb{R}^2$, it holds that

$$D_{12}f(x) = D_{21}f(x) \text{ for all } x \in U.$$

Let $x_0 \in U$ be an arbitrarily chosen point in U. Choose a sufficiently small open square $K \subseteq \mathbb{R}^2$ centred in x_0 such that $K \subseteq U$.

For points $x = (u, v) \in K$ and increments $h = (h_1, h_2) \in \mathbb{R}^2$ such that $x + h \in K$, we define a function $g = g(x, h)$ by

$$g(x, h) = f(u + h_1, v + h_2) - f(u + h_1, v) - f(u, v + h_2) + f(u, v),$$

and the following two functions in one real variable

$$\varphi(u) = f(u, v + h_2) - f(u, v)$$
$$\psi(v) = f(u + h_1, v) - f(u, v).$$

Then it is easy to see that

$$g(x, h) = \varphi(u + h_1) - \varphi(u)$$
$$= \psi(v + h_2) - \psi(v).$$

Now by the classical Mean Value Theorem, Theorem 5.6.1, there exists a number $\zeta_1 \in [0, 1]$ such that

$$g(x, h) = \varphi(u + h_1) - \varphi(u) = h_1 D_1 \varphi(u + \zeta_1 h_1)$$
$$= h_1 [D_1 f(u + \zeta_1 h_1, v + h_2) - D_1 f(u + \zeta_1 h_1, v)].$$

Again by applying Theorem 5.6.1 but this time to the partial derivative $D_1 f(x)$ considered as a function of its second variable v, there exists another number $\zeta_2 \in [0, 1]$ such that

$$D_1 f(u + \xi_1 h_1, v + h_2) - D_1 f(u + \zeta_1 h_1, v) = h_2 (D_{12} f(u + \zeta_1 h_1, v + \zeta_2 h_2)).$$

Altogether we get

$$g(x, h) = h_1 h_2 (D_{12} f(u + \zeta_1 h_1, v + \zeta_2 h_2)).$$

By similar applications of Theorem 5.6.1 to the function

$$g(x, h) = \psi(v + h_2) - \psi(v),$$

we obtain two numbers $\zeta_3, \zeta_4 \in [0, 1]$ such that

$$g(x, h) = h_2 h_1 (D_{21} f(u + \zeta_4 h_1, v + \zeta_3 h_2)).$$

By cancelling the factor $h_1 h_2$ in the two expressions for $g(x, h)$ we get

$$D_{12} f(u + \zeta_1 h_1, v + \zeta_2 h_2) = D_{21} f(u + \zeta_4 h_1, v + \zeta_3 h_2).$$

By continuity of the second order derivatives we get in the limit, when the increment h approaches zero, that $D_{12} f(x) = D_{21} f(x)$ at $x = (u, v)$. In particular, $D_{12} f(x_0) = D_{21} f(x_0)$. $\square$

Proof. (Theorem 6.6.1) If $f : U \to F$ is differentiable of class C^2, we have to prove that $D^2 f$ is a symmetric, bilinear mapping for all $x \in U$, i.e.

$$D^2 f(x)(h_1, h_2) = D^2 f(x)(h_2, h_1)$$

for arbitrary vectors $h_1, h_2 \in E$.

We embark on proving this.

Special case

First consider the case $F = \mathbb{R}$.

Let $x_0 \in U$ be an arbitrary point in U. Let $h_1, h_2 \in E$ be two arbitrarily chosen but then fixed vectors.

Define the function $g : K \to \mathbb{R}$ in two real variables $u, v \in \mathbb{R}$ by

$$g(u, v) = f(x_0 + uh_1 + vh_2), \quad (u, v) \in K,$$

where $K = [-a, a] \times [-a, a] \subseteq \mathbb{R}^2$ is a sufficiently small square in $\mathbb{R}^2$ such that $x_0 + uh_1 + vh_2 \in U$.

Then g is a differentiable function of two real variables of class C^2.

Using the Chain Rule we get

$$D_1 g(u, v) = Df(x_0 + uh_1 + vh_2)(h_1)$$
$$D_2 g(u, v) = Df(x_0 + uh_1 + vh_2)(h_2),$$

and next

$$D_{12} g(u, v) = D^2 f(x_0 + uh_1 + vh_2)(h_1, h_2)$$
$$D_{21} g(u, v) = D^2 f(x_0 + uh_1 + vh_2)(h_2, h_1).$$

From this follows by Theorem 6.6.2 that

$$D^2 f(x_0)(h_1, h_2) = D_{12} g(0, 0) = D_{21} g(0, 0) = D^2 f(x_0)(h_2, h_1),$$

proving the special case.

General case

For a general Banach space F, the theorem follows from the special case by making use of the fact (Theorem 6.5.2) that $D^2(\varphi \circ f) = \varphi \circ D^2 f$ for any continuous linear functional $\varphi : F \to \mathbb{R}$, combined with an application of the fact that for any non-zero vector $y \in F$ there exists a continuous linear functional $\varphi : F \to \mathbb{R}$ for which $\varphi(y) \neq 0$. The latter fact is an immediate corollary to the famous Hahn-Banach Theorem, see e.g. my book [Functional Analysis-Entering Hilbert Space, 2nd edition, Corollary 1.9.2].

The proof of Theorem 6.6.1 for an arbitrary $r \geq 2$ can be done in complete analogy to the proof for $r = 2$ by introducing r real variables in the help function g. $\square$

6.7   Taylor's formula

Let $f : U \to F$ be a differentiable mapping of class C^n, $n \geq 1$, defined in an open subset $U \subseteq E$ of the Banach space E and with values in the Banach space F. Then we have the following result, known as Taylor's formula.

Theorem 6.7.1 (Taylor's formula, integral remainder). *Let $x_0 \in U$ be an arbitrary point in U, and let $B_\rho(x_0) \subseteq U$ be an open ball in U with center x_0 and radius $\rho > 0$.*

For each $h \in E$ with norm $\|h\| < \rho$, we have the expansion

$$f(x_0+h) = f(x_0)+Df(x_0)(h)+\frac{D^2 f(x_0)}{2}(h^{[2]})+\cdots+\frac{D^{n-1}f(x_0)}{(n-1)!}(h^{[n-1]})+R(h),$$

where $(h^{[r]})$ stands for $(h, h, \ldots, h)$ (r times), and

$$R(h) = \frac{1}{(n-1)!} \int_0^1 (1-t)^{n-1} D^n f(x_0 + th)(h^{[n]}) dt.$$

Proof. In the following, the vector $h \in E$ with $\|h\| < \rho$ is kept fixed.
    Define the function

$$\sigma : [0,1] \to B_\rho(x_0) \subseteq E \quad \text{by} \quad \sigma(t) = x_0 + th, \ t \in [0,1].$$

Note that the first derivative of σ is constant, namely

$$\sigma'(t) = D\sigma(t)(1) = h, \ t \in [0,1].$$

The function $g = f \circ \sigma : [0,1] \to F$ is given by $g(t) = f(x_0 + th)$.

The functions σ and g, are actually defined in a slightly larger open interval containing $[0,1]$, and they are both differentiable of class C^n.

Using the Chain Rule, we find the first derivative of g to be

$$g'(t) = Dg(t)(1) = Df(\sigma(t)) \circ D\sigma(t)(1) = Df(\sigma(t))(h) = (Df \circ \sigma(t))(h).$$

When $n \geq 2$, we find by applying similar calculations to the function $Df \circ \sigma$, that the second derivative of g is given by

$$g^{(2)}(t) = D[(Df \circ \sigma(t))(h)] \circ D\sigma(t)(1) = (D^2 f \circ \sigma(t))(h^{[2]}).$$

For $2 \leq r \leq n$ we find by induction the rth derivative of g to be

$$g^{(r)}(t) = D[D^{r-1}f \circ \sigma(t))(h^{[r-1]})] \circ D\sigma(t)(1) = (D^r f \circ \sigma(t))(h^{[r]}).$$

Now consider the function $T : [0,1] \to F$ defined by

$$T(t) = g(t) + (1-t)g'(t) + \frac{(1-t)^2 g^{(2)}(t)}{2} + \cdots + \frac{(1-t)^{n-1} g^{(n-1)}(t)}{(n-1)!} .$$

By a direct calculation we get

$$\frac{dT}{dt} = \frac{(1-t)^{n-1}g^{(n)}(t)}{(n-1)!} \, .$$

Then by Theorem 5.6.4 we get the equation

$$T(1) = T(0) + \int_0^1 \frac{(1-t)^{n-1}g^{(n)}(t)}{(n-1)!} dt \, .$$

By inserting the values

$$g^{(n)}(t) = D^n f(x_0 + th)(h^{[n]})$$

$$T(1) = g(1) = f(x_0 + h)$$

$$T(0) = f(x_0) + Df(x_0)(h) + \frac{D^2 f(x_0)}{2}(h^{[2]}) + \cdots + \frac{D^{n-1}f(x_0)}{(n-1)!}(h^{[n-1]}) \, ,$$

in the equation, we get Taylor's formula with integral remainder. $\square$

By rewriting the integral remainder, we get a version of Taylor's formula more suitable for asymptotic analysis of the function $f : U \to F$.

Theorem 6.7.2 (Taylor's formula, asymptotic remainder). *Let $x_0 \in U$ be an arbitrary point in U, and let $B_\rho(x_0) \subseteq U$ be an open ball in U with center x_0 and radius $\rho > 0$.*

For each $h \in E$ with norm $\|h\| < \rho$, we have the expansion

$$f(x_0+h) = f(x_0)+Df(x_0)(h)+\frac{D^2 f(x_0)}{2}(h^{[2]})+\cdots+\frac{D^n f(x_0)}{n!}(h^{[n]})+\varepsilon(h)\|h\|^n,$$

where $\varepsilon(h) \to 0$ for $h \to 0$.

Proof. The integral remainder in the expansion of $f : U \to F$ in the ball $B_\rho(x_0)$ is given by the function

$$R(h) = \frac{1}{(n-1)!} \int_0^1 (1-t)^{n-1} D^n f(x_0 + th)(h^{[n]}) dt.$$

If we define the functions

$$R_1(h) = \frac{1}{(n-1)!} \int_0^1 (1-t)^{n-1} D^n f(x_0)(h^{[n]}) dt$$

$$R_2(h) = \frac{1}{(n-1)!} \int_0^1 (1-t)^{n-1} (D^n f(x_0 + th) - D^n f(x_0))(h^{[n]}) dt \, ,$$

we can write

$$R(h) = R_1(h) + R_2(h) \, .$$

We find

$$R_1(h) = \frac{1}{(n-1)!} \int_0^1 (1-t)^{n-1} D^n f(x_0)(h^{[n]}) dt$$

$$= \frac{D^n f(x_0)}{(n-1)!}(h^{[n]}) \int_0^1 (1-t)^{n-1} dt = \frac{D^n f(x_0)}{n!}(h^{[n]})$$

Define the function $\varepsilon = \varepsilon(h)$ by

$$\varepsilon(h) = \begin{cases} \frac{1}{(n-1)!} \int_0^1 \frac{(1-t)^{n-1}}{||h||^n}(D^n f(x_0 + th) - D^n f(x_0))(h^{[n]}) dt & h \neq 0 \\ 0 & h = 0. \end{cases}$$

Then

$$R(h) = \frac{D^n f(x_0)}{n!}(h^{[n]}) + \varepsilon(h)||h||^n ,$$

and it only remains to prove that $\varepsilon(h) \to 0$ for $h \to 0$.

Using properties of norms of integrals and of the operator norm for continuous n-linear mappings developed in Section 6.1, we get for $h \neq 0$ the following estimates

$$||\varepsilon(h)|| \leq \frac{1}{(n-1)!} \int_0^1 \frac{(1-t)^{n-1}}{||h||^n}||(D^n f(x_0 + th) - D^n f(x_0))(h^{[n]})|| dt$$

$$\leq \frac{1}{(n-1)!} \int_0^1 \frac{(1-t)^{n-1}}{||h||^n}||D^n f(x_0 + th) - D^n f(x_0)|| \, ||h||^n dt$$

$$= \frac{1}{(n-1)!} \int_0^1 (1-t)^{n-1} ||D^n f(x_0 + th) - D^n f(x_0)|| dt.$$

Since the mapping $f : U \to F$ is differentiable of class C^n, the n-linear mapping $D^n f : U \to L^n(E, F)$ is continuous.

For any $\varepsilon > 0$, we can therefore pick a δ, with $0 < \delta \leq \rho$, such that

$$||D^n f(x_0 + th) - D^n f(x_0)|| \leq \varepsilon, \quad \text{for all} \quad t \in [0, 1], \ h \in E, \ ||h|| \leq \delta .$$

Then we can continue our estimates of $||\varepsilon(h)||$ as follows

$$||\varepsilon(h)|| \leq \frac{1}{(n-1)!} \int_0^1 (1-t)^{n-1} \varepsilon \, dt = \frac{\varepsilon}{n!} .$$

This proves

$$\forall \varepsilon > 0 \ \exists \delta > 0 \ \forall h \in E : ||h|| \leq \delta \implies ||\varepsilon(h)|| \leq \varepsilon ,$$

which shows that $\varepsilon(h) \to 0$ for $h \to 0$.

Altogether, Taylor's formula with asymptotic remainder is proved. $\square$

6.8   The Inverse Function Theorem

Let $U \subseteq E$ and $V \subseteq F$ be open sets in Banach spaces E and F.

For a differentiable mapping $f : U \to V$ of class C^1, we defined in Section 5.9 the notions of *diffeomorphism*, and *local diffeomorphism* at a point $x_0 \in U$. These definitions have exact analogues for differentiable mappings of class C^r, such that we have the corresponding notions of *diffeomorphism* and *local diffeomorphism of class C^r* for every $1 \leq r \leq \infty$.

A diffeomorphism $f : U \to V$ of class C^r can be characterized as a differentiable mapping of class C^r, for which there exists a differentiable mapping $g : V \to U$ of class C^r such that $g \circ f = 1_U$ and $f \circ g = 1_V$, where 1_U and 1_V denote the identity mappings in U and V, respectively. Indeed, when such a mapping g exists, Lemma 3.7.1 shows that f is bijective with inverse $f^{-1} = g$.

If $f : U \to V$ is a diffeomorphism with inverse $g = f^{-1} : V \to U$, and $x_0 \in U$ is a point in U with image $y_0 = f(x_0) \in V$, we get by the Chain Rule

$$Dg(y_0) \circ Df(x_0) = 1_E \quad \text{and} \quad Df(x_0) \circ Dg(y_0) = 1_F,$$

where 1_E and 1_F are the identity toplinear isomorphisms in E and F, respectively. This shows that $Df(x_0) : E \to F$ is a toplinear isomorphism with toplinear inverse $Dg(y_0) : F \to E$, and hence

$$Df(x_0)^{-1} = Dg(y_0) = Df^{-1}(f(x_0)).$$

The general Inverse Function Theorem for higher order differentiable mappings can now be stated as follows.

Theorem 6.8.1 (The Inverse Function Theorem). *Let E and F be Banach spaces, and let $f : U \to F$ be a mapping defined in an open set $U \subseteq E$, which is differentiable of class C^r, $1 \leq r \leq \infty$. Assume that $x_0 \in U$ is a point in U at which the differential $Df(x_0) : E \to F$ of f is a toplinear isomorphism. Then f is a local diffeomorphism of class C^r at $x_0 \in U$.*

In other words: There exist open neighbourhoods $U(x_0) \subseteq U$ of x_0 and $V(f(x_0)) \subseteq V$ of $f(x_0)$ such that f maps $U(x_0)$ diffeomorphically of class C^r onto $V(f(x_0))$, i.e. the restriction $f|_{U(x_0)}$ of f to $U(x_0)$ defines a diffeomorphism $f|_{U(x_0)} : U(x_0) \to V(f(x_0))$ of class C^r; cf. Figure 5.6, page 143.

For the proof of Theorem 6.8.1, we need an extension of Theorem 5.8.3, on differentiability of the inverse transformation $\tau : Gl(E) \to Gl(E)$ in the general linear group

$$Gl(E) = \{L \in L(E,E) \mid L : E \to E \text{ toplinear isomorphism}\}.$$

Theorem 6.8.2. *Let E be a Banach space. Then we have:*

(i)   *$Gl(E)$ is an open set in $L(E, E)$.*

(ii) *The mapping $\tau : Gl(E) \to Gl(E)$, which assigns the inverse toplinear isomorphism $\tau(L) = L^{-1} \in Gl(E)$ to a toplinear isomorphism $L \in Gl(E)$, is differentiable of class C^∞. The differential of τ is given by*

$$D\tau(L)H = -L^{-1} \circ H \circ L^{-1}$$

for $L \in Gl(E)$ and $H \in L(E, E)$.

Proof. (Theorem 6.8.2) In Theorem 5.8.3 we proved that the inverse transformation τ is differentiable of class C^1 and that the differential map

$$D\tau : Gl(E) \to L(L(E, E), L(E, E))$$

is defined by

$$D\tau(L)H = -L^{-1} \circ H \circ L^{-1}, \text{ for } L \in Gl(E), \ H \in L(E, E).$$

By arguments similar to those employed in the proof of Theorem 5.8.3, we can prove that $D\tau$ is differentiable of class C^1, and hence that τ is differentiable of class C^2. By recurrence, we get that τ is differentiable of class C^∞. $\square$

Proof. (Theorem 6.8.1) The proof of Theorem 6.8.1 follows the proof of Theorem 5.9.3 verbatim right up to the end.

Continuing with the final paragraph in the proof of Theorem 5.9.3, the proof of Theorem 6.8.1 is completed as follows.

Since f is differentiable of class C^1, the differential Df is continuous. Hence Dg is also continuous, being the composition of continuous mappings, and then g is differentiable of class C^1.

Now assume that f is differentiable of class C^2. Then Df is differentiable of class C^1. Since also g is differentiable of class C^1, the chain rule shows that $Dg = \tau \circ Df \circ g$ is differentiable of class C^1. But that means that g is differentiable of class C^2.

In general for $r \geq 2$, it follows by induction that if f is differentiable of class C^r, then Df, and hence also $Dg = \tau \circ Df \circ g$, is differentiable of class C^{r-1}, showing that $g = f^{-1}$ is differentiable of class C^r.

This completes the proof of Theorem 6.8.1. $\square$

Exercises and Further Results

Exercise 6.1. Let E and F be normed vector spaces, and let $f : U \to F$ be a mapping defined in an open set U in E.

Prove that if the mapping $f : U \to F$ is r times differentiable in U, and the rth derivative $D^r f : U \to L^r(E, F)$ of f is s times differentiable in U, then the mapping f is $r + s$ times differentiable in U, and the derivative of order $r + s$ is given by $D^{r+s} f = D^s(D^r f)$.

Exercise 6.2. Let E and F be normed vector spaces.

Define a *quadratic* mapping $f : E \to F$ by

$$f(x) = B(x, x) \quad \text{for} \quad x \in E,$$

where $B : E \times E \to F$ is a continuous bilinear mapping.

Prove that f is differentiable of class C^∞, and determine the first derivative.

Exercise 6.3. Let $f = f(x_1) : \mathbb{R} \to \mathbb{R}$ and $g = g(x_2) : \mathbb{R} \to \mathbb{R}$ be two real-valued differentiable functions of class C^2 of one real variable.

Define the real-valued function $h : \mathbb{R}^2 \to \mathbb{R}$ of two real variables by

$$h(x_1, x_2) = f(x_1 + g(x_2)) \quad \text{for} \quad (x_1, x_2) \in \mathbb{R}^2.$$

Prove that h is differentiable in $\mathbb{R}^2$ and find formulas for all partial derivatives of first and second order in terms of derivatives of f and g.

Use this to verify the relation

$$(D_1 h)(D_{21}^2 h) = (D_2 h)(D_{11}^2 h).$$

Exercise 6.4. Let U be an open set in the normed vector space E. A real-valued function $f : U \to \mathbb{R}$ is said to have a *local maximum (minimum)* at a point $x_0 \in U$ if there exists an open neighbourhood $N \subseteq U$ of x_0 such that $f(x) \leq f(x_0)$ $(f(x) \geq f(x_0))$ for all $x \in N$.

1) Suppose that the function $f : U \to \mathbb{R}$ is differentiable at the point $x_0 \in U$. Prove that for each fixed $h \in E$, there exists an $r > 0$ such that the function $g(t) = f(x_0 + th)$ is defined for $t \in]-r, r[$ and is differentiable at 0 with derivative $g'(0) = Df(x_0)(h)$.

2) Suppose that the function $f : U \to \mathbb{R}$ is differentiable at the point $x_0 \in U$ and that f has a local maximum (minimum) at $x_0 \in U$. Prove that the differential of f at x_0 is zero, i.e. $Df(x_0) = 0$.

Exercise 6.5. Let $f : U \to \mathbb{R}$ be a real-valued function defined in an open set U in the Banach space E. Suppose f is differentiable of class $C^r, r \geq 2$.

By Theorem 6.6.1, the second derivative $D^2 f(x)$ of f is a symmetric, bilinear mapping for all $x \in U$, i.e. for arbitrary vectors $h, k \in E$ we have

$$D^2 f(x)(h, k) = D^2 f(x)(k, h).$$

A symmetric, bilinear mapping is also referred to as a *quadratic form*.

A quadratic form $Q = Q(h, k)$ in two variables $h, k \in E$ is called *positive definite*, if $Q(h, h) > 0$ for all $h \neq 0$.

The particular quadratic form $D^2 f(x)$ is called the *hessian* of f at $x \in U$. Let $x_0 \in U$ be an arbitrary point in U.

1) Prove that if $f : U \to \mathbb{R}$ has a local minimum at $x_0 \in U$, then the differential $Df(x_0) = 0$, and $D^2 f(x_0)(h, h) \geq 0$ for all $h \in E$.

[Hint: Taylor's formula with asymptotic remainder may be helpful.]

2) Prove that $f : U \to \mathbb{R}$ has a local minimum at $x_0 \in U$, if the differential $Df(x_0) = 0$, and the hessian $D^2 f(x_0)$ is positive definite.

3) Formulate corresponding results for the existence of a local maximum.

4) Interpret the above results on local minima and maxima for real-valued functions in case $E = \mathbb{R}^n$ is the n-dimensional real number space.

Exercise 6.6. Let $[a, b]$ be a closed and bounded interval in $\mathbb{R}$. Denote by $C^1([a, b], \mathbb{R}^n)$ the vector space of differentiable curves $x : [a, b] \to \mathbb{R}^n$ in $\mathbb{R}^n$ of class C^1. Equip $C^1([a, b], \mathbb{R}^n)$ with the norm

$$\|x\|_1 = \sup \left\{ \|x(t)\| + \|x'(t)\| \mid t \in [a, b] \right\},$$

in which $\| \cdot \|$ is the maximum norm in $\mathbb{R}^n$.

For an arbitrary open set U in $\mathbb{R} \times \mathbb{R}^n \times \mathbb{R}^n$, we denote by $\tilde{U}$ the subset of curves $x \in C^1([a, b], \mathbb{R}^n)$, in which $(t, x(t), x'(t)) \in U$ for all $t \in [a, b]$.

1) Show that $\tilde{U}$ is an open set in $C^1([a, b], \mathbb{R}^n)$.

Now let U be an open set in $\mathbb{R} \times \mathbb{R}^n \times \mathbb{R}^n$, considered with coordinates $(t, q, p) \in \mathbb{R} \times \mathbb{R}^n \times \mathbb{R}^n$, and let $L = L(t, q, p) : U \to \mathbb{R}$ be a differentiable function of class C^1.

Define the function $f : \tilde{U} \to \mathbb{R}$ by

$$f(x) = \int_a^b L\big(t, x(t), x'(t)\big) dt \quad \text{for } x \in \tilde{U} .$$

2) Show that $f : \tilde{U} \to \mathbb{R}$ is differentiable in $x \in \tilde{U}$ with the differential

$$Df(x)h = \int_a^b DL\big(t, x(t), x'(t)\big) \cdot \big(0, h(t), h'(t)\big) dt \quad \text{for } h \in C^1([a, b], \mathbb{R}^n).$$

In the following, the curves $x \in \tilde{U}$ and $h \in C^1([a, b], \mathbb{R}^n)$ are kept fixed.

3) Show that there exists an $\varepsilon > 0$, such that the curve $x + \lambda h$ belongs to $\tilde{U}$, for all $\lambda \in \,]-\varepsilon, \varepsilon\,[$, and use this to define a function $g : \,]-\varepsilon, \varepsilon\,[\,\to \mathbb{R}$ by

$$g(\lambda) = f(x + \lambda h) \quad \text{for} \quad \lambda \in \,]-\varepsilon, \varepsilon\,[\,.$$

4) Show that g is differentiable at $\lambda = 0$ with the differential quotient

$$g'(0) = \int_a^b DL\big(t, x(t), x'(t)\big) \cdot \big(0, h(t), h'(t)\big)\, dt$$

$$= \int_a^b \Big[\sum_{i=1}^n \Big(\frac{\partial L}{\partial q_i} h_i + \frac{\partial L}{\partial p_i} h_i{}'\Big)\Big]\, dt\ .$$

Here, as well as in 5), the partial derivatives of L shall be taken at the points $(t, x(t), x'(t)) \in U$ and the functions $h_i, h_i{}'$ at $t \in [a, b]$.

5) Now assume that $h(a) = h(b) = 0$ and that $L = L(t, q, p) : U \to \mathbb{R}$ is a differentiable function of class C^2.

Using integration by parts, first show that

$$\int_a^b \frac{\partial L}{\partial p_i} h_i{}'\, dt = -\int_a^b \frac{d}{dt}\Big(\frac{\partial L}{\partial p_i}\Big) h_i\, dt\ ,$$

and next that

$$g'(0) = \int_a^b \Big[\sum_{i=1}^n \Big(\frac{\partial L}{\partial q_i} - \frac{d}{dt}\Big(\frac{\partial L}{\partial p_i}\Big)\Big) h_i\Big]\, dt\ .$$

The system of equations

$$\frac{\partial L}{\partial q_i} - \frac{d}{dt}\Big(\frac{\partial L}{\partial p_i}\Big) = 0 \quad, i = 1, \ldots, n\ ,$$

is called the *Euler-Lagrange equations* for the above function $f = f(x)$ defined by L. These equations are fundamental in the calculus of variations.

6) Show that the differentiable curve $x : [a, b] \to \mathbb{R}^n$ in $\tilde{U}$ is a stationary point of $f : \tilde{U} \to \mathbb{R}$, i.e. $Df(x) = 0$, if and only if

$$\frac{\partial L}{\partial q_i}\big(t, x(t), x'(t)\big) = \frac{d}{dt}\Big(\frac{\partial L}{\partial p_i}\big(t, x(t), x'(t)\big)\Big)\ ,$$

for all $i = 1, \ldots, n$.

Exercise 6.7. In *Lagrangian mechanics* one studies the dynamics of a mechanical system by studying the paths $q = q(t)$ of the evolving mechanical system over a time interval $[a, b]$, by choosing a convenient set of independent parameters $q \in \mathbb{R}^n$ that completely characterize the possible configurations of the mechanical system. Along each path, the mechanical system has a kinetic energy T and a potential energy V. To find a solution to the underlying mechanical problem, one looks for paths where the total kinetic energy and the total potential energy are in balance. For this one introduces the *Lagrangian* $L = T - V$ and the *action integral*

$$S(q(t)) = \int_a^b L\big(t, q(t), q'(t)\big)\, dt.$$

The stable motions of the mechanical system are now found by solving the Euler-Lagrange equations for the action integral; cf. Exercise 6.6.

1) Consider a mechanical system described by a single real variable $q(t)$ which develops smoothly in a time interval $[a, b]$. Suppose the energy function for the mechanical system is kinetic energy $T(t) = \frac{1}{2} m q'(t)^2$ and potential energy $V(t) = \frac{1}{2} k q(t)^2$, where m and k are positive constants.

Determine the Euler-Lagrange equation for the system and solve it.

2) More generally, consider a point mass m moving smoothly in time in 3-dimensional Euclidean space $\mathbb{R}^3$, with potential energy $V(x(t))$ at the point $x(t) \in \mathbb{R}^3$. The kinetic energy of the particle is $T(t) = \frac{1}{2} m \|x'(t)\|^2$.

Show that the system of Euler-Lagrange equations in this case take the form of Newton's second law:

$$m x'' = -\nabla V(x).$$

Exercise 6.8. Let $U \subseteq E$ and $V \subseteq F$ be open sets in the Banach spaces E and F. Let $f : U \to V$ be a differentiable mapping of class $C^r, r \geq 1$. Assume that

(i) $f : U \to V$ is bijective.

(ii) $Df(x) : E \to F$ is a toplinear isomorphism for all $x \in U$.

Show that $f : U \to V$ is a diffeomorphism of class C^r.

Exercise 6.9. Let E_1, E_2 and F be normed vector spaces. We consider, as usual, $E_1 \times E_2$ as a normed vector space with the maximum norm.

Let $U \subseteq E_1 \times E_2$ be an open set, and let $(x_1, x_2) \in U$ be an arbitrarily chosen point in U, kept fixed in the following.

Consider a mapping $f : U \to F$ which is differentiable at $(x_1, x_2) \in U$.

The restriction $f|(E_1 \times \{x_2\}) \cap U$ of f to $(E_1 \times \{x_2\}) \cap U$ can be considered as a mapping of an open set in E_1 into F. This mapping is differentiable at x_1, and therefore has a continuous linear mapping $D_{E_1} f(x_1, x_2) : E_1 \to F$ as differential. We call $D_{E_1} f(x_1, x_2)$ the *partial differential* of f along E_1 at $(x_1, x_2) \in U$. We have in other words,

$$D_{E_1} f(x_1, x_2) = D(f|(E_1 \times \{x_2\}) \cap U)(x_1).$$

Similarly, we can define the partial differential $D_{E_2} f(x_1, x_2)$ of f along E_2 at $(x_1, x_2) \in U$.

1) In case $E_1 = \mathbb{R}^n$, $E_2 = \mathbb{R}^m$ and $F = \mathbb{R}^k$, find the matrix descriptions of the partial differentials $D_{E_1} f(x_1, x_2)$ and $D_{E_2} f(x_1, x_2)$ with respect to the canonical bases in $\mathbb{R}^n$, $\mathbb{R}^m$ and $\mathbb{R}^k$.

2) Show that the differential $Df(x_1, x_2) : E_1 \times E_2 \to F$ of $f : U \to F$ at $(x_1, x_2) \in U$ is given by

$$DF(x_1, x_2) \cdot (h_1, h_2) = D_{E_1} f(x_1, x_2) \cdot h_1 + D_{E_2} f(x_1, x_2) \cdot h_2 ,$$

for $(h_1, h_2) \in E_1 \times E_2$.

Or, with the obvious matrix notation,

$$DF(x_1, x_2)(h_1, h_2) = [D_{E_1} f(x_1, x_2) \quad D_{E_2} f(x_1, x_2)] \begin{bmatrix} h_1 \\ h_2 \end{bmatrix} .$$

Exercise 6.10 (The Implicit Function Theorem). Let X, Y, and Z be Banach spaces. (For simplicity, you may content yourself with considering $X = \mathbb{R}^n$, $Y = \mathbb{R}^m$ and $Z = \mathbb{R}^m$.)

Let $U \subseteq X \times Y$ be an open set, and let $f : U \to Z$ be a differentiable mapping of class $C^r, r \geq 1$.

Let (x_0, y_0) be a point in U, such that $f(x_0, y_0) = 0$, and assume that the partial differential $D_Y f(x_0, y_0) : Y \to Z$ of f along Y at $(x_0, y_0) \in U$ is a toplinear isomorphism.

Define the mapping $\Phi : U \to X \times Z$ by $\Phi(x, y) = \big(x, f(x, y)\big)$. Then Φ is clearly differentiable of class C^r, and the differential of Φ at $(x_0, y_0) \in U$ is a continuous linear mapping

$$D\Phi(x_0, y_0) : X \times Y \to X \times Z .$$

1) Show that the differential of Φ at $(x_0, y_0) \in U$ is given by

$$D\Phi(x_0, y_0) \cdot (h, k) = \big(h, D_X f(x_0, y_0) \cdot h + D_Y f(x_0, y_0) \cdot k\big)$$

for $(h, k) \in X \times Y$. Or, with the obvious matrix notation,

$$D\Phi(x_0, y_0)(h, k) = \begin{bmatrix} 1_X & 0 \\ D_X f(x_0, y_0) & D_Y f(x_0, y_0) \end{bmatrix} \begin{bmatrix} h \\ k \end{bmatrix} ,$$

where 1_X denotes the identity mapping in X.

2) Show that $D\Phi(x_0, y_0)$ is a toplinear isomorphism.

<u>Hint:</u> In matrix notation, the inverse linear mapping is given by

$$\begin{bmatrix} 1_X & 0 \\ -(D_Y f(x_0, y_0))^{-1} \circ D_X f(x_0, y_0) & (D_Y f(x_0, y_0))^{-1} \end{bmatrix} .$$

3) Show that there exist an open neighbourhood $W \subseteq X$ of $x_0 \in X$, an open neighbourhood $V \subseteq U \subseteq X \times Y$ of $(x_0, y_0) \in U$, and a differentiable mapping $g : W \to Y$ of class C^r, such that the following holds:

$$\left[\, (x, y) \in V \quad \text{and} \quad f(x, y) = 0 \,\right] \iff \left[\, x \in W \quad \text{and} \quad y = g(x) \,\right] .$$

Further show that the differential of g at $x \in W$ is given by

$$Dg(x) = -D_Y f\big(x, g(x)\big)^{-1} \circ D_X f\big(x, g(x)\big) .$$

Exercise 6.10. Let $f : U \to F$ be a mapping defined in an open subset $U \subseteq E$ of the Banach space E and with values in the Banach space F.

Prove that $f : U \to F$ is differentiable of class C^n, $n \geq 1$, if and only if for every $x \in U$, there exist r-linear mappings $f^{(r)}(x) \in L^r(E, F)$, $r = 1, \ldots, n$, such that for small vectors $h \in E$ it holds that

$$f(x+h) = f(x) + f^{(1)}(x)(h) + \frac{f^{(2)}(x)}{2}(h^{[2]}) + \cdots + \frac{f^{(n)}(x)}{n!}(h^{[n]}) + \varepsilon(h)\|h\|^n,$$

where $(h^{[r]})$ stands for $(h, h, \ldots, h)$ (r times), and $\varepsilon(h) \to 0$ for $h \to 0$.

Exercise 6.11. Let E be a Banach space with norm $\|\cdot\|$.

In this exercise we shall introduce a mapping in the Banach space $L(E, E)$ of bounded linear operators in E with the operator norm, which corresponds to the classical exponential function from real analysis. As usual, I denotes the identity operator in $L(E, E)$.

1) Show that for any bounded linear operator $A \in L(E, E)$, one gets a bounded linear operator $\exp(A) \in L(E, E)$ by the definition

$$\exp(A) = I + \sum_{n=1}^{\infty} \frac{A^n}{n!} .$$

The operator $\exp(A) \in L(E, E)$ is known as the *exponential* of the operator $A \in L(E, E)$.

2) For a fixed bounded linear operator $A \in L(E, E)$, define the mapping $\varphi = \varphi(t) : \mathbb{R} \to L(E, E)$ by

$$\varphi(t) = \exp(tA) \quad \text{for} \quad t \in \mathbb{R}.$$

Prove that $\varphi = \varphi(t)$ is differentiable of class C^∞, and that it is the unique solution to the Initial Value Problem (Exercise 5.22, page 152) in $L(E, E)$ consisting of the differential equation (i) with initial value (ii):

$$\text{(i)} \ \frac{dx}{dt} = Ax \qquad \text{(ii)} \ x(0) = I.$$

3) Determine Taylor's formula with asymptotic remainder for the mapping $\varphi = \varphi(t)$ at $t = 0$.

Chapter 7

Differentiable Manifolds

In a mathematical model, the local behaviour of a system can often be described by a set of parameters forming an open subset of a Euclidean space. As far as the global behaviour of the system is concerned a more general kind of object may be needed. Of particular importance in this connection are the so-called *manifolds*. Classically, manifolds occur as curves and surfaces in 3-dimensional space, and in higher dimensions typically as solution sets for nonlinear equations. The configuration spaces for mechanical systems are often manifolds. Many important infinite dimensional manifolds in the calculus of variations arise as mapping spaces.

In this chapter we provide an introduction to the theory of differentiable manifolds. We start out with a study of the solution sets for nonlinear equations in finitely many variables. By viewing such solution sets as embedded manifolds in Euclidean spaces one obtains a good understanding of the Lagrange multiplier method in the study of extremal problems subject to given constraints. Next we introduce the general notion of a differentiable manifold modelled on a Banach space, and differentiable mappings between such manifolds. At each point of a differentiable Banach manifold there is a tangent space, which is homeomorphic to the model Banach space, and which locally provides a linear approximation to the manifold. Using these tangent spaces we can define the differential of a differentiable mapping and prove the Chain Rule in this context.

We briefly describe the tangent bundle of a differentiable Banach manifold. Thereby we can define vector fields as sections in the tangent bundle and study differentiability of vector fields. We include an introduction to the Lie algebra of smooth vector fields on a smooth manifold. For finite dimensional manifolds, we also include a short study of immersions and embeddings, and of the basic properties of the transversality of mappings.

7.1 Solution sets for nonlinear equations

Let $\Omega \subseteq \mathbb{R}^{n+k}$ be an open set in $\mathbb{R}^{n+k}$ where $n, k \geq 1$. Let

$$g_1, \ldots, g_k : \Omega \to \mathbb{R}$$

be real-valued differentiable functions of class C^r, $r \geq 1$. These real-valued functions can be collected in the usual way into a vector-valued function

$$G = (g_1, \ldots, g_k) : \Omega \to \mathbb{R}^k.$$

In this section we shall examine the structure of the *zero-set* (the set of zeros) of G; i.e. the structure of the subset $M \subseteq \Omega$ given by

$$M = G^{-1}(0) = \left\{ x = (x_1, \ldots, x_{n+k}) \in \Omega \mid g_1(x) = \cdots = g_k(x) = 0 \right\}.$$

We consider M with the topology induced from $\mathbb{R}^{n+k}$.

Since G is continuous and $\{0\}$ is a closed set in $\mathbb{R}^k$, the zero-set M is a closed set in Ω. Conversely, a theorem from 1934 of the American mathematician Hassler Whitney (1907–1989), which we shall not discuss further here, states that an arbitrary closed subset M of $\mathbb{R}^{n+k}$ is the zero-set for a real-valued function of class C^∞.

In general, therefore, we can not conclude much about M just from the fact that it is a zero-set.

However, assume now that 0 is a *regular value* of G. By definition, this means that the Jacobian matrix

$$\mathbf{D}G(x) = \left[\frac{\partial g_i}{\partial x_j}(x) \right] \quad \text{has rank } k \text{ for all } x \in M,$$

i.e. that the $k \times (n + k)$-matrix $\mathbf{D}G(x)$ contains k linearly independent columns for all $x \in M$, or equivalently, that the k rows in the Jacobian matrix are linearly independent for all $x \in M$. Note that by a permutation of the coordinates in $\mathbb{R}^{n+k}$, we can, for every fixed $x_0 \in M$, obtain that the last k columns in $\mathbf{D}G(x_0)$ are linearly independent.

Then we can prove

Assertion 7.1.1. *Let $x_0 \in M$ be an arbitrary point in M. Then there exists an open neighbourhood $V \subseteq \Omega$ of x_0 and a diffeomorphism $\Phi : V \to V'$ of class C^r, which maps V onto an open neighbourhood V' of $0 \in \mathbb{R}^{n+k}$, such that $\Phi(x_0) = 0$ and*

$$G \circ \Phi^{-1}(x'_1, \ldots, x'_{n+k}) = (x'_{n+1}, \ldots, x'_{n+k}).$$

Proof. Let $K : V \to V_0$ be a diffeomorphism which maps an open neighbourhood $V \subseteq \Omega$ of x_0 onto an open neighbourhood V_0 of $0 \in \mathbb{R}^{n+k}$ such that $K(x_0) = 0$ and such that the $k \times k$-matrix

$$\left[\frac{\partial(g_i \circ K^{-1})}{\partial x_{n+j}}(0) \right],$$

for $i, j = 1, \ldots, k$, has linearly independent columns and hence non-zero determinant. The diffeomorphism $K : V \to V_0$ can be constructed as a parallel translation followed by a permutation of the coordinates in $\mathbb{R}^{n+k}$.

Now define the mapping $\Psi : V_0 \to \mathbb{R}^{n+k}$ by

$$\Psi(x) = \Psi(x_1, \ldots, x_{k+n}) = (x_1, \ldots, x_n, g_1 \circ K^{-1}(x), \ldots, g_k \circ K^{-1}(x)).$$

The determinant of the differential $D\Psi(0)$ can be computed by the Jacobian matrix $\mathbf{D}\Psi(0)$ and one easily finds that

$$\det \mathbf{D}\Psi(0) = \det \left[\frac{\partial(g_i \circ K^{-1})}{\partial x_{n+j}}(0) \right] \neq 0.$$

Hence the differential $D\Psi(0)$ is an isomorphism and by the Inverse Function Theorem we can therefore shrink V_0 such that Ψ is a diffeomorphism of class C^r. By shrinking V accordingly, the composition $\Phi = \Psi \circ K$ satisfies the requirements in the assertion. $\square$

In the following we shall, without explicitly writing it, identify $\mathbb{R}^n$ with the subspace $\mathbb{R}^n \times \{0\} \subset \mathbb{R}^{n+k}$.

Let $U = M \cap V$ and $U' = V' \cap \mathbb{R}^n$. Then U is an open neighbourhood of $x_0 \in M$, when M is equipped with the topology induced from $\mathbb{R}^{n+k}$, and similarly U' is an open neighbourhood of 0 in $\mathbb{R}^n = \mathbb{R}^n \times \{0\} \subset \mathbb{R}^{n+k}$. Note that Φ defines a homeomorphism $\varphi : U \to U'$. See Figure 7.1.

If we perform this construction at every point $x \in M$, the set M will be covered by open sets $U \subseteq M$ with associated homeomorphisms $\varphi : U \to U'$ onto open sets $U' \subseteq \mathbb{R}^n$. We call such a pair (U, φ) a *chart* on M. A collection of charts on M covering M is called an *atlas* on M. A topological space M, where any point has an open neighbourhood homeomorphic to an open subset of $\mathbb{R}^n$ is called an n-dimensional *topological manifold*.

Since the homeomorphism φ was obtained as a restriction of the diffeomorphism Φ of class C^r, it is clear that if two charts $(U_\alpha, \varphi_\alpha)$ and (U_β, φ_β) on M overlap (i.e. $U_\alpha \cap U_\beta \neq \emptyset$), then it holds that

$$\varphi_\beta \circ \varphi_\alpha^{-1} : \varphi_\alpha(U_\alpha \cap U_\beta) \to \varphi_\beta(U_\alpha \cap U_\beta)$$

is a diffeomorphism of class C^r between open sets in $\mathbb{R}^n$. Therefore we say that M is an n-dimensional *differentiable manifold* of class C^r.

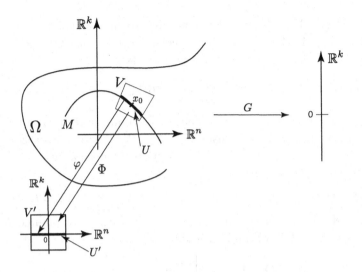

Fig. 7.1 The set M is a zero-set for the vector-valued function $G : \Omega \to \mathbb{R}^k$, and x_0 is a point in M. The homeomorphism φ connects a neighbourhood U around x_0 in M to a neighbourhood U' around 0 in $\mathbb{R}^n$.

If the functions $g_1, \ldots, g_k$ are polynomials, the corresponding zero-sets are called *algebraic varieties*.

In the situation described above, M is closely associated to $\mathbb{R}^{n+k}$. In fact, M is a so-called submanifold of $\mathbb{R}^{n+k}$.

With $m = n + k$ for $k \geq 1$, we have the formal definition.

Definition 7.1.2. A subset M of $\mathbb{R}^m$ with the induced topology is called an n-dimensional *submanifold* of $\mathbb{R}^m$ of class C^r, $r \geq 1$, if there for every $x \in M$ exists a diffeomorphism $\Phi : V \to V'$ of class C^r which maps an open neighbourhood V of x in $\mathbb{R}^m$ onto an open neighbourhood V' of 0 in $\mathbb{R}^m$ such that $\Phi(M \cap V) = \mathbb{R}^n \cap V'$.

Example 7.1.3. Define $g : \mathbb{R}^{n+1} \to \mathbb{R}$ by

$$g(x_1, \ldots, x_{n+1}) = \sum_{i=1}^{n+1} x_i{}^2 - 1.$$

Then $M = g^{-1}(0)$ is exactly the *unit sphere* S^n in $\mathbb{R}^{n+1}$. Since

$$\left(\frac{\partial g}{\partial x_1}, \ldots, \frac{\partial g}{\partial x_{n+1}} \right) = (2x_1, \ldots, 2x_{n+1}) \neq 0 \quad \text{on} \quad M,$$

it follows that $M = S^n$ is an n-dimensional submanifold of $\mathbb{R}^{n+1}$, in fact, an algebraic variety. ◄

Example 7.1.4. Define $g_1, g_2 : \mathbb{R}^4 \to \mathbb{R}$ by

$$g_1(x_1, x_2, x_3, x_4) = \sum_{i=1}^{4} x_i^2 - 1$$

$$g_2(x_1, x_2, x_3, x_4) = x_1^2 + x_2^2 - \frac{1}{2}.$$

Consider $M = \{x \in \mathbb{R}^4 \mid g_1(x) = g_2(x) = 0\}$. The Jacobian matrix

$$\left[\frac{\partial g_i}{\partial x_j}\right] = \begin{bmatrix} 2x_1 & 2x_2 & 2x_3 & 2x_4 \\ 2x_1 & 2x_2 & 0 & 0 \end{bmatrix}$$

has rank 2 on M, since $g_2(x) = 0$ gives $x_1^2 + x_2^2 = \frac{1}{2}$, which combined with $g_1(x) = 0$ gives $x_3^2 + x_4^2 = \frac{1}{2}$, such that both $(x_1, x_2) \neq (0, 0)$ and $(x_3, x_4) \neq (0, 0)$ on M.

It follows that M is a 2-dimensional submanifold of $\mathbb{R}^4$ contained in the 3-dimensional sphere S^3.

Note that M is determined by

$$M = \left\{(x_1, x_2, x_3, x_4) \in \mathbb{R}^4 = \mathbb{R}^2 \times \mathbb{R}^2 \middle| x_1^2 + x_2^2 = \frac{1}{2} \text{ and } x_3^2 + x_4^2 = \frac{1}{2}\right\}.$$

This shows that M is the product space of a circle in the $x_1 x_2$-plane and a circle in the $x_3 x_4$-plane, and thus M is topologically a *torus*. In the literature the torus described here is known as the *Clifford torus* in S^3. (It has interesting geometrical properties and is among others a so-called *minimal surface* in S^3, i.e. it displays minimal area relative to the boundary on small pieces of the surface.) ◄

Now let $c : I_\varepsilon \to \mathbb{R}^{n+k}$ be a differentiable curve defined in an open interval $I_\varepsilon = \,] -\varepsilon, \varepsilon\, [$ about $0 \in \mathbb{R}$, such that $c(0) = x_0$ and $c(I_\varepsilon) \subseteq M$. We write $c : I_\varepsilon \to M$.

For any $i = 1, \ldots, k$, the function $g_i \circ c(t)$ is constant, and therefore

$$0 = (g_i \circ c)'(0) = \langle \nabla g_i(x_0), c'(0) \rangle,$$

where

$$\nabla g_i(x_0) = \left(\frac{\partial g_i}{\partial x_1}(x_0), \ldots, \frac{\partial g_i}{\partial x_{n+k}}(x_0)\right)$$

is the *gradient* of g_i at x_0,

$$c'(0) = (c_1'(0), \ldots, c_{n+k}'(0))$$

is the *tangent vector* to the curve c at x_0, and

$\langle \cdot, \cdot \rangle$ denotes the canonical inner product in $\mathbb{R}^{n+k}$.

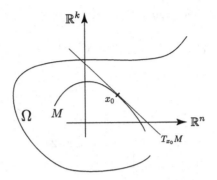

Fig. 7.2   The tangent space $T_{x_0}M$ to M at the point x_0.

Let

$$T_{x_0}M = \big\{\text{vectors } v \in \mathbb{R}^{n+k} \,\big|\, v = c'(0) \text{ for a diff. curve } c : I_\varepsilon \to M\big\}.$$

Then we have just proved that

$$v \in T_{x_0}M \quad \Rightarrow \quad \langle \nabla g_i(x_0), v \rangle = 0 \quad \text{for all} \quad i = 1, \dots, k.$$

We shall now prove the opposite implication. Suppose therefore that the vector $v \in \mathbb{R}^{n+k}$ satisfies the conditions $\langle \nabla g_i(x_0), v \rangle = 0$ for all $i = 1, \dots, k$. This is equivalent to the condition $DG(x_0)v = 0$ for the differential $DG(x_0)$, since $\nabla g_1(x_0), \dots, \nabla g_k(x_0)$ are the rows in the Jacobian matrix $\mathbf{D}G(x_0)$. Put

$$D\Phi(x_0)v = (v'_1, \dots, v'_{n+k}) \in \mathbb{R}^{n+k},$$

whereby

$$v = D\Phi^{-1}(0)(v'_1, \dots, v'_{n+k}).$$

Since $G \circ \Phi^{-1}(x'_1, \dots, x'_{n+k}) = (x'_{n+1}, \dots, x'_{n+k})$ and $DG(x_0)v = 0$, we get

$$\begin{aligned}
DG(x_0)v &= DG(x_0) \circ D\Phi^{-1}(0)(v'_1, \dots, v'_{n+k}) \\
&= D(G \circ \Phi^{-1})(0)(v'_1, \dots, v'_{n+k}) \\
&= (v'_{n+1}, \dots, v'_{n+k}) \\
&= (0, \dots, 0).
\end{aligned}$$

From this follows that

$$D\Phi(x_0)v = (v'_1, \dots, v'_n, 0, \dots, 0).$$

Now define the curve $c : I_\varepsilon \to M$, for a sufficiently small $\varepsilon > 0$, by

$$\Phi(c(t)) = (tv'_1, \dots, tv'_n, 0, \dots, 0).$$

Then $c(0) = x_0$, and
$$D\Phi(x_0)(c'(0)) = (v_1', \dots, v_n', 0, \dots, 0).$$
Since $D\Phi(x_0)$ is an isomorphism, we conclude that $v = c'(0) \in T_{x_0}M$.

Altogether we have proved that
$$T_{x_0}M = \{v \in \mathbb{R}^{n+k} | \langle \nabla g_i(x_0), v \rangle = 0, \ i = 1, \dots, k\}.$$
Conclusion: $T_{x_0}M$ *is an n-dimensional linear subspace of $\mathbb{R}^{n+k}$, and the orthogonal complement is spanned by* $\nabla g_1(x_0), \dots, \nabla g_k(x_0)$.

The description of the space $T_{x_0}M$ as tangent vectors for differentiable curves $c : I_\varepsilon \to M$ at $c(0) = x_0$ makes it reasonable to call $T_{x_0}M$ the *tangent space* of M at $x_0 \in M$. If $T_{x_0}M$ is translated from 0 to x_0, we get an n-dimensional affine subspace in $\mathbb{R}^{n+k}$, which geometrically is tangential to M at $x_0 \in M$; cf. Figure 7.2.

We shall now demonstrate how this geometrical way of thinking about sets of solutions of systems of nonlinear equations can be useful, for instance when investigating extremum problems for functions subject to given constraints.

Let $\Omega \subseteq \mathbb{R}^m$ be an open set in $\mathbb{R}^m$. Let $g_1, \dots, g_k : \Omega \to \mathbb{R}$ be differentiable functions of class C^r, $r \geq 1$, and let $x_0 \in \Omega$ be a point in Ω, such that $g_1(x_0) = \dots = g_k(x_0) = 0$. Assume that the Jacobian matrix
$$[D_j g_i(x_0)] = \left[\frac{\partial g_i}{\partial x_j}(x_0) \right] \quad \text{has rank } k.$$
A function $f : \Omega \to \mathbb{R}$ is said to have a *local maximum (minimum)* at $x_0 \in \Omega$ subject to the constraints $g_1(x) = \dots = g_k(x) = 0$, if there exists an open neighbourhood U_{x_0} of $x_0 \in \Omega \subseteq \mathbb{R}^m$, such that for all $x \in U_{x_0}$ for which $g_1(x) = \dots = g_k(x) = 0$ we have that $f(x) \leq f(x_0)$ $(f(x) \geq f(x_0))$. A local maximum or minimum is called a *local extremum*.

Theorem 7.1.5 (Lagrange multipliers). *Whenever the differentiable function $f : \Omega \to \mathbb{R}$ has a local extremum at $x_0 \in \Omega$ subject to the constraints $g_1(x) = \dots = g_k(x) = 0$, there exist real numbers $\lambda_1, \dots, \lambda_k \in \mathbb{R}$ such that*
$$\nabla f(x_0) = \lambda_1 \nabla g_1(x_0) + \dots + \lambda_k \nabla g_k(x_0) \,,$$
or written coordinate-wise
$$\frac{\partial f}{\partial x_j}(x_0) = \lambda_1 \frac{\partial g_1}{\partial x_j}(x_0) + \dots + \lambda_k \frac{\partial g_k}{\partial x_j}(x_0) \,,$$
for all $j = 1, \dots, m$.

A *system of real numbers $\lambda_1, \dots, \lambda_k$ satisfying these equations is called a system of* Lagrange multipliers.

Proof. Since $[D_j g_i(x_0)]$ has rank k, we can shrink Ω to an open neighbourhood of x_0, which we again for simplicity denote by Ω, where the Jacobian matrix $[D_j g_i(x)]$ has rank k in all points. Then

$$M = \{x \in \Omega \mid g_i(x) = 0, \ i = 1, \ldots, k\}$$

is an $(m - k)$-dimensional submanifold of $\mathbb{R}^m$.

Let $c : I_\varepsilon \to M$ be a differentiable curve such that $c(0) = x_0$. If f has a local extremum at x_0 subject to the constraints $g_1(x) = \cdots = g_k(x) = 0$, then $f \circ c : I_\varepsilon \to \mathbb{R}$ has a local extremum at $0 \in I_\varepsilon$. Thus $\frac{d}{dt}(f \circ c)(0) = 0$, or by the chain rule, equivalently, $\langle \nabla f(x_0), c'(0) \rangle = 0$. This shows that $\nabla f(x_0)$ is contained in the orthogonal complement of $T_{x_0} M$, which for its part, as we just proved, is spanned by $\nabla g_1(x_0), \ldots, \nabla g_k(x_0)$. This completes the proof. $\square$

Lagrange multipliers were introduced in 1755 by the French mathematician Joseph-Louis Lagrange (1736–1813). They give a necessary condition for having a local extremum subject to given constraints. The result is used to determine an extremum subject to constraints, in contexts, where for other reasons one knows that such an extremum exists. Introducing the Lagrange multipliers as extra variables we get altogether $k + m$ equations in $k + m$ variables to be solved, and this situation is apt for methods from linear algebra.

7.2 Manifolds

The notion of a manifold makes sense also without a surrounding Euclidean space, as we shall see now. It is useful to introduce manifolds in this more general setting even in finite dimensions, and to define manifolds of infinite dimension we are forced to free ourselves from a surrounding Euclidean space.

To develop the more general notion of a manifold, the first thing to be noticed is that a manifold locally has to look like an n-dimensional Euclidean space $\mathbb{R}^n$, or more generally, a normed vector space E. This is made precise in the next definition.

Definition 7.2.1. A *topological manifold*, modelled on the normed vector space E, is a Hausdorff space M such that every point $p \in M$ has an open neighbourhood U homeomorphic to an open set in E.

If $E = \mathbb{R}^n$, the space M is said to be an *n-dimensional* topological manifold, and the dimension is often noted by an upper index: M^n.

A pair (U, φ) consisting of an open set $U \subset M$ and a homeomorphism $\varphi : U \to \varphi(U)$ of U onto an open set $\varphi(U) \subset E$ is called a *local coordinate system*, or a local *chart*, on M. In the finite dimensional case we can make local computations in the open set U in M by transferring the coordinates $(x_1, \ldots, x_n) \in \varphi(U) \subset \mathbb{R}^n$ from $\varphi(U)$ to U using φ.

If two local coordinate systems $(U_\alpha, \varphi_\alpha)$ and (U_β, φ_β) on M have an overlap, i.e. if $U_\alpha \cap U_\beta \neq \emptyset$ (cf. Figure 7.3), then the change of coordinates

$$\varphi_\beta \circ \varphi_\alpha^{-1} : \varphi_\alpha(U_\alpha \cap U_\beta) \to \varphi_\beta(U_\alpha \cap U_\beta),$$

considered as a mapping between open sets in E, is interesting.

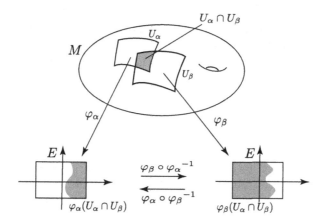

Fig. 7.3   A differentiable manifold M admits an atlas of local charts $\mathcal{A} = \{(\mathcal{U}_\alpha, \varphi_\alpha)\}_{\alpha \in \mathcal{I}}$ with differentiable exchange of coordinates between overlapping local charts.

In order to do any kind of mathematical analysis on M, it is necessary that M can be covered with local coordinate systems such that the change of coordinates between different systems of coordinates is differentiable in a suitable sense. This is made precise by introducing the notion of an atlas.

Definition 7.2.2.   Let M be a topological manifold modelled on the normed vector space E. An *atlas* of class C^r, $r \geq 1$, on M is a collection of local coordinate systems $\mathcal{A} = \{(\mathcal{U}_\alpha, \varphi_\alpha)\}_{\alpha \in \mathcal{I}}$ on M such that

(1) $\{U_\alpha\}_{\alpha \in I}$ is a covering of M.

(2) For every pair of indices $\alpha, \beta \in I$ for which $U_\alpha \cap U_\beta \neq \emptyset$, the change of coordinates

$$\varphi_\beta \circ \varphi_\alpha^{-1} : \varphi_\alpha(U_\alpha \cap U_\beta) \to \varphi_\beta(U_\alpha \cap U_\beta)$$

is a diffeomorphism of class C^r between open sets in E.

Remark 7.2.3. Since φ_α and φ_β are homeomorphisms of open sets in M onto open sets in E, the sets $\varphi_\alpha(U_\alpha \cap U_\beta)$ and $\varphi_\beta(U_\alpha \cap U_\beta)$ will automatically be open sets in E. Hence it is, strictly speaking, not necessary to explicitly include this condition in (2).

Furthermore it is sufficient in (2) to require that the mapping $\varphi_\beta \circ \varphi_\alpha^{-1}$ is differentiable of class C^r for all pairs of indices $\alpha, \beta \in I$. The reason being that also

$$\varphi_\alpha \circ \varphi_\beta^{-1} : \varphi_\beta(U_\alpha \cap U_\beta) \to \varphi_\alpha(U_\alpha \cap U_\beta)$$

will then be differentiable of class C^r, and since $\varphi_\alpha \circ \varphi_\beta^{-1}$ is the inverse to $\varphi_\beta \circ \varphi_\alpha^{-1}$, it follows that $\varphi_\beta \circ \varphi_\alpha^{-1}$ is a diffeomorphism of class C^r. ◁

Corresponding to the name atlas, the local coordinate systems $(U_\alpha, \varphi_\alpha)$ in an atlas $\mathcal{A}$ on M are called the *charts* in the atlas, and we say that two charts satisfying condition (2) in Definition 7.2.2 have an *overlap of class C^r*.

Definition 7.2.4. Two atlases $\mathcal{A}'$ and $\mathcal{A}''$ of class C^r on M are said to be *compatible* if the union $\mathcal{A}' \cup \mathcal{A}''$ of all the charts in $\mathcal{A}'$ and $\mathcal{A}''$ together form an atlas of class C^r on M.

To examine whether two atlases $\mathcal{A}'$ and $\mathcal{A}''$ are compatible it is enough to check that an arbitrarily chosen chart in $\mathcal{A}'$ has an overlap of class C^r with an arbitrarily chosen chart in $\mathcal{A}''$ (possibly an empty overlap).

Making use of the chain rule, it is easy to prove the following lemma.

Lemma 7.2.5. *Compatibility is an equivalence relation in the set of atlases of class C^r on M.*

Definition 7.2.6. An equivalence class of compatible atlases of class C^r on a topological manifold M is called a *differentiable structure* of class C^r on M. Any atlas in a differentiable structure is called an *atlas for the differentiable structure*.

Theorem 7.2.7. *Every differentiable structure of class C^r on a topological manifold M contains a unique maximal element (ordering by inclusion), which is called the* maximal atlas *for the differentiable structure.*

Proof. Let $\mathcal{A}$ be an arbitrary atlas for the differentiable structure on M. Take $\mathcal{A}^*$ to be the set of local coordinate systems on M that overlap of class C^r with every chart in $\mathcal{A}$ (possibly empty overlap). Since the charts in $\mathcal{A}$ cover M, it is easy to prove using the chain rule that two added local coordinate systems in $\mathcal{A}^*$ overlap of class C^r. Then it is clear that $\mathcal{A}^*$ is an atlas of class C^r on M which is compatible with $\mathcal{A}$. Furthermore, $\mathcal{A}^*$ is obviously the unique maximal element in the equivalence class of atlases on M defining the differentiable structure under consideration. $\square$

Following these preparations we can now define the notion of a differentiable manifold.

Definition 7.2.8. A *differentiable manifold* of class C^r, $r \geq 1$, modelled on the normed vector space E, is a topological manifold M modelled on E equipped with the maximal atlas for a differentiable structure of class C^r.

Chasing the definition backwards we get:

A *differentiable manifold* of class C^r, modelled on the normed vector space E, is a Hausdorff space M together with a collection of local coordinate systems $\mathcal{A} = \{(U_\alpha, \varphi_\alpha)\}_{\alpha \in I}$ satisfying the following conditions:

(0) For every $\alpha \in I$, the map $\varphi_\alpha : U_\alpha \to \varphi_\alpha(U_\alpha)$ is a homeomorphism of the open set U_α in M onto the open set $\varphi_\alpha(U_\alpha)$ in E.

(1) $\{U_\alpha\}_{\alpha \in I}$ is a covering of M.

(2) For every pair of indices $\alpha, \beta \in I$ for which $U_\alpha \cap U_\beta \neq \emptyset$, the change of coordinates
$$\varphi_\beta \circ \varphi_\alpha^{-1} : \varphi_\alpha(U_\alpha \cap U_\beta) \to \varphi_\beta(U_\alpha \cap U_\beta)$$
is a diffeomorphism of class C^r between open sets in E.

(3) $\mathcal{A}$ is maximal with respect to the property in (2).

We shall also admit differentiability class $r = 0$ in the above. In this case differentiability just has to be substituted by continuity and diffeomorphism by homeomorphism. A differentiable manifold of class C^0 is in other words nothing else but a topological manifold.

Example 7.2.9. An arbitrary open set U in a normed vector space E is a differentiable manifold of class C^∞ (often called a *smooth manifold*) modelled on E. An atlas for the *usual* differentiable structure on U can be constructed with just one chart, namely $(U, 1_U)$, where 1_U is the identity mapping of U.

More generally, an arbitrary open set U in a differentiable manifold M of class C^r modelled on E, is in itself a differentiable manifold of class C^r modelled on E. We get an atlas for the *induced* differentiable structure on U by shrinking the charts in an atlas on M to U. ◄

Example 7.2.10. Let $\Omega \subseteq \mathbb{R}^{n+k}$ be an open set in $\mathbb{R}^{n+k}$, and let $G : \Omega \to \mathbb{R}^k$ be a differentiable mapping of class C^r, $r \geq 1$. Assume that $M = G^{-1}(0)$ is non-empty and that the Jacobian matrix $\mathbf{D}G(x)$ has rank k for all $x \in M$. Then we have proved in Section 7.1 that M is an n-dimensional differentiable manifold of class C^r.

In particular, we have seen in Example 7.1.3 that the n-sphere

$$S^n = \left\{ (x_1, \ldots, x_{n+1}) \in \mathbb{R}^{n+1} \,\middle|\, \sum_{i=1}^{n+1} x_i^2 = 1 \right\}$$

can be given a differentiable structure by this method.

Corresponding to the $n+1$ coordinate axes in $\mathbb{R}^{n+1}$, the n-sphere S^n is divided into $2n + 2$ open hemispheres (defined repectively by the conditions $x_i > 0$ and $x_i < 0$ for $i = 1, \ldots, n+1$). By projecting these hemispheres orthogonally into the corresponding coordinate hyperplanes (given by $x_i = 0$), we get $2n + 2$ charts on S^n, which form an atlas for the differentiable structure on S^n.

An atlas with only two charts can be obtained by taking as maps in the charts the stereographic projection from each one of the two points in a pair of antipodal points on S^n onto the tangent plane to S^n at the other point in the pair. Since S^n is compact, the n-sphere cannot be covered by a single chart, and hence 2 is the minimal number of charts in an atlas for the differentiable structure on S^n. ◄

Example 7.2.11. Let E and F be normed vector spaces. Then the space $L(E, F)$ of continuous linear mappings $L : E \to F$ is a normed vector space and hence a differentiable manifold of class C^∞. If $E = F$ and E is a Banach space, then we know from Theorem 6.8.2 that the set of toplinear isomorphisms $Gl(E)$ in E is an open set in $L(E, E)$. Hence also $Gl(E)$ is a differentiable manifold of class C^∞.

If $E = \mathbb{R}^n$ and $F = \mathbb{R}^m$, then $L(\mathbb{R}^n, \mathbb{R}^m)$ can be identified with $\mathbb{R}^{n \cdot m}$ via matrix representation of linear mappings. In this case, $Gl(\mathbb{R}^n)$ can be identified with an open set in $\mathbb{R}^{n^2}$, and hence $Gl(\mathbb{R}^n)$ is an n^2-dimensional differentiable manifold of class C^∞. ◄

Example 7.2.12. Let M and N be differentiable manifolds of class C^r modelled on normed vector spaces E and F, respectively. Suppose that $\mathcal{A}_M = \{(U_\alpha, \varphi_\alpha)\}_{\alpha \in I}$ and $\mathcal{A}_N = \{(V_\beta, \psi_\beta)\}_{\beta \in J}$ are atlases for the differentiable structures on M and N, respectively. It is easy to verify that if $M \times N$ is given the product topology, then $\mathcal{A}_{M \times N} = \{(U_\alpha \times V_\beta, \varphi_\alpha \times \psi_\beta)\}_{(\alpha, \beta) \in I \times J}$ is an atlas for a differentiable structure of class C^r on $M \times N$. It is also easy to verify that if $\mathcal{A}'_M$ is compatible with $\mathcal{A}_M$, and $\mathcal{A}'_N$ is compatible with $\mathcal{A}_N$, then $\mathcal{A}'_{M \times N}$ is compatible with $\mathcal{A}_{M \times N}$. Hence we get a well defined differentiable structure of class C^r on $M \times N$, which we call *the product structure.* ◄

Example 7.2.13. Let $C(S^1, S^2)$ denote the space of continuous mappings of the circle S^1 into the 2-sphere S^2; cf. Figure 7.4.

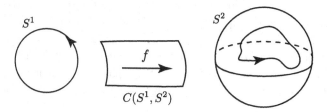

Fig. 7.4   The space of all continuous mappings from the circle S^1 into the sphere S^2.

We equip $C(S^1, S^2)$ with the topology determined by the metric

$$D(f, g) = \sup_{x \in S^1} d(f(x), g(x)) ,$$

where $d(f(x), g(x))$ is the shortest distance along a great circle arc on S^2 from $f(x)$ to $g(x)$.

For a fixed continuous mapping $f : S^1 \to S^2$, we define

$$E_f = \left\{ v : S^1 \to \mathbb{R}^3 \mid v \text{ is continuous}; \langle v(x), f(x) \rangle = 0 \text{ for all } x \in S^1 \right\} .$$

We can think of v as a vector field on S^2 along f. The space E_f is a Banach space with norm

$$\|v\| = \sup_{x \in S^1} \|v(x)\| ,$$

where $\|v(x)\|$ is the Euclidean norm of $v(x) \in \mathbb{R}^3$.

Consider

$$U'_f = \left\{ v \in E_f \mid \|v\| < \pi \right\} ,$$

and define

$$p_f : U'_f \to C(S^1, S^2)$$

by

$$p_f(v)(x) = \begin{array}{l} \text{The point at distance } \|v(x)\| \text{ from } f(x) \text{ along} \\ \text{the great circle arc determined by } f(x) \text{ and } v(x), \\ \text{oriented in the direction of } v(x). \end{array}$$

Then p_f is a homeomorphism of the open set U_f' in E_f onto an open set U_f in $C(S^1, S^2)$. The inverse mapping to p_f, which we shall denote by $\varphi_f = p_f^{-1}$, therefore defines a local coordinate system (U_f, φ_f) on $C(S^1, S^2)$ about $f \in C(S^1, S^2)$.

It can be proved that with these local coordinate systems, $C(S^1, S^2)$ gets the structure of a differentiable manifold of class C^∞ modelled on Banach spaces.

The method indicated for construction of a differentiable structure on a space of continuous mappings $C(X, M)$ was suggested by the American mathematician James Eells (1926–2007) in 1952 and is very general. It works when X is a compact topological space and M is a Riemannian manifold (e.g. a submanifold of a Euclidean space), substituting great circle arcs on S^2 in the above construction by geodesic curves on M. ◄

7.3 Differentiable mappings

Let M and N be differentiable manifolds of class C^r, $r \geq 1$, modelled on normed vector spaces E and F, respectively.

It goes without saying that when we consider charts on M, respectively N, in the following, we refer to charts in the maximal atlas for the differentiable structure on M, respectively N.

Definition 7.3.1. A continuous mapping $f : M \to N$ is said to be *differentiable (of class C^r)* if for any pair of charts (U, φ) on M and (V, ψ) on N it holds that the mapping

$$\psi \circ f \circ \varphi^{-1} : \varphi(f^{-1}(V))) \to \psi(V)$$

in Figure 7.5 is differentiable (of class C^r) in the usual sense as a mapping between open sets in normed vector spaces.

Remark 7.3.2. Using the chain rule it is easy to prove that it suffices to verify differentiability for the charts in two arbitrarily chosen atlases for the differentiable structures on M and N. ◁

It is also easy to prove the following

Theorem 7.3.3. *Let $f_1 : M_1 \to M_2$ and $f_2 : M_2 \to M_3$ be differentiable mappings of class C^r between differentiable manifolds of class C^r. Then the composition of mappings $f_2 \circ f_1 : M_1 \to M_3$ is differentiable of class C^r.*

Finally, we remark that the identity mapping $1_M : M \to M$ on a differentiable manifold M of class C^r is a differentiable mapping of class C^r;

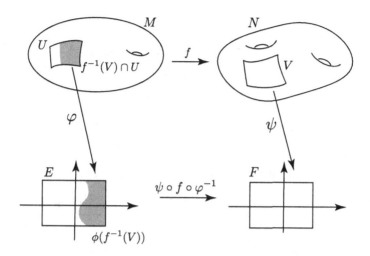

Fig. 7.5 A map f between two manifolds is differentiable if and only if the associated map $\psi \circ f \circ \varphi^{-1}$ between E and F is differentiable in the usual sense of mappings between open sets in normed vector spaces.

indeed, this is equivalent to the differentiable overlap of the charts in an atlas on M.

7.4 Tangent spaces

Let M be a differentiable manifold of class C^r, $r \geq 1$, modelled on the normed vector space E. Consider a point $p \in M$, which is kept fixed in the following.

For any $\epsilon > 0$, let $I_\epsilon =]-\epsilon, \epsilon[$ denote the open interval in $\mathbb{R}$ with center $0 \in \mathbb{R}$ and radius $\epsilon > 0$. We shall study differentiable curves $c : I_\epsilon \to M$ with $c(0) = p$.

Definition 7.4.1. Two differentiable curves $c_i : I_{\epsilon_i} \to M$, $i = 1, 2$, with $c_i(0) = p$ are called *tangential* if

$$\frac{d}{dt}(\varphi \circ c_1)(0) = \frac{d}{dt}(\varphi \circ c_2)(0)$$

for every chart (U, φ) on M about $p \in M$.

According to Definition 7.4.1, two curves in M are in other words tangential at $p \in M$ if in every chart (U, φ) on M about $p \in M$ they are transferred into curves in E with common tangent vector at $\varphi(p) \in E$; cf. Figure 7.6.

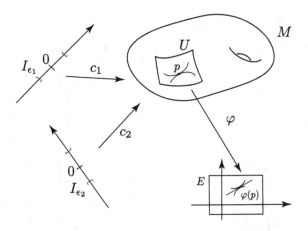

Fig. 7.6   Two curves, c_1 and c_2 are tangential at a point $p \in M$ if in any chart they are mapped into curves in E with a common tangent at the point $\varphi(p) \in E$.

Remark 7.4.2. If $(U_\alpha, \varphi_\alpha)$ and (U_β, φ_β) are two charts on M about $p \in M$ and $c : I_\epsilon \to M$ is a differentiable curve in M with $c(0) = p$ it follows by the chain rule that

$$\frac{d}{dt}(\varphi_\beta \circ c)(0) = D(\varphi_\beta \circ \varphi_\alpha^{-1})(\varphi_\alpha(p))\left(\frac{d}{dt}(\varphi_\alpha \circ c)(0)\right).$$

In Definition 7.4.1 we therefore only have to show that the tangent condition $\frac{d}{dt}(\varphi \circ c_1)(0) = \frac{d}{dt}(\varphi \circ c_2)(0)$ holds in a single chart (U, φ) about $p \in M$ to make sure that it holds in every chart. ◁

We get immediately the following result.

Lemma 7.4.3. *The relation 'tangential' between differentiable curves through $p \in M$ is an equivalence relation.*

Motivated by geometrical ideas borrowed from surfaces in 3-dimensional space we can then define tangent vectors to a differentiable manifold as follows.

Definition 7.4.4. A *tangent vector* to M at $p \in M$ is an equivalence class of tangential curves in M through $p \in M$.

We denote a tangent vector at $p \in M$ with symbols such as v_p, X_p, or $X(p)$ and call p the *foot* of the tangent vector. If $c : I_\epsilon \to M$ is a representative for the tangent vector v_p we write $v_p = cls(c)$.

Let T_pM denote the set of tangent vectors to M at $p \in M$. We call T_pM the *tangent space* to M at $p \in M$.

Now let (U, φ) be a chart on M about $p \in M$. Then we can define a mapping

$$T(\varphi, p) : T_p M \to E,$$

by associating to $v_p = cls(c) \in T_p M$ the vector

$$T(\varphi, p)(v_p) = \frac{d}{dt}(\varphi \circ c)(0).$$

Lemma 7.4.5. *The mapping $T(\varphi, p) : T_p M \to E$ is bijective. If $(U_\alpha, \varphi_\alpha)$ and (U_β, φ_β) are any two charts on M about $p \in M$, then*

$$D(\varphi_\beta \circ \varphi_\alpha^{-1})(\varphi_\alpha(p)) \circ T(\varphi_\alpha, p) = T(\varphi_\beta, p),$$

i.e. the following diagram of mappings is commutative

$$
\begin{array}{ccc}
T_p M & = & T_p M \\
{\scriptstyle T(\varphi_\alpha, p)} \downarrow & & \downarrow {\scriptstyle T(\varphi_\beta, p)} \\
E & \xrightarrow[\;D(\varphi_\beta \circ \varphi_\alpha^{-1})(\varphi_\alpha(p))\;]{} & E \; .
\end{array}
$$

Proof. (i) To prove that $T(\varphi, p) : T_p M \to E$ is injective we observe that if $T(\varphi, p)(cls(c_1)) = T(\varphi, p)(cls(c_2))$ then $\frac{d}{dt}(\varphi \circ c_1)(0) = \frac{d}{dt}(\varphi \circ c_2)(0)$. But then it follows by Remark 7.4.2 that the curves c_1 and c_2 are tangential at $p \in M$ and hence that $cls(c_1) = cls(c_2)$, proving that $T(\varphi, p)$ is injective.

(ii) To prove that $T(\varphi, p) : T_p M \to E$ is surjective, we need to prove that every $h \in E$ is in the image of $T(\varphi, p)$. But this is easy, since given an arbitrary $h \in E$, we can, for a sufficiently small $\epsilon > 0$, define the curve $c : I_\epsilon \to M$ by $c(t) = \varphi^{-1}(\varphi(p) + th)$, and for this curve $c(0) = p$ and $T(\varphi, p)(cls(c)) = \frac{d}{dt}(\varphi \circ c)(0) = h$.

(iii) That the diagram is commutative follows by the computation below making use of the chain rule:

$$
\begin{aligned}
D(\varphi_\beta \circ \varphi_\alpha^{-1})&(\varphi_\alpha(p)) \circ T(\varphi_\alpha, p)(cls(c)) \\
&= D(\varphi_\beta \circ \varphi_\alpha^{-1})(\varphi_\alpha(p))\left(\frac{d}{dt}(\varphi_\alpha \circ c)(0)\right) \\
&= \frac{d}{dt}(\varphi_\beta \circ \varphi_\alpha^{-1} \circ \varphi_\alpha \circ c)(0) = \frac{d}{dt}(\varphi_\beta \circ c)(0) \\
&= T(\varphi_\beta, p)(cls(c)) \; .
\end{aligned}
$$

This completes the proof of the lemma. $\square$

A (real) vector space E equipped with a topology is called a *topological vector space* if the mappings defined by addition of vectors and multiplication by scalars,

$$+ : E \times E \to E$$

$$\cdot : \mathbb{R} \times E \to E \ ,$$

are continuous. As an example, a normed vector space E, equipped with the topology induced by the norm, is a topological vector space. Here we shall only consider such topological vector spaces E, where the topology can be defined by a norm. Topological vector spaces with this property are said to be *normalizable*. The reason for introducing this new notion of a topological vector space is that in certain situations we do not wish to be stuck with a specific norm. If E and F are topological vector spaces, we can consider continuous linear mappings $L : E \to F$. As in Definition 5.8.1, a linear isomorphism $L : E \to F$, such that both L and the inverse mapping L^{-1} are continuous, is called a *toplinear isomorphism*.

Theorem 7.4.6. *Let M be a differentiable manifold of class C^r, $r \geq 1$, modelled on the normed vector space E. Then the tangent space T_pM to M at $p \in M$ can be equipped in a unique way with the structure of a topological vector space such that $T(\varphi, p) : T_pM \to E$ is a toplinear isomorphism for every chart (U, φ) on M about $p \in M$.*

Proof. Let (U, φ) be an arbitrary chart on M about $p \in M$. For real scalars $\lambda, \mu \in \mathbb{R}$ and tangent vectors $v_p, w_p \in T_pM$, we define $\lambda v_p + \mu w_p \in T_pM$ as the unique tangent vector characterized by the equation

$$T(\varphi, p)(\lambda v_p + \mu w_p) = \lambda T(\varphi, p)(v_p) + \mu T(\varphi, p)(w_p) \ .$$

By this definition the vector space structure in E is transferred to T_pM via $T(\varphi, p)$. By the very definition, we have forced the bijective mapping $T(\varphi, p)$ to be an isomorphism.

It is easy to prove that this vector space structure in T_pM is independent of the chart chosen about $p \in M$, since for any two charts $(U_\alpha, \varphi_\alpha)$ and (U_β, φ_β) about $p \in M$, we get $D(\varphi_\beta \circ \varphi_\alpha^{-1})(\varphi_\alpha(p)) \circ T(\varphi_\alpha, p) = T(\varphi_\beta, p)$ by Lemma 7.4.5, and since $D(\varphi_\beta \circ \varphi_\alpha^{-1})(\varphi_\alpha(p))$ is a linear isomorphism.

Since $D(\varphi_\beta \circ \varphi_\alpha^{-1})(\varphi_\alpha(p))$ is a homeomorphism, we can also in a consistent manner lift the topology from E to T_pM by forcing any, and thereby every, of the bijective mappings $T(\varphi, p)$ to be a homeomorphism. (We cannot, however, transfer the norm in E in a consistent manner to T_pM, since $D(\varphi_\beta \circ \varphi_\alpha^{-1})(\varphi_\alpha(p))$ does not necessarily preserve the norm in E.)

With the above vector space structure and topology in T_pM, clearly T_pM is a topological vector space and $T(\varphi, p)$ is a toplinear isomorphism for every chart (U, φ) about $p \in M$. $\square$

In the following we shall always consider T_pM as a topological vector space with the structure given by Theorem 7.4.6.

Now let $f : \Omega \to \mathbb{R}$ be a differentiable function defined in an open neighbourhood $\Omega \subseteq M$ of $p \in M$, and let $v_p \in T_pM$ be an arbitrary tangent vector to M at p. Then we define the *directional derivative* of f along v_p at $p \in M$ by

$$v_p[f] = \frac{d}{dt}(f \circ c)(0) \ ,$$

where the differentiable curve $c : I_\epsilon \to M$ is a representative for v_p.

In an arbitrary chart (U, φ) about $p \in M$ we have

$$v_p[f] = \frac{d}{dt}(f \circ c)(0) = \frac{d}{dt}(f \circ \varphi^{-1} \circ \varphi \circ c)(0)$$

$$= D(f \circ \varphi^{-1})(\varphi(p))(\frac{d}{dt}(\varphi \circ c)(0)$$

$$= D(f \circ \varphi^{-1})(\varphi(p)) \circ T(\varphi, p)(v_p) \ .$$

This rewriting shows that the definition of $v_p[f]$ is independent of the choice of curve $c : I_\epsilon \to M$ representing $v_p \in T_pM$. The rewriting also shows that for tangent vectors $v_p, w_p \in T_pM$ and scalars $\lambda, \mu \in \mathbb{R}$ it holds that

$$(\lambda v_p + \mu w_p)[f] = \lambda v_p[f] + \mu w_p[f] \ .$$

Let $\mathcal{E}(M, p)$ denote the set of germs of differentiable functions $f : \Omega \to \mathbb{R}$ defined in open neighbourhoods Ω of $p \in M$. Technically speaking, an element in $\mathcal{E}(M, p)$ is an equivalence class of differentiable functions defined in open neighbourhoods of $p \in M$, where two such functions define the *same germ* if they agree in a common neighbourhood of $p \in M$. The idea behind considering the *germ* of a function at $p \in M$, rather than the function itself, is that we are free then to shrink the neighbourhood on which the function is defined as we see fit. And since our objective is to perform differentiations of functions at the point p, it is only the local behaviour of the functions in neighbourhoods as small as we wish that matters anyhow. In the following we shall, from a notational point of view, identify a representative for a germ with the germ itself, as it is common to do, and we shall think of a germ as an ordinary function with the built-in freedom to shrink domains.

For germs $f, g \in \mathcal{E}(M, p)$ and scalars $\lambda, \mu \in \mathbb{R}$, it is clear how to define differentiable germs $\lambda f + \mu g \in \mathcal{E}(M, p)$ and $f \cdot g \in \mathcal{E}(M, p)$ by the obvious pointwise defined operations in the intersection of domains of definition of functions representing the germs. Equipped with these operations, $\mathcal{E}(M, p)$ is turned into a real algebra. (Loosely speaking, a real *algebra* is a real vector space equipped with a product such that all sensible rules for addition and product of elements in the algebra and multiplication with scalars are satisfied.)

A tangent vector $v_p \in T_p M$ defines a function

$$v_p : \mathcal{E}(M, p) \to \mathbb{R} \;,$$

which to $f \in \mathcal{E}(M, p)$ associates the number $v_p[f] \in \mathbb{R}$.

Elementary computations show that

(i) $v_p[\lambda f + \mu g] = \lambda v_p[f] + \mu v_p[g]$

(ii) $v_p[f \cdot g] = v_p[f] \cdot g + f \cdot v_p[g]$

for all $f, g \in \mathcal{E}(M, p)$ and $\lambda, \mu \in \mathbb{R}$.

A function $v_p : \mathcal{E}(M, p) \to \mathbb{R}$ with the above properties (i) and (ii) is called a *derivation* in $\mathcal{E}(M, p)$.

Finally in this section, we restrict to the finite dimensional case.

Let M be an n-dimensional differentiable manifold of class C^r, $r \geq 1$, and let (U, φ) be a chart with coordinates $x = (x_1, \ldots, x_n) \in \varphi(U) \subseteq \mathbb{R}^n$, about $p \in M$. Suppose p has the coordinates $x^0 = (x_1^0, \ldots, x_n^0) \in \varphi(U)$. Then define the *parameter curves* through p in Figure 7.7 as the curves determined by

$$\varphi(c_i(t)) = (x_1^0, \ldots, x_i^0 + t, \ldots, x_n^0) \;, \quad i = 1, \ldots, n \;.$$

The parameter curves define tangent vectors at $p \in M$, denoted by

$$\frac{\partial}{\partial x_1}(p), \ldots, \frac{\partial}{\partial x_n}(p) \;.$$

In other words,

$$\frac{\partial}{\partial x_i}(p) = cls(c_i) \;.$$

This notation is well chosen, since

$$\frac{\partial}{\partial x_i}(p)[f] = \frac{d}{dt}(f \circ c_i)(0) = \frac{\partial}{\partial x_i}(f \circ \varphi^{-1})(\varphi(p))$$

for every differentiable germ $f \in \mathcal{E}(M, p)$.

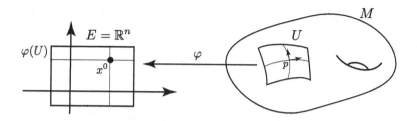

Fig. 7.7 Parameter curves in a chart on a finite dimensional differentiable manifold.

Theorem 7.4.7. *In every chart* (U, φ) *about* $p \in M$, *the corresponding tangent vectors*

$$\left\{ \frac{\partial}{\partial x_1}(p), \ldots, \frac{\partial}{\partial x_n}(p) \right\}$$

form a basis for T_pM.

Proof. Consider the isomorphism $T(\varphi, p) : T_pM \to \mathbb{R}^n$. Since

$$T(\varphi, p)\left(\frac{\partial}{\partial x_i}(p)\right) = \frac{d}{dt}(\varphi \circ c_i)(0) = (0, \ldots, 1, \ldots, 0) ,$$

it follows that $T(\varphi, p)$ maps $\left\{\frac{\partial}{\partial x_1}(p), \ldots, \frac{\partial}{\partial x_n}(p)\right\}$ onto the canonical basis $\{e_1, \ldots, e_n\}$ in $\mathbb{R}^n$. Then the matter is clear. $\square$

The tangent vectors $\frac{\partial}{\partial x_1}(p), \ldots, \frac{\partial}{\partial x_n}(p)$ are defined at every point p in the domain of definition U for the chart (U, φ) on M. The symbols $\frac{\partial}{\partial x_1}, \ldots, \frac{\partial}{\partial x_n}$ define in other words n vector fields on U. These local vector fields are called the *coordinate vector fields* in the chart (U, φ) on M.

7.5   The tangent bundle

Let M be a differentiable manifold of class C^r, $r \geq 1$, modelled on the normed vector space E.

Let TM denote the set of all tangent vectors v_p to M at all points $p \in M$. In symbols

$$TM = \bigcup_{p \in M} T_pM.$$

Define the map $\pi : TM \to M$ by $\pi(v_p) = p$.

The triple (TM, π, M) is called the *tangent bundle* of M. In this context, TM is called the *total space*, π the *projection*, and M the *base space* in the tangent bundle.

As we shall show now, the total space TM in the tangent bundle of M can in a natural way be equipped with a topology and a differentiable structure.

Let $\mathcal{A} = \{(U_\alpha, \varphi_\alpha)\}_{\alpha \in I}$ be an atlas for the differentiable structure on M. Then we get an induced system $\tilde{\mathcal{A}} = \{(\tilde{U}_\alpha, \tilde{\varphi}_\alpha)\}_{\alpha \in I}$ of pairs $(\tilde{U}_\alpha, \tilde{\varphi}_\alpha)$ associated with TM, by setting

$$\tilde{U}_\alpha = \pi^{-1}(U_\alpha) \, ,$$

and defining

$$\tilde{\varphi}_\alpha : \tilde{U}_\alpha \to \varphi_\alpha(U_\alpha) \times E \, ,$$

by

$$\tilde{\varphi}_\alpha(v_p) = \big(\varphi_\alpha(p), T(\varphi_\alpha, p)(v_p)\big) \, .$$

Here $T(\varphi_\alpha, p) : T_p M \to E$ is the toplinear isomorphism introduced in Section 7.4, defined using the chart $(U_\alpha, \varphi_\alpha)$ about $p \in U_\alpha$. Note that $\tilde{\varphi}_\alpha$ maps $\tilde{U}_\alpha$ bijectively onto the open set $\varphi_\alpha(U_\alpha) \times E$ in $E \times E$.

If $U_\alpha \cap U_\beta \neq \emptyset$, clearly $\tilde{U}_\alpha \cap \tilde{U}_\beta \neq \emptyset$. Also note that

$$\tilde{\varphi}_\beta \circ \tilde{\varphi}_\alpha^{-1} : \tilde{\varphi}_\alpha(\tilde{U}_\alpha \cap \tilde{U}_\beta) \to \tilde{\varphi}_\beta(\tilde{U}_\alpha \cap \tilde{U}_\beta),$$

maps the open set

$$\tilde{\varphi}_\alpha(\tilde{U}_\alpha \cap \tilde{U}_\beta) = \varphi_\alpha(U_\alpha \cap U_\beta) \times E \quad \text{in} \quad E \times E$$

bijectively onto the open set

$$\tilde{\varphi}_\beta(\tilde{U}_\alpha \cap \tilde{U}_\beta) = \varphi_\beta(U_\alpha \cap U_\beta) \times E \quad \text{in} \quad E \times E.$$

For every $p \in U_\alpha \cap U_\beta$, it follows by Lemma 7.4.5 that

$$D(\varphi_\beta \circ \varphi_\alpha^{-1})(\varphi_\alpha(p)) \circ T(\varphi_\alpha, p) = T(\varphi_\beta, p),$$

which implies that

$$\tilde{\varphi}_\beta \circ \tilde{\varphi}_\alpha^{-1}(x, h) = \big(\varphi_\beta \circ \varphi_\alpha^{-1}(x), D(\varphi_\beta \circ \varphi_\alpha^{-1})(x)(h)\big)$$

for $(x, h) \in \tilde{\varphi}_\alpha(\tilde{U}_\alpha \cap \tilde{U}_\beta) = \varphi_\alpha(U_\alpha \cap U_\beta) \times E \subseteq E \times E$.

This expression for $\tilde{\varphi}_\beta \circ \tilde{\varphi}_\alpha^{-1}$ shows that it is a diffeomorphism of class C^{r-1} between open sets in $E \times E$. (As usual class C^0 shall be interpreted as continuity and a diffeomorphism of class C^0 as a homeomorphism.)

We conclude from the above that the system of pairs $\tilde{\mathcal{A}} = \{(U_\alpha, \varphi_\alpha)\}_{\alpha \in I}$ has the properties:

(1) $\{\tilde{U}_\alpha\}_{\alpha \in I}$ is a covering of TM.

(2) For every pair of indexes $\alpha, \beta \in I$ for which $\tilde{U}_\alpha \cap \tilde{U}_\beta \neq \emptyset$, the map

$$\tilde{\varphi}_\beta \circ \tilde{\varphi}_\alpha^{-1} : \tilde{\varphi}_\alpha(\tilde{U}_\alpha \cap \tilde{U}_\beta) \to \tilde{\varphi}_\beta(\tilde{U}_\alpha \cap \tilde{U}_\beta)$$

is a diffeomorphism of class C^{r-1} between open sets in $E \times E$.

Then it is easy to prove the following:

Theorem 7.5.1. *There is a well defined topology on TM such that TM is a topological manifold modelled on the normed vector space $E \times E$, and such that TM has a uniquely defined differentiable structure of class C^{r-1} for which $\tilde{\mathcal{A}}$ is an atlas of class C^{r-1} on TM, if $\mathcal{A}$ is an atlas of class C^r for the differentiable structure on M.*

Proof. Via $\tilde{\varphi}_\alpha$ we can lift the system of open sets in $\varphi_\alpha(U_\alpha) \times E$ to $\tilde{U}_\alpha$. In an overlap $\tilde{U}_\alpha \cap \tilde{U}_\beta$, the maps $\tilde{\varphi}_\alpha$ and $\tilde{\varphi}_\beta$ dictate the same open sets, since $\tilde{\varphi}_\beta \circ \tilde{\varphi}_\alpha^{-1}$ is a diffeomorphism. This defines a topology on TM as asserted in the theorem, which is Hausdorff. It is easy to see that compatible atlases on M determine the same topology on TM. Equipped with this topology, TM is a topological manifold modelled on $E \times E$, and it is also clear that $\tilde{\mathcal{A}}$ is an atlas of class C^{r-1} on TM and that compatible atlases on M will induce compatible atlases on TM. This completes the proof of Theorem 7.5.1. $\square$

In the following TM is considered as a differentiable manifold of class C^{r-1} equipped with the differentiable structure given by Theorem 7.5.1.

Let (U, φ) be a chart on M and let $(\tilde{U}, \tilde{\varphi})$ be the corresponding chart on TM. Then the diagram below is *commutative*, i.e. all compositions of mappings obtained by following the direction of arrows between two spaces in the diagram coincide.

$$
\begin{array}{ccccc}
TM \hookleftarrow \tilde{U} & \xrightarrow{\ \tilde{\varphi}\ } & \varphi(U) \times E & \hookrightarrow & E \times E \\
\downarrow{\scriptstyle \pi} \quad \downarrow{\scriptstyle \pi} & & \downarrow{\scriptstyle proj_1} & \downarrow{\scriptstyle proj_1} & \\
M \hookleftarrow U & \xrightarrow[\ \varphi\ ]{} & \varphi(U) & \hookrightarrow & E
\end{array}
$$

Here $proj_1$ denotes the projection map onto the first factor of the product space in question and $\hookrightarrow$ and $\hookleftarrow$ are inclusion maps. It follows from the diagram that $\pi : TM \to M$ is a differentiable mapping of class C^{r-1}.

Altogether we have now proved that if M is a differentiable manifold of class C^r, $r \geq 1$, modelled on E, then TM is a differentiable manifold of class C^{r-1} modelled on $E \times E$, and $\pi : TM \to M$ is a differentiable mapping of class C^{r-1}.

For every $p \in M$, the space $\pi^{-1}(p) = T_pM$ is a topological vector space isomorphic to E, and furthermore, we have seen in the above that locally $\pi : TM \to M$ looks like the projection map $proj_1 : \varphi(U) \times E \to \varphi(U)$ in a coordinate neighbourhood $U \subseteq M$. Such a *locally trivial* bundle of vector spaces is called a *vector bundle*. In this setting, the vector space $\pi^{-1}(p)$ is called the *fibre* in the vector bundle. Vector bundles are important objects in many connections. In theoretical mechanics for example, the total space TM in the tangent bundle of M occurs as the phase space of a mechanical system with configuration space M.

Finally, we specialize once again to the finite dimensional case.

Let M be an n-dimensional differentiable manifold of class C^r, $r \geq 1$, and let (U, φ) be a chart on M with coordinates $(x_1, \ldots, x_n) \in \varphi(U) \subseteq \mathbb{R}^n$. Corresponding to this chart we have the chart $(\tilde{U}, \tilde{\varphi})$ on TM, where

$$\tilde{\varphi} : \tilde{U} = \pi^{-1}(U) \to \varphi(U) \times \mathbb{R}^n \subseteq \mathbb{R}^n \times \mathbb{R}^n = \mathbb{R}^{2n}$$

is defined by

$$\tilde{\varphi}(v_p) = \big(\varphi(p), T(\varphi, p)(v_p)\big) .$$

As conclusion we get that TM is a $2n$-dimensional differentiable manifold of class C^{r-1}.

For $p \in U$, the set of tangent vectors $\{\frac{\partial}{\partial x_1}(p), \ldots, \frac{\partial}{\partial x_n}(p)\}$ is a basis for T_pM, and hence we can write every tangent vector $v_p \in T_pM$ in a unique way as a linear combination

$$v_p = \sum_{i=1}^{n} h_i \frac{\partial}{\partial x_i}(p) .$$

In the proof of Theorem 7.4.7, we have earlier proved that $T(\varphi, p)$ maps $\{\frac{\partial}{\partial x_1}(p), \ldots, \frac{\partial}{\partial x_n}(p)\}$ into the canonical basis $\{e_1, \ldots, e_n\}$ for $\mathbb{R}^n$. Hence

$$\tilde{\varphi}(v_p) = (x, h) ,$$

where $x = \varphi(p) \in \varphi(U) \subseteq \mathbb{R}^n$, and $h = (h_1, \ldots, h_n) \in \mathbb{R}^n$ is the set of coordinates of v_p in the basis $\{\frac{\partial}{\partial x_1}(p), \ldots, \frac{\partial}{\partial x_n}(p)\}$ for T_pM.

Now let (V, ψ) be a second chart on M with coordinates $(y_1, \ldots, y_n)$ from $\psi(V) \subseteq \mathbb{R}^n$. We assume that the overlap between (U, φ) and (V, ψ) is of class C^r. Then the overlap between $(\tilde{U}, \tilde{\varphi})$ and $(\tilde{V}, \tilde{\psi})$ is of class C^{r-1}. In the overlap we have

$$\tilde{\psi} \circ \tilde{\varphi}^{-1}(x, h) = \big(\psi \circ \varphi^{-1}(x), D(\psi \circ \varphi^{-1})(x)(h)\big) ,$$

for $x \in \varphi(U) \subseteq \mathbb{R}^n$, and $h \in \mathbb{R}^n$. Or, in coordinates:

$$\tilde{\psi} \circ \tilde{\varphi}^{-1}(x_1, \ldots, x_n, h_1, \ldots, h_n)$$

$$= \left(y_1(x_1, \ldots, x_n), \ldots, y_n(x_1, \ldots, x_n), \sum_{j=1}^{n} \frac{\partial y_1}{\partial x_j}(x)h_j, \ldots, \sum_{j=1}^{n} \frac{\partial y_n}{\partial x_j}(x)h_j \right),$$

where
$$y_1 = y_1(x_1, \ldots, x_n)$$
$$\vdots$$
$$y_n = y_n(x_1, \ldots, x_n)$$

is the coordinate description of the diffeomorphism

$$\psi \circ \varphi^{-1} : \varphi(U \cap V) \to \psi(U \cap V) .$$

7.6 Vector fields

Let M be a differentiable manifold of class C^r, $r \geq 1$, modelled on the normed vector space E.

A vector field X on M is an assignment of a tangent vector X_p to M at every point $p \in M$. If the tangent vector X_p varies differentiably with p, the vector field is said to be differentiable. These vague definitions can be made precise by exploiting that the total space TM in the tangent bundle of M is a differentiable manifold of class C^{r-1}.

Definition 7.6.1. A *vector field* on M is a mapping $X : M \to TM$ such that $\pi \circ X = 1_M$. If the mapping X is differentiable of class C^{r-1}, then X is said to be a *differentiable vector field* of class C^{r-1} on M.

The condition $\pi \circ X = 1_M$ ensures that $X_p = X(p)$ is a tangent vector to M at p. (Note that we usually write X_p instead of the more traditional $X(p)$ in this connection.) In other words, the mapping X picks out exactly one tangent vector $X_p \in T_pM$ in every fibre $\pi^{-1}(p) = T_pM$ of the tangent bundle of M.

As mentioned earlier, a tangent bundle is an example of the more general notion of a *vector bundle*. In this context, a vector field X is called a *section* in the tangent bundle.

Let (U, φ) be a chart on M and let $(\tilde{U}, \tilde{\varphi})$ be the corresponding chart on TM. If $X : M \to TM$ is a vector field on M then it is easy to see that we get a commutative diagram

$$
\begin{array}{ccc}
TM \hookleftarrow \tilde{U} & \xrightarrow{\tilde{\varphi}} & \varphi(U) \times E \hookrightarrow E \times E \\
\end{array}
$$

in which $X^{\varphi}(x) = \big(x, h \circ \varphi^{-1}(x)\big)$, $x \in \varphi(U)$, for a uniquely defined mapping $h : U \to E$. We call X^{φ}, or h, the *local representative* of X.

It is easy to see that X is a vector field of class C^{r-1} on M if and only if the local representative X^{φ}, or equivalently h, is differentiable of class C^{r-1}, in the usual sense of a mapping between open sets in normed vector spaces, for every chart (U, φ) on M.

In the remaining part of this section we assume that M is n-dimensional.

In a chart (U, φ) on M with coordinates $(x_1, \ldots, x_n) \in \varphi(U) \subseteq \mathbb{R}^n$, a vector field X on M has a unique decomposition

$$
X(p) = \sum_{i=1}^{n} h_i(p) \frac{\partial}{\partial x_i}(p) \qquad \text{for all} \qquad p \in U .
$$

The collection of functions $h_1, \ldots, h_n : U \to \mathbb{R}$ determines the local representative $h : U \to \mathbb{R}^n$ of X in the chart (U, φ). The function h_i is called the ith *coordinate function* of the vector field X in the chart (U, φ).

Now let $f : \Omega \to \mathbb{R}$ be a differentiable function defined in an open set $\Omega \subseteq M$. For every vector field X on M we can define a function $X[f] : \Omega \to \mathbb{R}$ by the definition $X[f](p) = X_p[f]$ for $p \in \Omega$. For obvious reasons we call $X[f]$ the *directional derivative* of f in Ω in the direction of the vector field X.

Let $\varphi = (\varphi_1, \ldots, \varphi_n) : U \to \mathbb{R}^n$ be the coordinate functions of φ in the chart (U, φ) on M. For an arbitrary vector field X on M we have the following computation at the point $p \in U$:

$$
X[\varphi_i](p) = X_p[\varphi_i] = \sum_{j=1}^{n} h_j(p) \frac{\partial}{\partial x_j}(p)[\varphi_i]
$$

$$
= \sum_{j=1}^{n} h_j(p) \frac{\partial}{\partial x_j}(\varphi_i \circ \varphi^{-1})(\varphi(p))
$$

$$
= h_i(p) .
$$

From this computation it follows that $X[\varphi_i] = h_i$, which shows that $X[\varphi_i]$ is just the ith coordinate function of X in the chart (U, φ) on M.

Theorem 7.6.2. *Let M be an n-dimensional differentiable manifold of class C^r, $r \geq 1$. A vector field X on M is differentiable of class C^{r-1} if and only if one of the following equivalent conditions is satisfied:*

V.1 $X : M \to TM$ *is differentiable of class C^{r-1}.*

V.2 *The coordinate functions $h_1, \ldots, h_n : U \to \mathbb{R}$ of X are differentiable of class C^{r-1} in every chart (U, φ) on M.*

V.3 *If $f : \Omega \to \mathbb{R}$ is a differentiable function of class C^r defined in the open set $\Omega \subseteq M$, then $X[f] : \Omega \to \mathbb{R}$ is a differentiable function of class C^{r-1}.*

Proof. Since the condition (V.1) is the definition of differentiability of the vector field X, we only have to prove equivalence of the three conditions.

The equivalence of (V.1) and (V.2) follows by observing that

$$\tilde{\varphi} \circ X \circ \varphi^{-1}(x) = \big(x, h \circ \varphi^{-1}(x)\big) , \ x \in \varphi(U) \subseteq \mathbb{R}^n,$$

where $h = (h_1, \ldots, h_n) : U \to \mathbb{R}^n$ is the local representative of X in (U, φ).

The equivalence of (V.2) and (V.3) follows using that in any chart (U, φ) on M, the function $X[\varphi_i] = h_i$ is the ith coordinate function of X. $\square$

Finally in this section, we consider the smooth case. As usual *smooth* is taken as differentiability of class C^∞.

Let M be an n-dimensional smooth manifold and let $C^\infty(M)$ denote the set of smooth functions $f : M \to \mathbb{R}$. Then $C^\infty(M)$ is a real algebra with the sum $f + g$ and the product $f \cdot g$ of smooth functions $f, g \in C^\infty(M)$ defined by the usual pointwise defined operations.

A smooth vector field X on M defines a mapping

$$X : C^\infty(M) \to C^\infty(M),$$

which to $f \in C^\infty(M)$ associates the function $X[f] \in C^\infty(M)$.

This mapping is a so-called *derivation* in $C^\infty(M)$, i.e. for all $\lambda, \mu \in \mathbb{R}$ and $f, g \in C^\infty(M)$ it satisfies the conditions:

Linearity: $\qquad X[\lambda f + \mu g] = \lambda X[f] + \mu X[g]$

Leibniz' formula: $\quad X[f \cdot g] = X[f] \cdot g + f \cdot X[g]$.

It can be proved that every derivation in $C^\infty(M)$ in a unique way corresponds to a smooth vector field on M. There is in other words a bijective correspondence between the set of smooth vector fields on M and the set of derivations in $C^\infty(M)$. In fact, at each point $p \in M$, the tangent space T_pM and the set of derivations (directional derivatives) in the set $\mathcal{E}(M,p)$ of germs of smooth functions at $p \in M$ both have natural structures as n-dimensional vector spaces, and the map which to a tangent vector $v_p \in T_pM$ associates the corresponding directional derivative is an isomorphism. Furthermore, in a chart (U, φ) on M, the coordinate vector fields $\{\frac{\partial}{\partial x_1}(p), \ldots, \frac{\partial}{\partial x_n}(p)\}$ provide a basis in both spaces; cf. Exercise 7.10, page 231.

Now let X and Y be smooth vector fields on M and define

$$[X, Y] : C^\infty(M) \to C^\infty(M)$$

by the formula

$$[X, Y][f] = X[Y[f]] - Y[X[f]]$$

for $f \in C^\infty(M)$.

A simple calculation shows that $[X, Y]$ is a derivation in $C^\infty(M)$, and hence $[X, Y]$ determines a smooth vector field on M, which is known as the *Lie product* of X and Y.

Let $\mathcal{X}(M)$ denote the space of smooth vector fields on M. Then $\mathcal{X}(M)$ can in a natural way be equipped with the structure of a real vector space by defining the linear combination $\lambda X + \mu Y \in \mathcal{X}(M)$, for $\lambda, \mu \in \mathbb{R}$ and $X, Y \in \mathcal{X}(M)$, by $(\lambda X + \mu Y)(p) = \lambda X(p) + \mu Y(p)$, for all $p \in M$. Elementary computations establish that the Lie product in $\mathcal{X}(M)$ satisfies the following rules for all $\lambda, \mu \in \mathbb{R}$ and $X, Y, Z \in \mathcal{X}(M)$:

Bilinearity:
$$[\lambda X + \mu Y, Z] = \lambda[X, Z] + \mu[Y, Z]$$
$$[X, \lambda Y + \mu Z] = \lambda[X, Y] + \mu[X, Z].$$

Anti-commutativity: $[X, Y] = -[Y, X]$.

Jacobi's identity: $[X, [Y, Z]] + [Z, [X, Y]] + [Y, [Z, X]] = 0$.

A real vector space with a product satisfying the above rules is called a *Lie algebra*. The space of smooth vector fields $\mathcal{X}(M)$ on a smooth finite dimensional manifold M is in other words a real Lie algebra with the Lie product $[\cdot, \cdot] : \mathcal{X}(M) \times \mathcal{X}(M) \to \mathcal{X}(M)$ as the product.

7.7 Induced mappings

Let M and N be differentiable manifolds of class C^r, $r \geq 1$, modelled on normed vector spaces E and F, respectively.

Let $f : M \to N$ be a differentiable mapping. For every $p \in M$ we intend to define a 'linear approximation' $T_p f : T_p M \to T_{f(p)} N$ to f at p. The definition is secured by the following observation.

Lemma 7.7.1. *For every pair of charts* $(U_\alpha, \varphi_\alpha)$ *and* (U_β, φ_β) *about* $p \in M$ *and every pair of charts* (V_α, ψ_α) *and* (V_β, ψ_β) *about* $f(p) \in N$, *the following diagram is commutative,*

$$
\begin{array}{ccccccc}
T_p M & \xrightarrow{T(\varphi_\alpha, p)} & E & \xrightarrow{D(\psi_\alpha \circ f \circ \varphi_\alpha^{-1})(\varphi_\alpha(p))} & F & \xleftarrow{T(\psi_\alpha, f(p))} & T_{f(p)} N \\
\Big\| & \scriptstyle D(\varphi_\beta \circ \varphi_\alpha^{-1})(\varphi_\alpha(p)) \Big\downarrow & & & \Big\downarrow \scriptstyle D(\psi_\beta \circ \psi_\alpha^{-1})(\psi_\alpha(f(p))) & & \Big\| \\
T_p M & \xrightarrow[T(\varphi_\beta, p)]{} & E & \xrightarrow[D(\psi_\beta \circ f \circ \varphi_\beta^{-1})(\varphi_\beta(p))]{} & F & \xleftarrow[T(\psi_\beta, f(p))]{} & T_{f(p)} N \; .
\end{array}
$$

Proof. The two outer rectangles are commutative by Lemma 7.4.5. That the middle rectangle is commutative follows immediately by the following computation using the chain rule:

$$
\begin{aligned}
D(\psi_\beta &\circ f \circ \varphi_\beta^{-1})(\varphi_\beta(p)) \circ D(\varphi_\beta \circ \varphi_\alpha^{-1})(\varphi_\alpha(p)) \\
&= D(\psi_\beta \circ f \circ \varphi_\alpha^{-1})(\varphi_\alpha(p)) \\
&= D(\psi_\beta \circ \psi_\alpha^{-1})(\psi_\alpha(f(p))) \circ D(\psi_\alpha \circ f \circ \varphi_\alpha^{-1})(\varphi_\alpha(p)) \; .
\end{aligned}
$$

From the commutativity of the three rectangles follows the commutativity of the diagram as a whole. $\square$

Theorem 7.7.2. *Let* $f : M \to N$ *be a differentiable mapping. For every point* $p \in M$, *there exists a uniquely defined continuous linear mapping* $T_p f : T_p M \to T_{f(p)} N$ *such that the following diagram is commutative for every pair of charts* (U, φ) *about* $p \in M$ *and* (V, ψ) *about* $f(p) \in N$:

$$
\begin{array}{ccc}
T_p M & \xrightarrow{T_p f} & T_{f(p)} N \\
\scriptstyle T(\varphi, p) \Big\downarrow & & \Big\downarrow \scriptstyle T(\psi, f(p)) \\
E & \xrightarrow[D(\psi \circ f \circ \varphi^{-1})(\varphi(p))]{} & F \; .
\end{array}
$$

In other words:

$$
T(\psi, f(p)) \circ T_p f = D(\psi \circ f \circ \varphi^{-1})(\varphi(p)) \circ T(\varphi, p) \; .
$$

Proof. The topology and vector space structure on $T_p M$, respectively $T_{f(p)} N$, is defined by requiring $T(\varphi, p)$, respectively $T(\psi, f(p))$, to be a toplinear isomorphism. Hence we are forced to define $T_p f$ as the composition

$$
T_p f = T(\psi, f(p))^{-1} \circ D(\psi \circ f \circ \varphi^{-1})(\varphi(p)) \circ T(\varphi, p)
$$

for a pair of charts (U, φ) about $p \in M$ and (V, ψ) about $f(p) \in N$. By Lemma 7.7.1, the definition is independent of the choice of charts. This proves the theorem. $\square$

The linear mapping $T_p f : T_p M \to T_{f(p)} N$ is called the *differential*, or the *tangential mapping*, or the *induced mapping* of f at $p \in M$.

We shall now provide a geometrical interpretation of $T_p f$.

Theorem 7.7.3. *Let $f : M \to M$ be a differentiable mapping and let $p \in M$. The differential $T_p f : T_p M \to T_{f(p)} N$ has the properties:*

(1) *If $v_p \in T_p M$ is represented by the differentiable curve $c : I_\epsilon \to M$, then the tangent vector $T_p f(v_p) \in T_{f(p)} N$ is represented by the differentiable curve $f \circ c : I_\epsilon \to N$.*

(2) *If $v_p \in T_p M$ is considered as a derivation in $\mathcal{E}(M, p)$, then the tangent vector $T_f(v_p) \in T_{f(p)} N$ operates as a derivation in $\mathcal{E}(N, f(p))$ by the rule*

$$T_f(v_p)[g] = v_p[g \circ f] \quad \text{for} \quad g \in \mathcal{E}(N, f(p)) .$$

Proof. (1) Let $c : I_\epsilon \to M$ be a differentiable curve representing $v_p \in T_p M$. For charts (U, φ) about $p \in M$ and (V, ψ) about $f(p) \in N$, we then get:

$$\begin{aligned}
T(\psi, f(p))\big(T_p f(v_p)\big) &= D(\psi \circ f \circ \varphi^{-1})(\varphi(p))\big(T(\varphi, p)(v_p)\big) \\
&= D(\psi \circ f \circ \varphi^{-1})(\varphi(p))\big(\frac{d}{dt}(\varphi \circ c)(0)\big) \\
&= \frac{d}{dt}(\psi \circ f \circ \varphi^{-1} \circ \varphi \circ c)(0) \\
&= \frac{d}{dt}\big(\psi \circ (f \circ c)\big)(0) .
\end{aligned}$$

This computation shows exactly that the differentiable curve $f \circ c : I_\epsilon \to N$ is a representative of $T_p f(v_p)$.

(2) Again let $v_p \in T_p M$ be represented by $c : I_\epsilon \to M$. Then from (1) we know that $T_p f(v_p)$ is represented by $f \circ c : I_\epsilon \to N$. Therefore we get

$$T_p f(v_p)[g] = \frac{d}{dt}\big(g \circ (f \circ c)\big)(0) = \frac{d}{dt}\big((g \circ f) \circ c\big)(0) = v_p[g \circ f] .$$

This completes the proof of the theorem. $\square$

The first assertion in the following theorem is the chain rule in the context of differentiable manifolds.

Theorem 7.7.4. *Let $f : M \to N$ and $g : N \to L$ be differentiable mappings between differentiable manifolds. As usual, 1_M denotes the identity mapping of M. Then for every $p \in M$ it holds that:*

(1) $T_p(g \circ f) = T_{f(p)}g \circ T_p f$

(2) $T_p 1_M = 1_{T_p M}$.

Proof. Let the tangent vector $v_p \in T_p M$ be represented by the differentiable curve $c : I_\epsilon \to M$. Then the tangent vector $T_{f(p)}g(T_p f(v_p)) \in T_{g(f(p))}L$ is represented by the curve $g \circ (f \circ c) = (g \circ f) \circ c$, which on the other hand also is a representative of the tangent vector $T_p(g \circ f)(v_p)$. This proves (1). Since $c = 1_M \circ c$, the curve $c : I_\epsilon \to M$ is a representative of both v_p and $T_p 1_M(v_p)$. This proves (2). $\square$

Definition 7.7.5. A differentiable mapping (of class C^r) $f : M \to N$ between differentiable manifolds (of class C^r) is called a *diffeomorphism* (of class C^r) if there exists a differentiable mapping (of class C^r) $g : N \to M$ such that $g \circ f = 1_M$ and $f \circ g = 1_N$. Two differentiable manifolds (of class C^r) M and N are said to be *diffeomorphic* (of class C^r) if there exists a diffeomorphism (of class C^r) $f : M \to N$.

We remark that according to Lemma 3.7.1, the conditions $g \circ f = 1_M$ and $f \circ g = 1_N$ ensure that f is a bijective mapping with inverse mapping g. A diffeomorphism is in other words nothing else but a differentiable mapping $f : M \to N$ with a differentiable inverse mapping $g : N \to M$.

Theorem 7.7.6. *Let M and N be differentiable manifolds modelled on normed vector spaces E and F, respectively. If M and N are diffeomorphic, then E and F are toplinear isomorphic. In particular, if two finite dimensional differentiable manifolds M and N are diffeomorphic, then they have the same dimension.*

Proof. Let $f : M \to N$ and $g : N \to L$ be differentiable mappings such that $g \circ f = 1_M$ and $f \circ g = 1_N$. According to Theorem 7.7.4, we then have for any point $p \in M$:

$$T_{f(p)}g \circ T_p f = 1_{T_p M} \quad \text{and} \quad T_p f \circ T_{f(p)}g = 1_{T_{f(p)} N} .$$

This proves that $T_p f : T_p M \to T_{f(p)} N$ is a toplinear isomorphism with inverse $T_{f(p)}g : T_{f(p)}N \to T_p M$. Then the theorem follows. $\square$

We shall now extend the above notions to the setting of tangent bundles.

For an arbitrary differentiable mapping $f : M \to N$ of class C^r, we can define a mapping $Tf : TM \to TN$ between the total spaces in the tangent bundles of M and N by the definition

$$Tf(v_p) = T_p f(v_p) \quad \text{for all} \quad v_p \in T_p M \quad \text{and all} \quad p \in M .$$

The mapping Tf is in other words defined by the condition that the restriction of Tf to $T_p M$ for every $p \in M$, is nothing else but the continuous linear mapping $T_p f : T_p M \to T_{f(p)} N$. Clearly we get a commutative diagram

$$
\begin{array}{ccc}
TM & \xrightarrow{\;Tf\;} & TN \\
\pi_M \downarrow & & \downarrow \pi_N \\
M & \xrightarrow[\;f\;]{} & N ,
\end{array}
$$

in which π_M and π_N are the projections in the tangent bundles of M and N, respectively.

Note that Tf maps the fibre $T_p M$ over $p \in M$ in the tangent bundle of M linearly into the fibre $T_{f(p)} N$ over $f(p) \in N$ in the tangent bundle of N. In the context of vector bundles such a mapping is called a *bundle map*.

The total spaces TM and TN in the tangent bundles of M and N are differentiable manifolds of class C^{r-1}. (For $r = 1$ topological manifolds.) Hence the following theorem is pleasing.

Theorem 7.7.7. *If $f : M \to N$ is a differentiable mapping of class C^r, then $Tf : TM \to TN$ is a differentiable mapping of class C^{r-1}. (Continuous for $r = 1$.)*

Proof. Let (U, φ) and (V, ψ) be charts on M, respectively N, chosen so that $f(U) \subseteq V$. Let $(\tilde{U}, \tilde{\varphi})$ and $(\tilde{V}, \tilde{\psi})$ be the corresponding charts on TM, respectively TN. Then it is sufficient to prove that $\tilde{\psi} \circ Tf \circ \tilde{\varphi}^{-1}$ is differentiable of class C^{r-1}.

The definition of the differentials $T_p f : T_p M \to T_{f(p)} N$ for $p \in U$ shows directly that we get a commutative diagram

$$
\begin{array}{ccc}
\tilde{U} & \xrightarrow{\;Tf\;} & \tilde{V} \\
\tilde{\varphi} \downarrow & & \downarrow \tilde{\psi} \\
U \times E & \xrightarrow[\;\tilde{f}\;]{} & V \times F ,
\end{array}
$$

where $\tilde{f}(x, h) = \big(\psi \circ f \circ \varphi^{-1}(x), D(\psi \circ f \circ \varphi^{-1})(x)h\big)$.

By the construction of the differentiable structures on TM and TN, the chart mappings $\tilde{\varphi}$ and $\tilde{\psi}$ are diffeomorphisms of class C^{r-1}. Hence the above diagram shows immediately that $\tilde{\psi} \circ Tf \circ \tilde{\varphi}^{-1}$ is differentiable of class C^{r-1}. This completes the proof. $\square$

The mapping $Tf : TM \to TN$ is also called the *differential*, or the *tangential mapping*, or the *induced mapping* of f.

From Theorem 7.7.4 we get immediately

Theorem 7.7.8. *Let $f : M \to N$ and $g : N \to L$ be differentiable mappings between differentiable manifolds. Then*

(1) $T(g \circ f) = Tg \circ Tf$

(2) $T1_M = 1_{TM}$.

7.8 Immersions, submersions, embeddings, submanifolds

In this section *all manifolds are finite dimensional.*

Let M^n and N^m be differentiable manifolds of class C^r, $r \geq 1$, of dimension n and m, respectively. By laying down conditions on the induced linear mappings $T_p f : T_p M \to T_{f(p)} N$ for $p \in M$, one can define certain differentiable mappings of particular interest, namely immersions and submersions.

We say that the differentiable mapping $f : M^n \to N^m$ has *rank k* at a point $p \in M$ if the linear mapping $T_p f : T_p M \to T_{f(p)} N$ has rank k. If this is the case, we write $\mathrm{rk}_p f = k$. Note that $0 \leq \mathrm{rk}_p f \leq \min\{n, m\}$. If (U, φ) is a chart on M about $p \in M$, and (V, ψ) is a chart on N about $f(p) \in N$, then we have the commutative diagram

$$
\begin{array}{ccc}
T_p M & \xrightarrow{\;\;\;T_p f\;\;\;} & T_{f(p)} N \\[2mm]
{\scriptstyle T(\varphi, p)} \big\downarrow & & \big\downarrow {\scriptstyle T(\psi, f(p))} \\[2mm]
\mathbb{R}^n & \xrightarrow[D(\psi \circ f \circ \varphi^{-1})(\varphi(p))]{} & \mathbb{R}^m \;.
\end{array}
$$

Since $T(\varphi, p)$ and $T(\psi, f(p))$ are isomorphisms, it follows that the rank of $T_p f$ is equal to the rank of the classical differential $D(\psi \circ f \circ \varphi^{-1})(\varphi(p))$, which on the other hand is determined by the rank of the Jacobian matrix $\mathbf{D}(\psi \circ f \circ \varphi^{-1})(\varphi(p))$.

Definition 7.8.1. An *immersion* is a differentiable mapping $f : M^n \to N^m$ that satisfies one of the following equivalent conditions:

I.1 $T_p f : T_p M \to T_{f(p)} N$ is injective for all $p \in M$.

I.2 $\mathrm{rk}_p f = n$ for all $p \in M$.

I.3 For every point $p \in M$ there exist charts (U, φ) on M about $p \in M$ and (V, ψ) on N about $f(p) \in N$ in which the coordinate description $\psi \circ f \circ \varphi^{-1}$ of f has the form

$$\psi \circ f \circ \varphi^{-1}(x_1, \ldots, x_n) = (x_1, \ldots, x_n, 0, \ldots, 0)$$

for $(x_1, \ldots, x_n) \in \varphi(f^{-1}(V) \cap U) \subseteq \mathbb{R}^n$.

Naturally, we have to prove that the conditions (I.1), (I.2) and (I.3) are equivalent. It is trivial that the conditions (I.1) and (I.2) are equivalent. It is also clear that (I.3) implies (I.2). Hence it only remains to prove that (I.2) implies (I.3). Since we can construct new charts on M^n and N^m by composing some initially given chart mappings φ and ψ on M^n, respectively N^m, by local diffeomorphisms in $\mathbb{R}^n$, respectively $\mathbb{R}^m$, it suffices to prove the following:

Assertion 7.8.2. *Suppose $n \leq m$ and let $G = (g_1, \ldots, g_m) : \Omega \to \mathbb{R}^m$ be a differentiable mapping of class C^r, $r \geq 1$, defined in an open neighbourhood Ω of $0 \in \mathbb{R}^n$. Suppose that $G(0) = 0$ and that $\mathrm{rk}_0 G = n$. Then there exists a local diffeomorphism $H = (h_1, \ldots, h_m)$ between open neighbourhoods of $0 \in \mathbb{R}^m$ such that*

$$H \circ G(x_1, \ldots, x_n) = (x_1, \ldots, x_n, 0, \ldots, 0)$$

for $(x_1, \ldots, x_n)$ in a suitable open neighbourhood of $0 \in \mathbb{R}^n$.

Proof. Possibly after an initial permutation of the coordinates in $\mathbb{R}^m$ (which will define a local diffeomorphism in $\mathbb{R}^m$), we can assume that the first n rows in the Jacobian matrix $\mathbf{D}G(0)$ are linearly independent. Then define $K = (k_1, \ldots, k_m)$ in an open neighbourhood of $0 \in \mathbb{R}^m$ by

$$k_i(y_1, \ldots, y_m) = \begin{cases} g_i(y_1, \ldots, y_n) & \text{for } 1 \leq i \leq n \\ y_i + g_i(y_1, \ldots, y_n) & \text{for } n+1 \leq i \leq m. \end{cases}$$

It is easy to prove that

$$\det \mathbf{D}K(0) = \det \left[\frac{\partial g_i}{\partial y_j}(0) \right]_{i,j=1,\ldots,n} \neq 0.$$

Hence the differential $DK(0)$ is an isomorphism and by the Inverse Function Theorem we can then shrink the domain of definition for K so that it is a diffeomorphism of class C^r. It is now easy to prove that the inverse mapping $H = K^{-1}$ satisfies the condition in the assertion. $\square$

Using Assertion 7.1.1, or Theorem 8.2.2 in the next chapter, it is easy to prove the equivalence of (S.2) and (S.3) in the following definition.

Definition 7.8.3. A *submersion* is a differentiable mapping $f : M^n \to N^m$ that satisfies one of the following equivalent conditions:

S.1 $T_p f : T_p M \to T_{f(p)} N$ is surjective for all $p \in M$.

S.2 $\mathrm{rk}_p f = m$ for all $p \in M$.

S.3 For every point $p \in M$, there exist charts (U, φ) on M about $p \in M$ and (V, ψ) on N about $f(p) \in N$ in which the coordinate description $\psi \circ f \circ \varphi^{-1}$ of f has the form

$$\psi \circ f \circ \varphi^{-1}(x_1, \ldots, x_n) = (x_1, \ldots, x_m)$$

for $(x_1, \ldots, x_n) \in \varphi(f^{-1}(V) \cap U) \subseteq \mathbb{R}^n$.

The existence of a submersion $f : M^n \to N^m$ implies that for $q \in f(M)$, the set $M_q = \{p \in M | f(p) = q\}$ is an $(n - m)$-dimensional submanifold in M; cf. page 219. Clearly, the family of submanifolds M_q for $q \in f(M)$ exhaust all of M. We say that these submanifolds *foliate* M. We shall not pursue the study of submersions further here.

We shall, on the other hand, make a further study of immersions. From condition (I.3) it follows that an immersion is locally injective. It is, however, not necessarily globally injective.

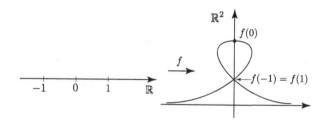

Fig. 7.8   Immersion with a double point.

Example 7.8.4. Define the mapping $f : \mathbb{R} \to \mathbb{R}^2$ by

$$f(t) = \left(t^3 - t, \ \frac{2}{1 + t^2}\right), \ t \in \mathbb{R} \ .$$

Since

$$f'(t) = \left(3t^2 - 1, \ \frac{-4t}{(1 + t^2)^2}\right) \neq (0, 0)$$

for all $t \in \mathbb{R}$, the mapping f is an immersion.

For $t_1, t_2 \in \mathbb{R}$ we have

$$f(t_1) = f(t_2) \iff \begin{cases} t_1^3 - t_1 = t_2^3 - t_2, & \text{i.e.} \quad t_1(t_1^2 - 1) = t_2(t_2^2 - 1) \\ \frac{2}{1+t_1^2} = \frac{2}{1+t_2^2}, & \text{i.e.} \quad t_1^2 = t_2^2 \,. \end{cases}$$

From this follows that if $t_1^2 = t_2^2 \neq 1$, then $t_1 = t_2$. But $t_1^2 = t_2^2 = 1$ also satisfy the equations. Hence it follows that f has the double point $f(-1) = f(1)$, and otherwise no multiple points; cf. Figure 7.8. ◄

Example 7.8.5. The pictures one sees of non-orientable surfaces in 3-space, such as the one in Figure 7.9 of the Klein bottle for instance, are immersions of the corresponding abstract surfaces that are not globally injective. ◄

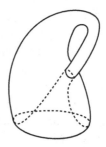

Fig. 7.9   Immersion of the Klein bottle in 3-space.

The following class of immersions is of particular interest.

Definition 7.8.6. An injective immersion $f : M^n \to N^m$ that maps M homeomorphically onto its image $f(M) \subseteq N$ (with the induced topology from N) is called an *embedding*.

Example 7.8.7. Let $M = \,]-2, 1[\, \subseteq \mathbb{R}$ and $N = \mathbb{R}^2$. Let $f : M \to N$ be the mapping defined in Example 7.8.4 but now only for $t \in M$; cf. Figure 7.10. From the investigation in Example 7.8.4, it follows that $f : M \to N$ is an injective immersion. However, f does not map M homeomorphically onto its image $f(M)$ in $\mathbb{R}^2$ with the induced topology from $\mathbb{R}^2$. This can be seen by noticing that the inverse mapping $f^{-1} : f(M) \to M$, which is defined as a mapping of sets, is not continuous, since $f(t) \to f(-1)$ as $t \to 1$. ◄

When M is compact, the image $f(M)$ cannot get unpleasant 'accumulation points of itself' as in Example 7.8.7.

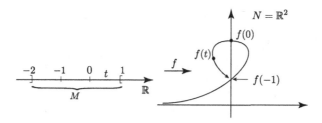

Fig. 7.10   An example of an injective immersion which is not an embedding.

Theorem 7.8.8. *Let $f : M^n \to N^m$ be an injective immersion and assume furthermore that M is compact. Then f is an embedding.*

Proof. We have to prove that $f^{-1} : f(M) \to M$ is continuous when $f(M)$ is equipped with the induced topology from N. With that in mind, let C be an arbitrary closed set in M. We have to prove that $(f^{-1})^{-1}(C) = f(C)$ is a closed set in $f(M)$. We argue as follows. Since M is compact, any closed set in M, in particular C, is compact. Hence $f(C)$ is a compact set in N, being the continuous image of the compact set C in M. Since N is a Hausdorff space, the set $f(C)$ is therefore closed in N. Then $f(C)$ is also closed in $f(M)$ which was to be proved. $\square$

Corresponding to the different types of immersions we can define similar types of submanifolds.

Definition 7.8.9. Let M^n and N^m be differentiable manifolds of class C^r, $r \geq 1$, and assume that M is a subset of N. Let $J : M \to N$ be the inclusion mapping of M into N. If J is an immersion of class C^r, the manifold M is said to be an *immersed submanifold* of class C^r in N. If J is an embedding of class C^r, the manifold M is said to be a *submanifold* of class C^r in N.

In differential geometry it is usually immersed submanifolds that are of interest. In differential topology the focus is most often on submanifolds.

Example 7.8.10. For $n \leq m$, the space $\mathbb{R}^n$ can be considered as a submanifold of $\mathbb{R}^m$ by the embedding $J(x_1, \ldots, x_n) = (x_1, \ldots, x_n, 0, \ldots, 0)$. ◀

As the following theorem shows, the local picture of a submanifold is like that described in Example 7.8.10; cf. Figure 7.11.

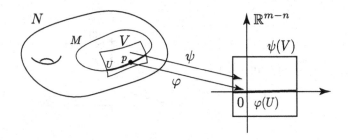

Fig. 7.11   Local picture of a submanifold.

Theorem 7.8.11. *Let M^n and N^m be differentiable manifolds of class C^r, $r \geq 1$, and assume that M is a subset of N. Then M is a submanifold of N of class C^r if and only if for each point $p \in M$ there exists a chart (V, ψ) on N about $p \in M \subseteq N$, and a chart (U, φ) on M about $p \in M$, such that the following conditions are satisfied:*

   (i) $\psi(p) = \varphi(p) = 0$. *(We say that the charts are* centred *at p.)*

   (ii) $U = M \cap V$.

  (iii) *If $\mathbb{R}^n$ is identified with the submanifold $\mathbb{R}^n \times \{0\}$ of $\mathbb{R}^m$, then the following diagram is commutative:*

$$
\begin{array}{ccc}
U & \hookrightarrow & V \\
{\scriptstyle\varphi}\downarrow & & \downarrow{\scriptstyle\psi} \\
\mathbb{R}^n & \hookrightarrow & \mathbb{R}^m .
\end{array}
$$

The proof is easily carried out by exploiting property (I.3) of an immersion (Definition 7.8.1) and is left to the reader. You should also note that Theorem 7.8.11 establishes the connection to the submanifolds in $\mathbb{R}^m$ defined in Definition 7.1.2.

7.9   Transversality

In this section *all manifolds are finite dimensional.*

Let M^n and N^m be differentiable manifolds of class C^r, $r \geq 1$, and of dimension n and m, respectively.

Let $f : M^n \to N^m$ be a differentiable mapping of class C^r. For every point $q \in N$ we put

$$M_q = \{p \in M | f(p) = q\} \ .$$

The set M_q is a closed set in M, possibly the empty set, since it is the preimage by f of the closed set $\{q\}$ in N. According to the theorem of Hassler Whitney mentioned on page 182, every closed subset of M is the zero-set for a differentiable real-valued function, and hence in general, we can say no more about M_q than it is a closed set. Proceeding in analogy with the method employed in Section 7.1, in particular by using Assertion 7.1.1, we can, however, prove the following result in case $n \geq m$: *If $M_q \neq \emptyset$ and* $\mathrm{rk}_p f = m$ *for all $p \in M_q$, then M_q is a submanifold of dimension $n - m$ in* M. An example is indicated in Figure 7.12.

Thus in general, the preimage M_q of a point $q \in N$ for $f : M \to N$ can be an arbitrary closed set in M. But with suitable regularity conditions for $f : M \to N$ satisfied in relation to the point $q \in N$, we get a submanifold.

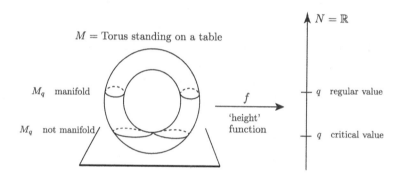

Fig. 7.12 f measures the height of the points in the torus over the table.

Let $f : M^n \to N^m$ be an arbitrary differentiable mapping of class C^r, $r \geq 1$, and now let W be a submanifold in N. Assume that $f^{-1}(W) \neq \emptyset$. Then we can ask whether $f^{-1}(W)$ is a submanifold of M. In general, this will not be the case, but with suitable regularity conditions for f satisfied in relation to W, the set $f^{-1}(W)$ will be a submanifold in M. A suitable regularity condition is *transversality*, which we are about to define. In this connection the *codimension* of W in N plays a significant role. We define this number by

$$\text{codimension } W = \text{dimension } N - \text{dimension } W \ .$$

Definition 7.9.1. Let $f : M^n \to N^m$ be a differentiable mapping of class C^r, $r \geq 1$, between differentiable manifolds of class C^r. Let W be a submanifold of class C^r in N of codimension k, and let $p \in M$ be a point in M. Then we say that f is *transversal* to W at $p \in M$ (denoted $f\pitchfork_p W$) if either, (a) $f(p) \notin W$, or (b) $f(p) \in W$ and one of the following equivalent conditions is satisfied:

T.1 $T_{f(p)}N = T_{f(p)}W + T_p f(T_p M)$.

T.2 If $\bar{\psi} : V \to \mathbb{R}^k$ is a differentiable mapping of class C^r defined in an open neighbourhood V of $f(p)$ in N^m, such that $W \cap V = \bar{\psi}^{-1}(0)$ and such that $\bar{\psi}$ has rank k at $f(p)$, then $\bar{\psi} \circ f : f^{-1}(V) \to \mathbb{R}^k$ has rank k at p.

T.3 If (V, ψ) is a chart on N^m centred at $f(p)$ (i.e. $\psi(f(p)) = 0$) with coordinates $(y_1, \ldots, y_m) \in \psi(V) \subseteq \mathbb{R}^m$, such that $W \cap V$ is defined by the equations $y_1 = \cdots = y_k = 0$, and (U, φ) is a chart on M^n centred at p (i.e. $\varphi(p) = 0$) with coordinates $(x_1, \ldots, x_n) \in \varphi(U) \subseteq \mathbb{R}^n$, then the matrix

$$\left[\frac{\partial(\psi_i \circ f \circ \varphi^{-1})}{\partial x_j}(0) \right]_{\substack{1 \leq i \leq k \\ 1 \leq j \leq n}} \qquad \text{has rank} \quad k .$$

T.4 Let (V, ψ) be a chart on N^m as in (T.3). Then there exists a chart (U, φ) on M^n centred at p with coordinates $(x_1, \ldots, x_n) \in \varphi(U) \subseteq \mathbb{R}^n$, such that f in these coordinates has the coordinate description

$$\psi \circ f \circ \varphi^{-1}(x_1, \ldots, x_n) = \big(x_1, \ldots, x_k, f_{k+1}(x_1, \ldots, x_n), \ldots, f_m(x_1, \ldots, x_n)\big) .$$

Proof of the equivalence of the conditions (T.1) – (T.4).

We prove the cycle: (T.1) $\Rightarrow$ (T.2) $\Rightarrow$ (T.3) $\Rightarrow$ (T.4) $\Rightarrow$ (T.1).

(T.1) $\Rightarrow$ (T.2).
Let $\bar{\psi} : V \to \mathbb{R}^k$ be a differentiable mapping of class C^r defined in an open neighbourhood V of $f(p)$ in N such that $W \cap V = \bar{\psi}^{-1}(0)$ and such that $\bar{\psi}$ has rank k at $f(p)$. We shall prove that $\bar{\psi} \circ f : f^{-1}(V) \to \mathbb{R}^k$ has rank k at p, when (T.1) is satisfied. Since $\bar{\psi} \circ f(p) = \bar{\psi}(f(p)) = 0$ we shall in other words prove that $T_p(\bar{\psi} \circ f) : T_p M \to T_0 \mathbb{R}^k$ is surjective. With this in mind, let $v \in T_0 \mathbb{R}^k$ be given. Since $\bar{\psi}$ has rank k at $f(p)$, the linear map $T_{f(p)}\bar{\psi} : T_{f(p)}N \to T_0 \mathbb{R}^k$ is surjective and hence we can choose $v' \in T_{f(p)}N$ such that $T_{f(p)}\bar{\psi}(v') = v$. According to (T.1) we can next choose $w \in T_{f(p)}W$ and $u \in T_p M$ such that $v' = w + T_p f(u)$. Since $\bar{\psi}$ is constant along $W \cap V$, we have $T_{f(p)}\bar{\psi}(w) = 0$. Then it follows immediately that $v = T_{f(p)}\bar{\psi}(v') = T_{f(p)}\bar{\psi}(T_p f(u)) = T_p(\bar{\psi} \circ f)(u)$. This proves that $T_p(\bar{\psi} \circ f)$ is surjective. $\square$

(T.2) $\Rightarrow$ (T.3).

Let $\bar{\psi} : V \to \mathbb{R}^k$ be the first k coordinate functions in a chart (V, ψ) on N^m as in (T.3). Then $\bar{\psi}$ has rank k at $f(p)$ and $W \cap V = \bar{\psi}^{-1}(0)$. Hence according to (T.2), $\bar{\psi} \circ f$ has rank k at p. But then equivalently, the matrix

$$\left[\frac{\partial(\psi_i \circ f \circ \varphi^{-1})}{\partial x_j}(0) \right]_{\substack{1 \le i \le k \\ 1 \le j \le n}} \qquad \text{has rank} \quad k \;,$$

which was to be proved. $\square$

(T.3) $\Rightarrow$ (T.4).

Let (V, ψ) be a chart on N^m centred at $f(p)$, such that $W \cap V$ is defined by the equations $y_1 = \cdots = y_k = 0$ in the coordinates $(y_1, \ldots, y_m)$ from $\psi(V) \subseteq \mathbb{R}^m$. Furthermore, let (U', φ') be an arbitrary chart on M^n centred at p with coordinates $(x'_1, \ldots, x'_n) \in \varphi'(U') \subseteq \mathbb{R}^n$. Possibly after an initial permutation of the coordinates $(x'_1, \ldots, x'_n)$ we can, according to (T.3), without loss of generality assume that the $k \times k$-matrix

$$\left[\frac{\partial(\psi_i \circ f \circ \varphi'^{-1})}{\partial x'_j}(0) \right]_{\substack{1 \le i \le k \\ 1 \le j \le k}} \qquad \text{has rank} \quad k \;.$$

Then define $\Phi : \varphi'(U') \to \mathbb{R}^n$ by

$$\Phi(x'_1, \ldots, x'_n)$$
$$= \big(\psi_1 \circ f \circ \varphi'^{-1}(x'_1, \ldots, x'_n), \ldots, \; \psi_k \circ f \circ \varphi'^{-1}(x'_1, \ldots, x'_n), \; x'_{k+1}, \ldots, \; x'_n \big).$$

It is easy to see that the Jacobian matrix for Φ at $\varphi'(p) = 0$ has rank n. The Inverse Function Theorem then shows that there exists an open neighbourhood $U \subseteq U'$ of p such that the restriction of $\Phi \circ \varphi'$ to U defines a chart (U, φ) on M^n centred at p with coordinates $(x_1, \ldots, x_n)$ from the open set $\varphi(U) \subseteq \mathbb{R}^n$. The chart (U, φ) satisfies the requirement in (T.4), since if we put $(x_1, \ldots, x_n) = \Phi(x'_1, \ldots, x'_n)$ we get

$$\psi \circ f \circ \varphi^{-1}(x_1, \ldots, x_n) = \psi \circ f \circ \varphi'^{-1} \circ \Phi^{-1}(x_1, \ldots, x_n)$$
$$= \psi \circ f \circ \varphi'^{-1}(x'_1, \ldots, x'_n)$$
$$= \big(\psi_1 \circ f \circ \varphi'^{-1}(x'_1, \ldots, x'_n), \ldots, \; \psi_k \circ f \circ \varphi'^{-1}(x'_1, \ldots, x'_n),$$
$$\ldots, \psi_m \circ f \circ \varphi'^{-1}(x'_1, \ldots, x'_n) \big)$$
$$= \big(x_1, \ldots, \; x_k, \; f_{k+1}(x_1, \ldots, x_n), \ldots, \; f_m(x_1, \ldots, x_n) \big) \;,$$

which was to be proved. $\square$

(T.4) $\Rightarrow$ (T.1).

Let (V, ψ) and (U, φ) be charts as in (T.4). Every tangent vector $v \in T_{f(p)}N$ has a unique decomposition as a linear combination:

$$v = \sum_{j=1}^{m} a_j \frac{\partial}{\partial y_j}(f(p)) \ , \ a_j \in \mathbb{R} \ , j = 1, \ldots, m \ .$$

Put

$$u = \sum_{i=1}^{k} a_i \frac{\partial}{\partial x_i} \in T_p M \ ,$$

and consider the tangent vector

$$v_1 = v - T_p f(u)$$

$$= \sum_{j=1}^{m} a_j \frac{\partial}{\partial y_j}(f(p)) - \sum_{i=1}^{k} a_i \left\{ \sum_{j=1}^{m} \frac{\partial}{\partial x_i}(\psi_j \circ f \circ \varphi^{-1})(0) \frac{\partial}{\partial y_j}(f(p)) \right\} \ .$$

From the form of $\psi \circ f \circ \varphi^{-1}$ it follows immediately that v_1 will be a linear combination of the tangent vectors $\frac{\partial}{\partial y_{k+1}}(f(p)), \ldots, \frac{\partial}{\partial y_m}(f(p))$. Since $W \cap V$ is defined by the equations $y_1 = \cdots = y_k = 0$, i.e. $W \cap V$ is described by the coordinates $y_{k+1}, \ldots, y_m$, it is clear that $v_1 \in T_{f(p)}W$. This proves that $T_{f(p)}N = T_{f(p)}W + T_p f(T_p M)$. $\square$

Remark 7.9.2. The condition (T.1) does not involve coordinates. From this follows that the conditions (T.2), (T.3) and (T.4), which are equivalent to (T.1), are independent of the choice of coordinate systems. $\triangleleft$

Example 7.9.3. In this example $M = \mathbb{R}$ and $N = \mathbb{R}^2$. We consider $W = \mathbb{R}$ as a submanifold of $N = \mathbb{R}^2$ by the usual embedding

$$W = \mathbb{R} = \mathbb{R} \times \{0\} \subset \mathbb{R}^2 = N \ .$$

For each of the following mappings $f : \mathbb{R} \to \mathbb{R}^2$ we look for transversality of the mapping in relation to $W = \mathbb{R} \times \{0\}$ at the point $p = 0$:

(1) $f(x) = (x, x + x^3)$
(2) $f(x) = (x, x^3)$
(3) $f(x) = (x, x^2 + 1)$.

With reference to Figure 7.13 we see that:

(1) The tangent to $f(M)$ at $f(p)$ cuts 'right through' the tangent to W at $f(p)$. Thus $f \pitchfork_p W$.

(2) The tangent to $f(M)$ at $f(p)$ coincides with the tangent to W at $f(p)$. Thus f is not transversal to W at p.

(3) $f(p) \notin W$. Thus $f \pitchfork_p W$. $\blacktriangleleft$

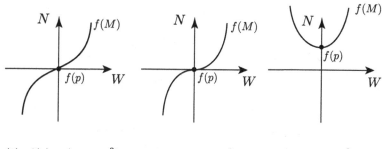

(1) $f(x) = (x, x + x^3)$ (2) $f(x) = (x, x^3)$ (3) $f(x) = (x, x^2 + 1)$

Fig. 7.13   Transversality of smooth mappings.

Note that $T_0 f(T_0\mathbb{R})$ for the mappings (1) and (2) in Example 7.9.3, is the tangent to the submanifold $f(\mathbb{R}) \subset \mathbb{R}^2$. In general, $f(M)$ is not a submanifold in N, but it supports 'healthy' geometrical intuition to think of $f(M)$ as a submanifold in N in connection with transversality questions.

Definition 7.9.4. Let $f : M^n \to N^m$ be a differentiable mapping of class C^r, $r \geq 1$, between differentiable manifolds of class C^r, and let W be a submanifold of class C^r in N.

(i) Let $A \subseteq M$ be a subset in M. We say that f is *transversal* to W along A, denoted $f \pitchfork_A W$, if $f \pitchfork_p W$ for all $p \in A$.

(ii) If f is transversal to W along all of M we say that f is *transversal* to W and write $f \pitchfork W$.

Example 7.9.5. Let $M = S^1$ be the unit circle and $N = \mathbb{R}^2$ the plane. Let $W = S^1 \subset \mathbb{R}^2 = N$ be the unit circle considered as a submanifold of the plane in the usual way. Let $f : S^1 \to \mathbb{R}^2$ be an immersion, or, in other words, a closed differentiable curve with velocity vector $\neq 0$. Then transversality of f with respect to W can be visualized geometrically: the map f is transversal to W exactly when the curve determined by f and the submanifold $W = S^1$ have linearly independent tangent vectors at all points of intersection. Some typical examples are shown in Figure 7.14. ◀

Remark 7.9.6. A slight perturbation of f in Figure 7.14 (1) will not destroy transversality. We say that *transversality is stable*. By an arbitrary small perturbation of f in Figure 7.14 (2), we can obtain transversality. In

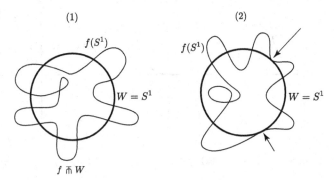

The mapping $f : S^1 \to \mathbb{R}^2$
is transversal to $W = S^1$

Non-transversal intersection at
the points indicated by an arrow

Fig. 7.14

other words: *transversality is typical*, or, as we say, *generic*; transversality
is 'almost always' the case. These loose remarks are crystallized and made
precise in a famous transversality theorem from 1954 of the French mathematician René Thom (1923–2002). With reference to Definition 7.9.4, let
$C^r(M^n, N^m)$ denote the space of differentiable mappings $f : M^n \to N^m$ of
class C^r equipped with a suitable C^k-topology; cf. Section 8.5, page 246.
Then Thom's Transversality Theorem can be formulated as follows: *The
set of differentiable mappings $f : M^n \to N^m$ transversal to W is open and
dense in* $C^r(M^n, N^m)$.

We shall not pursue this theorem further here. ◁

Proposition 7.9.7. *Let M^n and N^m be differentiable manifolds of class
C^r, $r \geq 1$, and let W be a submanifold of class C^r in N of codimension
$k > n$. Suppose $f : M^n \to N^m$ is a differentiable mapping of class C^r for
which $f \pitchfork W$. Then the image $f(M)$ of M by f is disjoint with W, i.e.
$f(M) \cap W = \emptyset$.*

Proof. We give an indirect proof. Assume therefore that $f(p) \in W$ for a
point $p \in M$. Then we have

$$\dim \big(T_{f(p)}W + T_p f(T_p M)\big) \leq \dim T_{f(p)}W + \dim T_p M$$
$$= m - k + n$$
$$< m = \dim T_{f(p)}N \;,$$

which contradicts condition (T.1) for transversality. $\square$

The following theorem contains the result about the structure of the pre-image of a submanifold by a differentiable mapping, which we mentioned as a motivation for introducing the notion of transversality.

Theorem 7.9.8. *Let $f : M^n \to N^m$ be a differentiable mapping of class C^r, $r \geq 1$, between differentiable manifolds of class C^r, and let W be a submanifold of class C^r in N of codimension k. Assume that $f^{-1}(W) \neq \emptyset$ and that $f \pitchfork W$. Then $f^{-1}(W)$ is a submanifold of class C^r in M of codimension k. (In particular: If $n = k$, then $f^{-1}(W)$ will contain only isolated points in M; cf. Figure 7.15.)*

Proof. Let $p \in f^{-1}(W)$. Choose a chart (U, φ) on M centred at p with coordinates $(x_1, \ldots, x_n) \in \varphi(U) \subseteq \mathbb{R}^n$, and a chart (V, ψ) on N centred at $f(p)$ with coordinates $(y_1, \ldots, y_m) \in \psi(V) \subseteq \mathbb{R}^m$, such that $W \cap V$ is defined by the equations $y_1 = \cdots = y_k = 0$, and such that $f(U) \subseteq V$, and for which $\psi \circ f \circ \varphi^{-1} : \varphi(U) \to \psi(V)$ has the form

$$\psi \circ f \circ \varphi^{-1}(x_1, \ldots, x_n) = \big(x_1, \ldots, x_k, f_{k+1}(x_1, \ldots, x_n), \ldots, f_m(x_1, \ldots, x_n)\big).$$

This can be obtained according to (T.4).

Then $f^{-1}(W) \cap U$ is defined by the equations $x_1 = \cdots = x_k = 0$ and hence described by the coordinates $x_{k+1}, \ldots, x_n$. This proves that $f^{-1}(W)$ is a submanifold of class C^r in M of codimension k. $\square$

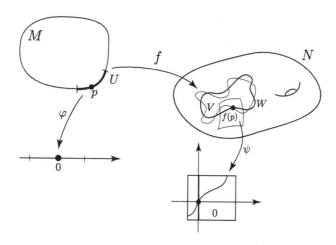

Fig. 7.15   Submanifolds defined by transversality.

There are numerous applications of the notion of transversality. Here we shall be content with outlining a single such application.

Let M be a compact differentiable manifold of class C^2, and let $\mathcal{X}_1(M)$ denote the set of differentiable vector fields of class C^1 on M.

First we shall equip $\mathcal{X}_1(M)$ with a topology. For simplicity we assume that M is a submanifold of class C^2 in a Euclidean space $\mathbb{R}^m$. (In reality, this is no restriction, due to an embedding theorem for abstract manifolds in Euclidean spaces proved by Whitney.) Let $J : M \to \mathbb{R}^m$ be the inclusion map and consider the commutative diagram

$$
\begin{array}{ccc}
TM & \xrightarrow{\;TJ\;} & T\mathbb{R}^m = \mathbb{R}^m \times \mathbb{R}^m \\
{\scriptstyle \pi_M}\downarrow & & \downarrow{\scriptstyle proj_1} \\
M & \xrightarrow{\;\;J\;\;} & \mathbb{R}^m \; .
\end{array}
$$

It is easy to prove that TJ is an embedding. A vector field $X \in \mathcal{X}_1(M)$ defines for each point $p \in M$ a linear mapping $T_pX : T_pM \to T_{X(p)}TM$. Via the differentials for the embeddings J and TJ, the tangent spaces T_pM and $T_{X(p)}TM$ can be identified with linear subspaces in $\mathbb{R}^m$ and $\mathbb{R}^m \times \mathbb{R}^m = \mathbb{R}^{2m}$, respectively, and thereby be given the structure of normed vector spaces. Let $||T_pX||$ denote the operator norm for the linear mapping T_pX with respect to these norms in T_pM and $T_{X(p)}TM$. Now put

$$
||X||_1 = \sup_{p \in M} \left\{ ||X_p|| + ||T_pX|| \right\} \; .
$$

It is easy to prove that $\mathcal{X}_1(M)$ is a vector space and that $|| \cdot ||$ is a norm in $\mathcal{X}_1(M)$. The topology on $\mathcal{X}_1(M)$ induced by the norm $|| \cdot ||_1$ is independent of the chosen embedding of M in a Euclidean space and is called the C^1-*topology* on $\mathcal{X}_1(M)$.

Now let $W \subseteq TM$ be the submanifold in TM consisting of the zero tangent vector at all points of M (the zero-section). Let $X \in \mathcal{X}_1(M)$. A zero for X is a point $p \in M$ such that $X(p) \in W$; cf. Figure 7.16.

It is easy to prove that if $X \pitchfork W$ then X has isolated zeros, and since M is compact therefore at most finitely many zeros. By Thom's Transversality Theorem, the set of vector fields $X \in \mathcal{X}_1(M)$ for which $X \pitchfork W$, is an open and dense subset $\mathcal{G}$ in $\mathcal{X}_1(M)$; i.e. transversality is stable and generic ($\mathcal{G}$ for 'generic'). Thereby we have the following theorem.

Theorem 7.9.9. *Let M be a compact differentiable manifold of class C^2, and let $\mathcal{X}_1(M)$ be the set of vector fields of class C^1 on M equipped with the C^1-topology. Then there exists an open and dense subset $\mathcal{G} \subseteq \mathcal{X}_1(M)$,*

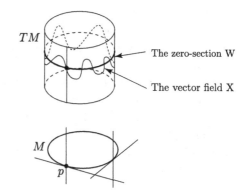

In Figure 7.16 we think of
M as the circle S^1. Then we
can identify TM with $S^1 \times \mathbb{R}$
by turning all the tangents to
vertical position.

Fig. 7.16 Generically smooth vector fields have isolated zeros.

such that every vector field $X \in \mathcal{G}$ has isolated, and thereby at most finitely many, zeros.

Whether there exists a vector field X on M without zeros depends on a topological invariant of M, called the *Euler-Poincaré characteristic*, denoted by $\chi(M)$. If M is homeomorphic to a polyhedron, $\chi(M)$ can be determined as the alternating sum $\chi(M) = \sum_{i=0}^{n}(-1)^i c_i$, where c_i denotes the number of i-dimensional faces in the polyhedron. A famous result from 1926 of Heinz Hopf says that M admits a vector field without zeros exactly when $\chi(M) = 0$. Among the orientable closed surfaces, the torus is the only surface with Euler-Poincaré characteristic 0. The torus is therefore the only closed orientable surface that admits a vector field without any zeros.

Exercises and Further Results

Exercise 7.1. Let M be a topological manifold modelled on the normed vector space E. Suppose M admits an atlas of class C^r, $r \geq 1$.

Prove that compatibility is an equivalence relation in the set of atlases of class C^r on M.

Exercise 7.2. Consider the n-sphere S^n in $\mathbb{R}^{n+1}$ defined by

$$S^n = \left\{ (x_1, \ldots, x_{n+1}) \in \mathbb{R}^{n+1} \,\middle|\, \sum_{i=1}^{n+1} x_i^2 = 1 \right\}.$$

1) Corresponding to the $n + 1$ coordinate axes in $\mathbb{R}^{n+1}$, the n-sphere S^n is divided into $2n + 2$ open hemispheres (defined repectively by the

conditions $x_i > 0$ and $x_i < 0$ for $i = 1, \ldots, n+1$). By projecting these hemispheres orthogonally into the corresponding coordinate hyperplanes (given by $x_i = 0$), we get a set of $2n + 2$ charts on S^n.

Prove that this set of charts is an atlas of class C^∞ for the differentiable structure on S^n defined in Example 7.1.3.

2) Prove that one can define an atlas of class C^∞ on S^n with only two charts by taking as maps in the charts the stereographic projection from each one of the two points $P_N = (0, \ldots, 0, 1), P_S = (0, \ldots, 0, -1) \in S^n$ onto the coordinate hyperplane in $\mathbb{R}^{n+1}$ defined by $x_{n+1} = 0$.

3) Determine the minimal number of charts in an atlas for the differentiable structure on S^n.

Exercise 7.3. The *real projective space* of dimension n, denoted by $\mathbb{RP}^n$, is the set of straight lines through the origin of $\mathbb{R}^{n+1}$. There is a canonical surjective mapping $\pi : S^n \to \mathbb{RP}^n$, which maps a point $x = (x_1, \ldots, x_{n+1})$ in S^n into the line through x. Each line in $\mathbb{RP}^n$ has exactly two preimages in S^n, namely a pair of antipodal points $x, -x \in S^n$.

Equip $\mathbb{RP}^n$ with the quotient topology induced from the topology on S^n by the mapping $\pi : S^n \to \mathbb{RP}^n$.

Prove that $\mathbb{RP}^n$ in a unique way can be turned into an n-dimensional differentiable manifold with a differentiable structure of class C^∞, such that the mapping π is a local diffeomorphism of class C^∞.

Exercise 7.4. Let M be the line with an extra origo defined in Example 2.9.7. Show that we can turn M into a 1-dimensional manifold, if we do not require manifolds to be Hausdorff spaces.

Exercise 7.5. Let M and N be differentiable manifolds of class C^r, $r \geq 1$, modelled on normed vector spaces E and F, respectively.

Let $\mathcal{A}_M = \{(U_\alpha, \varphi_\alpha)\}_{\alpha \in I}$ and $\mathcal{A}_N = \{(V_\beta, \psi_\beta)\}_{\beta \in J}$ be atlases for the differentiable structures on M and N, respectively.

1) Prove that if $M \times N$ is given the product topology, then

$$\mathcal{A}_{M \times N} = \{(U_\alpha \times V_\beta, \varphi_\alpha \times \psi_\beta)\}_{(\alpha, \beta) \in I \times J}$$

is an atlas for a differentiable structure of class C^r on $M \times N$.

2) Prove that if $\mathcal{A}'_M$ is compatible with $\mathcal{A}_M$, and $\mathcal{A}'_N$ is compatible with $\mathcal{A}_N$, then $\mathcal{A}'_{M \times N}$ is compatible with $\mathcal{A}_{M \times N}$.

3) Prove that we get a well defined differentiable structure of class C^r on $M \times N$ by taking products of atlases for the differentiable structures on M and N.

Exercise 7.6. Let $C(S^1, S^2)$ be the space of continuous mappings of the circle S^1 into the 2-sphere S^2. Convince yourself that the set of local coordinate systems on $C(S^1, S^2)$ constructed in Example 7.2.13, defines an atlas for a differentiable structure of class C^∞ on $C(S^1, S^2)$.

Exercise 7.7. Think of $\mathbb{R}^9$ as the product $\mathbb{R}^9 = \mathbb{R}^3 \times \mathbb{R}^3 \times \mathbb{R}^3$ with coordinates $(x, y, z) \in \mathbb{R}^9$, in which $x = (x_1, x_2, x_3)$, $y = (y_1, y_2, y_3)$ and $z = (z_1, z_2, z_3)$ are vectors in $\mathbb{R}^3$. Let $\langle \cdot, \cdot \rangle$ denote the standard inner product in $\mathbb{R}^3$.

Define the subset $M \subset \mathbb{R}^9$ by

$$M = \left\{ (x, y, z) \in \mathbb{R}^9 \;\middle|\; \begin{array}{l} \langle x, x \rangle = \langle y, y \rangle = \langle z, z \rangle = 1 \\ \langle x, y \rangle = \langle x, z \rangle = \langle y, z \rangle = 0 \end{array} \right\}.$$

1) Show that M is a compact, 3-dimensional, differentiable manifold of class C^∞.

2) A linear mapping $L : \mathbb{R}^3 \to \mathbb{R}^3$ such that $\langle L(v), L(w) \rangle = \langle v, w \rangle$ for all $v, w \in \mathbb{R}^3$ is called a linear isometry in $\mathbb{R}^3$. Let $\mathbf{O}(3)$ denote the space of linear isometries in $\mathbb{R}^3$ equipped with the induced topology when considered as a subset of $L(\mathbb{R}^3, \mathbb{R}^3)$.

Let $\{e_1, e_2, e_3\}$ denote the canonical basis in $\mathbb{R}^3$. Show that the mapping

$$\Phi : \mathbf{O}(3) \to M \quad \text{defined by} \quad \Phi(L) = (L(e_1), L(e_2), L(e_3))$$

is a homeomorphism.

3) Let $\mathbf{O}(n)$ denote the space of linear isometries in $\mathbb{R}^n$ equipped with the canonical inner product. Show that $\mathbf{O}(n)$ can be given the structure of a compact, differentiable manifold of class C^∞ and dimension

$$n^2 - n - \binom{n}{2} = \frac{n \cdot (n-1)}{2}.$$

With composition of isometries as multiplication, the manifold $\mathbf{O}(n)$ is also a group, and the group operations are differentiable. Such a group manifold is called a *Lie group* after the Norwegian mathematician Sophus Lie (1842–1899).

The particular Lie group $\mathbf{O}(n)$ is known as *the orthogonal group* in n-dimensional Euclidean space.

Exercise 7.8. Let M be an n-dimensional, differentiable manifold of class $C^r, r \geq 1$. Show that for every point $x \in M$ there exists an open neighbourhood U of $x \in M$ and n differentiable functions $\varphi_1, \ldots, \varphi_n : M \to \mathbb{R}$

of class C^r, such that the restriction of $\varphi = (\varphi_1, \ldots, \varphi_n) : M \to \mathbb{R}^n$ to $U \subseteq M$ defines a chart (U, φ) on M about $x \in M$.

Hint: You can make use of the *localization function* ρ defined below.

First define the function $\lambda : \mathbb{R} \to \mathbb{R}$ by

$$\lambda(x) = \begin{cases} \exp(-1/x^2) & \text{for } x > 0 \\ 0 & \text{for } x \leq 0 . \end{cases}$$

Then it is well known that $\lambda : \mathbb{R} \to \mathbb{R}$ is differentiable of class C^∞.

Now let a and b be arbitrary real numbers with $0 < a < b$. Then define the function $\rho : \mathbb{R}^n \to \mathbb{R}$ by

$$\rho(x) = \frac{\lambda(b^2 - \|x\|^2)}{\lambda(b^2 - \|x\|^2) + \lambda(\|x\|^2 - a^2)} ,$$

where $\| \cdot \|$ is the Euclidean norm in $\mathbb{R}^n$. Then $\rho : \mathbb{R}^n \to \mathbb{R}$ is differentiable of class C^∞, and

$$\rho(x) = \begin{cases} 1 & \text{for } \|x\| \leq a \\ 0 & \text{for } \|x\| \geq b . \end{cases}$$

Exercise 7.9. Let M be a compact, n-dimensional, differentiable manifold of class $C^r, r \geq 1$.

1) Show that there exist a finite number of open sets $U_1, \ldots, U_k$ in M that cover M, and with associated differentiable functions of class C^r,

$$\varphi_{ij} : M \to \mathbb{R} \quad \text{and} \quad \rho_i : M \to \mathbb{R} , \quad i = 1, \ldots, k , \ j = 1, \ldots, n ,$$

such that

    a) the restriction of $\varphi_i = (\varphi_{i1}, \ldots, \varphi_{in}) : M \to \mathbb{R}^n$
       to U_i defines a chart (U_i, φ_i) on M,

    b) $\rho_i(x) = 0$ for $x \notin U_i$,

    c) $\sum_{i=1}^{k} \rho_i(x) > 0$ for all $x \in M$.

2) In continuation of 1), define the mapping $f : M \to \mathbb{R}^{(n+1)k}$ by

$$f(x) = \big(\varphi_{11}(x), \ldots, \varphi_{1n}(x), \ldots, \varphi_{k1}(x), \ldots, \varphi_{kn}(x), \rho_1(x), \ldots, \rho_k(x)\big) ,$$

for $x \in M$.

Show that f is injective and a homeomorphism of M onto its image $f(M)$ in $\mathbb{R}^{(n+1)k}$ with the induced topology.

3) Show that f is an embedding of class C^r.

Exercise 7.10. Let M be an n-dimensional, differentiable manifold of class $C^r, r \geq 1$, and let $\mathcal{E}(M, p)$ denote the space of germs of differentiable functions $f : \Omega \to \mathbb{R}$ of class C^r defined in open neighbourhoods Ω of $p \in M$.

By a *derivation* in $\mathcal{E}(M, p)$ we understand a mapping $X_p : \mathcal{E}(M, p) \to \mathbb{R}$ which for all $\lambda, \mu \in \mathbb{R}$ and $f, g \in \mathcal{E}(M, p)$ satisfies:

$$\text{(i)} \quad X_p[\lambda f + \mu g] = \lambda X_p[f] + \mu X_p[g]$$

$$\text{(ii)} \quad X_p[f \cdot g] = X_p[f] \cdot g + f \cdot X_p[g] .$$

Let $\mathcal{D}er(M, p)$ denote the space of derivations in $\mathcal{E}(M, p)$.

1) Show that $\mathcal{D}er(M, p)$ can be equipped in a natural way with structure as a real vector space.

2) Show that the mapping $T_p M \to \mathcal{D}er(M, p)$, which maps the tangent vector $v_p \in T_p M$ into the associated derivation in $\mathcal{E}(M, p)$, is linear.

3) Now consider the situation in a chart (U, φ) on M around $p \in M$ only. Let $(x_1, \ldots, x_n) \in \varphi(U) \subseteq \mathbb{R}^n$ be the coordinates belonging to the chart, and assume that $\varphi(p) = (0, \ldots, 0)$.

Then show that every germ $f \in \mathcal{E}(M, p)$ can be written in the form $f = \sum_{i=1}^{n} x_i g_i$ in which $g_i \in \mathcal{E}(M, p)$ for $i = 1, \ldots, n$, and x_i is the germ at $p \in M$ of the ith coordinate function of $\varphi : U \to \mathbb{R}^n$; cf. Lemma 8.4.6.

Further show that

$$g_i(p) = \frac{\partial}{\partial x_i}(p)[f] .$$

4) (Continuation of 3)). Show that an arbitrary derivation X_p in $\mathcal{E}(M, p)$ can be written in a unique way as a linear combination

$$X_p = \sum_{i=1}^{n} a_i \frac{\partial}{\partial x_i}(p) , \quad a_i \in \mathbb{R}, \ i = 1, \ldots, n ,$$

and conclude from this that the linear mapping $T_p M \to \mathcal{D}er(M, p)$, defined in question 2), is an isomorphism.

Exercise 7.11. Let M be a finite dimensional smooth manifold.

A mapping $X : C^\infty(M) \to C^\infty(M)$ is called a *derivation* in $C^\infty(M)$, if it for all $\lambda, \mu \in \mathbb{R}$ and $f, g \in C^\infty(M)$ satisfies the conditions:

i) $X[\lambda f + \mu g] = \lambda X[f] + \mu X[g]$ and *ii)* $X[f \cdot g] = X[f] \cdot g + f \cdot X[g]$.

Let $\mathcal{D}er(M)$ denote the space of derivations in $C^\infty(M)$, and let $\mathcal{X}(M)$ denote the space of smooth vector fields on M.

1) Prove that $\mathcal{D}er(M)$ and $\mathcal{X}(M)$ are real vector spaces when equipped with the obvious pointwise defined vector space operations.

2) Prove that the mapping $\mathcal{X}(M) \to \mathcal{D}er(M)$, which to the smooth vector field $X \in \mathcal{X}(M)$ associates the derivation in $C^\infty(M)$ defined by X, is an isomorphism of real vector spaces.

Exercise 7.12. Let M and N be differentiable manifolds modelled on Banach spaces. Prove that if M is connected, then a differentiable mapping $f : M \to N$ has constant value if and only if the differential $T_p f : T_p M \to T_{f(p)} N$ equals 0 in all points $p \in M$.

Exercise 7.13. Prove that there does not exist an immersion of class C^1 of a compact, n-dimensional, differentiable manifold M^n into $\mathbb{R}^n$.

Exercise 7.14. Let $f : M \to N$ be a differentiable mapping of class C^r, $r \geq 1$, between finite dimensional differentiable manifolds M and N.

The graph $G(f)$ of f is by definition the subset of the product manifold $M \times N$ given by

$$G(f) = \{(p, f(p)) \mid p \in M\}.$$

1) Prove that $G(f)$ is a submanifold of $M \times N$.

2) Prove that for every $p \in M$, the tangent space

$$T_{(p, f(p))} G(f) \subseteq T_p M \times T_{f(p)} N$$

can be identified with the graph of the differential $T_p f : T_p M \to T_{f(p)} N$.

3) Prove that for any finite dimensional differentiable manifold M, the set of *diagonal* points $\Delta M = \{(p, p) \in M \times M \mid p \in M\}$ in $M \times M$, is a submanifold of the product manifold $M \times M$.

Exercise 7.15. Let $S_1^1 \times S_2^1$ be a smooth manifold, a *torus*, defined as the product of two copies S_1^1 and S_2^1 of the unit circle S^1 in the Euclidean plane.

Construct an embedding of $S_1^1 \times S_2^1$ into 3-dimensional Euclidean space $\mathbb{R}^3$ of class C^∞ by making use of the trigonometric functions cos and sin.

Exercise 7.16. A *spherical double pendulum* is a mechanical system consisting of two coupled spherical pendula in Euclidean 3-space $\mathbb{R}^3$.

Let $q_1 \in \mathbb{R}^3$ be the position unit vector of the first pendulum relative to its pivot O at the origo of $\mathbb{R}^3$, and let $q_2 \in \mathbb{R}^3$ be the position unit vector of the second pendulum relative to its pivot at the end of the first pendulum.

Show that the *configuration space* for a spherical double pendulum, which describes the possible positions of the double pendulum, can be identified with a 4-dimensional smooth manifold defined as the product of two copies of the unit sphere S^2 in $\mathbb{R}^3$.

Chapter 8

An Introduction to Singularity Theory

This chapter is an introduction to some fundamental concepts in the theory of singularities of smooth mappings and to parts of the mathematical foundations of the theory of elementary catastrophes created by the French mathematician René Thom in the 1960s. Emphasis is given to the study of the geometry of smooth germs, i.e. the study of how a smooth mapping crumples up a neighbourhood of a point.

First we introduce the notions of germs and equivalence of germs. Then as our first result, we provide normal forms for the regular germs corresponding to the local forms known from immersions and submersions.

Among the singular germs we determine normal forms for the germs of real-valued functions of one real variable of finite order. A particular study is made of the so-called Morse germs. The classification of these germs is of interest in its own right, since it can be considered as a geometrical description of the classical method of finding local extrema for functions of several real variables.

For applications to catastrophe theory, the Whitney C^k-topologies on function spaces are of special interest when formulating results about genericity of appropriate classes of mappings. Such results are then proved by transversality arguments. The last two sections of this chapter are devoted to these topics.

All spaces in this chapter are finite dimensional, and differentiable mappings are of class C^∞, i.e. smooth mappings.

8.1 Equivalence of germs

It will be convenient in this chapter to write vector-valued mappings with capital letters and their coordinate functions with small letters.

Let $F = (f_1, \ldots, f_m) : U \to \mathbb{R}^m$ be a smooth mapping defined in an open set $U \subseteq \mathbb{R}^n$ and let $x_0 \in U$ be a point in U. The Jacobian matrix $\mathbf{D}F(x_0)$ for the differential $DF(x_0) : \mathbb{R}^n \to \mathbb{R}^m$ of F at $x_0 \in U$ is then an $m \times n$-matrix. The rank of F at x_0, as before denoted by $\mathrm{rk}_{x_0} F$, is therefore an integer in the interval $0 \leq \mathrm{rk}_{x_0} F \leq \min\{n, m\}$.

Definition 8.1.1. We say that a point $x_0 \in U$ in the domain of definition $U \subseteq \mathbb{R}^n$ of a smooth mapping $F : U \to \mathbb{R}^m$ is a *regular point* of F, if the rank $\mathrm{rk}_{x_0} F = \min\{n, m\}$, and a *singular point*, if $\mathrm{rk}_{x_0} F < \min\{n, m\}$.

In singularity theory one investigates, among other things, the local behaviour of a smooth mapping $F : U \to \mathbb{R}^m$ in the neighbourhood of a singular point $x_0 \in U \subseteq \mathbb{R}^n$ of F. What is meant by this is most conveniently described by introducing the notion of a *germ* of a smooth mapping. In Chapter 7, page 199, we have earlier introduced the notion of germs of smooth, real-valued functions, and now we proceed in complete analogy to define germs of smooth, vector-valued mappings.

Let $x_0 \in \mathbb{R}^n$ be a fixed point in $\mathbb{R}^n$. Two smooth maps $F_1 : U_1 \to \mathbb{R}^m$ and $F_2 : U_2 \to \mathbb{R}^m$ defined in open neighbourhoods U_1 and U_2 of $x_0 \in \mathbb{R}^n$, are said to define the *same smooth germ*, if there exists an open neighbourhood U of $x_0 \in \mathbb{R}^n$ such that $U \subseteq U_1 \cap U_2$ and such that the restrictions of F_1 and F_2 to U agree in U, i.e. $F_1|U = F_2|U$. Clearly, this defines an equivalence relation in the set of smooth mappings into $\mathbb{R}^m$ defined in open neighbourhoods of $x_0 \in \mathbb{R}^n$. The equivalence classes are called *germs* of smooth mappings of $\mathbb{R}^n$ into $\mathbb{R}^m$ at $x_0 \in U \subseteq \mathbb{R}^n$. The germ of the smooth mapping $F : U \to \mathbb{R}^m$ at $x_0 \in U \subseteq \mathbb{R}^n$ and with $y_0 = F(x_0)$ is denoted by $F : (\mathbb{R}^n, x_0) \to (\mathbb{R}^m, y_0)$.

The idea behind considering the *germ* $F : (\mathbb{R}^n, x_0) \to (\mathbb{R}^m, y_0)$ of a mapping at $x_0 \in U \subseteq \mathbb{R}^n$, rather than the mapping itself, is that we are then free to shrink the neighbourhood on which the mapping is defined as we see fit. And since the objective in singularity theory is to study the local behaviour of smooth mappings, it is only the behaviour in neighbourhoods as small as we wish that matters anyhow. Accordingly, we shall think of a germ as an ordinary smooth mapping with the built-in freedom to shrink domains. Furthermore, we shall follow the tradition and use F to denote both the mapping itself and the germ of the mapping.

Suppose $F : (\mathbb{R}^n, x_0) \to (\mathbb{R}^m, y_0)$ and $G : (\mathbb{R}^m, y_0) \to (\mathbb{R}^k, z_0)$ are two smooth germs, where the germ of G is taken at the point $y_0 = F(x_0)$. Then we have a well defined *composed germ* $G \circ F : (\mathbb{R}^n, x_0) \to (\mathbb{R}^k, z_0)$. If a germ $F : (\mathbb{R}^n, x_0) \to (\mathbb{R}^n, y_0)$ has a smooth representative $F : U \to V$

that maps an open neighbourhood U of $x_0 \in \mathbb{R}^n$ diffeomorphically onto an open neighbourhood V of $y_0 \in \mathbb{R}^n$, then we get a well defined *inverse germ* $F^{-1} : (\mathbb{R}^n, y_0) \to (\mathbb{R}^n, x_0)$.

A smooth germ $F : (\mathbb{R}^n, x_0) \to (\mathbb{R}^m, y_0)$ has a well defined *differential* $DF(x_0) : \mathbb{R}^n \to \mathbb{R}^m$. Hence we can define the *rank* of the smooth germ F as the rank of the linear mapping $DF(x_0)$, which in turn can be computed from the Jacobian matrix. We denote the rank of the germ F by $\text{rk}_{x_0} F$.

A smooth germ $F : (\mathbb{R}^n, x_0) \to (\mathbb{R}^n, y_0)$, which has a smooth inverse germ $F^{-1} : (\mathbb{R}^n, y_0) \to (\mathbb{R}^n, x_0)$, is said to be the *germ of a diffeomorphism*.

From the Inverse Function Theorem (Theorem 6.8.1) for smooth mappings, we get immediately the following version for smooth germs.

Theorem 8.1.2 (Inverse Germs). *A germ $F : (\mathbb{R}^n, x_0) \to (\mathbb{R}^n, y_0)$ of a smooth mapping is the germ of a diffeomorphism if and only if the differential $DF(x_0) : \mathbb{R}^n \to \mathbb{R}^n$ is an isomorphism, or equivalently, $\text{rk}_{x_0} F = n$.*

By the geometry in a smooth germ $F : (\mathbb{R}^n, x_0) \to (\mathbb{R}^m, y_0)$ we understand, informally speaking, how F crumples up a neighbourhood of x_0 under the mapping into $\mathbb{R}^m$. Singularity theory is designed, among other things, to study this geometry. Since the qualitative geometry in a smooth germ is left unaffected by local shifts of coordinates in $\mathbb{R}^n$ and $\mathbb{R}^m$ by local diffeomorphisms, the smooth germs are divided into equivalence classes under the following notion of equivalence.

Definition 8.1.3. We say that two smooth germs $F_0 : (\mathbb{R}^n, x_0) \to (\mathbb{R}^m, y_0)$ and $F_1 : (\mathbb{R}^n, x_1) \to (\mathbb{R}^m, y_1)$ are *equivalent*, if there exist germs of diffeomorphisms $G : (\mathbb{R}^n, x_0) \to (\mathbb{R}^n, x_1)$ and $H : (\mathbb{R}^n, y_0) \to (\mathbb{R}^n, y_1)$, such that $H \circ F_0 = F_1 \circ G$, i.e. such that the following diagram is commutative

$$
\begin{array}{ccc}
(\mathbb{R}^n, x_0) & \xrightarrow{\ \ G\ \ } & (\mathbb{R}^n, x_1) \\
\Big\downarrow{\scriptstyle F_0} & & \Big\downarrow{\scriptstyle F_1} \\
(\mathbb{R}^m, y_0) & \xrightarrow[\ \ H\ \ ]{} & (\mathbb{R}^m, y_1) \ .
\end{array}
$$

In the applications of singularity theory to catastrophe theory, the so-called elementary catastrophes are singularities derived from smooth real-valued functions that occur as potential functions in the models of the phenomena under consideration. In the case $m = 1$, it is therefore appropriate not to allow local changes of coordinates in $\mathbb{R}$. This leads to a stronger notion of equivalence as follows.

Definition 8.1.4. We say that two smooth germs $f_0 : (\mathbb{R}^n, x_0) \to (\mathbb{R}, 0)$ and $f_1 : (\mathbb{R}^n, x_1) \to (\mathbb{R}, 0)$ are *right equivalent*, if there exists a germ of a diffeomorphism $G : (\mathbb{R}^n, x_0) \to (\mathbb{R}^n, x_1)$, such that $f_0 = f_1 \circ G$, i.e. such that the following diagram is commutative

$$
\begin{array}{ccc}
(\mathbb{R}^n, x_0) & \xrightarrow{\;G\;} & (\mathbb{R}^n, x_1) \\
\downarrow{\scriptstyle f_0} & & \downarrow{\scriptstyle f_1} \\
(\mathbb{R}, 0) & =\!\!=\!\!= & (\mathbb{R}, 0) \, .
\end{array}
$$

An arbitrary smooth germ will always be equivalent to a smooth germ of the type $F : (\mathbb{R}^n, 0) \to (\mathbb{R}^m, 0)$. This is readily seen by composing the germ with suitable parallel translations. It is therefore sufficient to consider germs of the type $F : (\mathbb{R}^n, 0) \to (\mathbb{R}^m, 0)$, when studying the geometry of a smooth germ.

The notion of equivalence in Definition 8.1.3 defines an equivalence relation in the set of smooth germs $F : (\mathbb{R}^n, 0) \to (\mathbb{R}^m, 0)$ according to which they are divided into classes of equivalent germs. A major problem in singularity theory is to find algebraic invariants that will distinguish between these equivalence classes. This is a most difficult problem solved around 1970 in works by the American mathematician John Mather, which we shall not discuss further here. Another problem consists in finding for any given equivalence class of smooth germs a particularly nice model (representative) for the equivalence class, a so-called *normal form*.

In the following three sections, we shall determine normal forms for certain equivalence classes of smooth germs $F : (\mathbb{R}^n, 0) \to (\mathbb{R}^m, 0)$ characterized by conditions on the partial derivatives of the mapping F at 0, that are independent of the representative for the equivalence class.

8.2 Regular germs

A smooth germ $F : (\mathbb{R}^n, 0) \to (\mathbb{R}^m, 0)$ is called a *regular germ*, if the rank of F at 0 is maximal, i.e. $\mathrm{rk}_0 F = \min\{n, m\}$. If $\mathrm{rk}_0 F < \min\{n, m\}$, the germ F is called a *singular germ*.

Making use of the chain rule it is easy to prove that if two smooth germs $(\mathbb{R}^n, 0) \to (\mathbb{R}^m, 0)$ are equivalent, then either both of them are regular, or both of them are singular.

In this section we shall determine normal forms for the regular germs $(\mathbb{R}^n, 0) \to (\mathbb{R}^m, 0)$. There are two cases corresponding to $n \leq m$, respectively $n \geq m$. The smooth germs E and S in the two theorems below are the normal forms for regular germs that we are looking for.

Theorem 8.2.1. *Suppose* $n \leq m$. *If* $F : (\mathbb{R}^n, 0) \to (\mathbb{R}^m, 0)$ *is a smooth germ with* $\mathrm{rk}_0 F = n$, *then* F *is equivalent to the germ* $E : (\mathbb{R}^n, 0) \to (\mathbb{R}^m, 0)$ *defined by* $E(x_1, \ldots, x_n) = (x_1, \ldots, x_n, 0, \ldots, 0)$.

Theorem 8.2.2. *Suppose* $n \geq m$. *If* $F : (\mathbb{R}^n, 0) \to (\mathbb{R}^m, 0)$ *is a smooth germ with* $\mathrm{rk}_0 F = m$, *then* F *is equivalent to the germ* $S : (\mathbb{R}^n, 0) \to (\mathbb{R}^m, 0)$ *defined by* $S(x_1, \ldots, x_n) = (x_1, \ldots, x_m)$.

In the proofs of the above theorems, we let $(x_1, \ldots, x_n) \in \mathbb{R}^n$ denote the coordinates in $\mathbb{R}^n$, and $(y_1, \ldots, y_m) \in \mathbb{R}^m$ the coordinates in $\mathbb{R}^m$. Furthermore, let $F = (f_1, \ldots, f_m) : (\mathbb{R}^n, 0) \to (\mathbb{R}^m, 0)$ be the coordinate description of a representative for the smooth germ F. The Jacobian matrix $\mathbf{D}F(0)$ for the germ F at $0 \in \mathbb{R}^n$ is well defined and is given by

$$
\mathbf{D}F(0) = \begin{bmatrix} \frac{\partial f_1}{\partial x_1}(0) & \cdots & \frac{\partial f_1}{\partial x_n}(0) \\ \vdots & & \vdots \\ \frac{\partial f_m}{\partial x_1}(0) & \cdots & \frac{\partial f_m}{\partial x_n}(0) \end{bmatrix} .
$$

Proof of Theorem 8.2.1. Since $n \leq m$ and $\mathrm{rk}_0 F =$ rank of $\mathbf{D}F(0) = n$, we can assume, possibly after an initial permutation of the coordinates in $\mathbb{R}^m$, that the first n rows in $\mathbf{D}F(0)$ are linearly independent. Since we are interested only in the equivalence class of F, it is permissible to perform such an initial permutation. The proof now follows the same pattern as the proof of Assertion 7.8.2.

Define the smooth germ $K = (k_1, \ldots, k_m) : (\mathbb{R}^m, 0) \to (\mathbb{R}^m, 0)$ by

$$
k_i(y_1, \ldots, y_m) = \begin{cases} f_i(y_1, \ldots, y_n) & \text{for } 1 \leq i \leq n \\ y_i + f_i(y_1, \ldots, y_n) & \text{for } n+1 \leq i \leq m . \end{cases}
$$

It is easy to prove that

$$
\det \mathbf{D}K(0) = \det \left[\frac{\partial f_i}{\partial y_j}(0) \right]_{i,j=1,\ldots,n} \neq 0 .
$$

Hence the differential $DK(0)$ is an isomorphism and by Theorem 8.1.2, K is then the germ of a diffeomorphism. Let $H : (\mathbb{R}^m, 0) \to (\mathbb{R}^m, 0)$ be the inverse germ of K. Then it is easy to prove that $H \circ F = E$, which completes the proof of Theorem 8.2.1. $\square$

Proof of Theorem 8.2.2. Since $n \geq m$ and $\mathrm{rk}_0 F =$ rank of $\mathbf{D}F(0) = m$, we can assume, possibly after an initial permutation of the coordinates in $\mathbb{R}^n$, that the first m columns in $\mathbf{D}F(0)$ are linearly independent. Since again we are interested only in the equivalence class of F, it is permissible to perform such an initial permutation. The proof now follows the same pattern as the proof of Assertion 7.1.1.

Define the smooth germ $K = (k_1, \ldots, k_n) : (\mathbb{R}^n, 0) \to (\mathbb{R}^n, 0)$ by

$$k_i(x_1, \ldots, x_n) = \begin{cases} f_i(x_1, \ldots, x_n) & \text{for } 1 \leq i \leq m \\ x_i & \text{for } m+1 \leq i \leq n \, . \end{cases}$$

It is easy to prove that

$$\det \mathbf{D}K(0) = \det \left[\frac{\partial f_i}{\partial x_j}(0) \right]_{i,j=1,\ldots,m} \neq 0 \, .$$

Hence the differential $DK(0)$ is an isomorphism and by Theorem 8.1.2, K is then the germ of a diffeomorphism. Let $G : (\mathbb{R}^n, 0) \to (\mathbb{R}^n, 0)$ be the inverse germ of K. Then it is easy to prove that $F \circ G = S$, which completes the proof of Theorem 8.2.2. $\square$

As a corollary to the proof of Theorem 8.2.2 we get the theorem.

Theorem 8.2.3. *Let $f : (\mathbb{R}^n, 0) \to (\mathbb{R}, 0)$ be a regular germ. Then f is right-equivalent to the germ $p_1 : (\mathbb{R}^n, 0) \to (\mathbb{R}, 0)$, represented by the projection of $\mathbb{R}^n$ onto the first coordinate axis.*

8.3   Germs $(\mathbb{R}, 0) \to (\mathbb{R}, 0)$ of finite order

Let $f : (\mathbb{R}, 0) \to (\mathbb{R}, 0)$ be a smooth germ. We say that the germ f has *finite order*, if for $k \geq 1$, at least one of the derivatives of higher order $f^{(k)}(0) \neq 0$.

If the smooth germ $f : (\mathbb{R}, 0) \to (\mathbb{R}, 0)$ has finite order, the least $k \geq 1$ for which $f^{(k)}(0) \neq 0$ is called the *order* of the germ f.

Lemma 8.3.1. *Let $f : (\mathbb{R}, 0) \to (\mathbb{R}, 0)$ be a smooth germ of finite order $k \geq 1$. Then a smooth germ $g : (\mathbb{R}, 0) \to (\mathbb{R}, 0)$, which is equivalent to f, will also have finite order k.*

Proof. Since g is equivalent to f there exist germs $\varphi, \psi : (\mathbb{R}, 0) \to (\mathbb{R}, 0)$ of diffeomorphisms such that $g = \psi \circ f \circ \varphi$.

By well known rules for differentiation we first get

$$g^{(1)} = \left(\psi^{(1)} \circ f \circ \varphi \right) \cdot \left(f^{(1)} \circ \varphi \right) \cdot \varphi^{(1)} \, ,$$

and next that

$$
\begin{aligned}
g^{(2)} = {} & \left(\psi^{(2)} \circ f \circ \varphi \right) \cdot \left(f^{(1)} \circ \varphi \right)^2 \cdot \left(\varphi^{(1)} \right)^2 \\
& + \left(\psi^{(1)} \circ f \circ \varphi \right) \cdot \left(f^{(2)} \circ \varphi \right) \cdot \left(\varphi^{(1)} \right)^2 \\
& + \left(\psi^{(1)} \circ f \circ \varphi \right) \cdot \left(f^{(1)} \circ \varphi \right) \cdot \varphi^{(2)} .
\end{aligned}
$$

We could go on this way to determine the derivatives of higher and higher order of g, but since $\varphi(0) = \psi(0) = 0$ and $f^{(i)}(0) = 0$ for all $i \leq k-1$, clearly $g^{(i)}(0) = 0$ for $i \leq k - 1$, and

$$
g^{(k)}(0) = \psi^{(1)}(0) \cdot f^{(k)}(0) \cdot \left(\varphi^{(1)}(0) \right)^k .
$$

The germs φ and ψ are germs of diffeomorphisms, and hence $\varphi^{(1)}(0) \neq 0$ and $\psi^{(1)}(0) \neq 0$. Since also $f^{(k)}(0) \neq 0$, it follows that $g^{(k)}(0) \neq 0$. This completes the proof that g has order k. $\square$

For every $k \geq 1$ there is an obvious germ $f_k : (\mathbb{R}, 0) \to (\mathbb{R}, 0)$ of order k, namely the germ represented by the smooth function $f_k(x) = x^k$. By Lemma 8.3.1, any two germs f_k and f_l are equivalent if and only if $k = l$.

The following theorem shows that f_k is a normal form for the equivalence class of smooth germs $f : (\mathbb{R}, 0) \to (\mathbb{R}, 0)$ of order k.

Theorem 8.3.2. *Let $f : (\mathbb{R}, 0) \to (\mathbb{R}, 0)$ be a smooth germ of finite order $k \geq 1$. Then f is equivalent to the germ $f_k : (\mathbb{R}, 0) \to (\mathbb{R}, 0)$.*

Proof. Let $f : I \to \mathbb{R}$ be a smooth function defined in an open interval I about $0 \in \mathbb{R}$, which represents the smooth germ $f : (\mathbb{R}, 0) \to (\mathbb{R}, 0)$ of finite order $k \geq 1$.

Assertion 8.3.3. *There exists a smooth function $g : I \to \mathbb{R}$ such that $f = f_k \cdot g$, or, equivalently, such that $f(x) = x^k \cdot g(x)$ for all $x \in I$.*

Proof of Assertion. Since $f(0) = 0$, it holds for all $x \in I$ that

$$
f(x) = \int_0^1 \frac{d}{dt} f(t \cdot x) dt = \int_0^1 f'(t \cdot x) \cdot x \, dt = x \cdot h_1(x),
$$

where

$$
h_1(x) = \int_0^1 f'(t \cdot x) dt .
$$

A well known rule for differentiation of a function defined by an integral shows that $h_1 : I \to \mathbb{R}$ is a smooth function. Note also that $h_1(0) = f'(0)$.

If $f'(0) \neq 0$, we can by applying the same technique to h_1 factorize one more x from f and write f in the form $f(x) = x^2 \cdot h_2(x)$, where $h_2 : I \to \mathbb{R}$ is a smooth function. An elementary computation shows that

$$h_2(0) = \frac{f^{(2)}(0)}{2!} .$$

Since $f^{(i)}(0) = 0$ for $i \leq k - 1$, we can continue in this way and eventually write f in the form $f(x) = x^k \cdot g(x)$, where $g = h_k : I \to \mathbb{R}$ is a smooth function. This proves the assertion. $\square$

Continuing the proof of Theorem 8.3.2, we work further on the factorization $f(x) = x^k \cdot g(x)$. An elementary computation shows that

$$g(0) = \frac{f^{(k)}(0)}{k!} .$$

We only worry about the germ at $0 \in \mathbb{R}$, and hence we can shrink the interval I as necessary. Since g is continuous, we can therefore assume that

$$\frac{g(x)}{g(0)} > 0 \quad \text{for all} \quad x \in I .$$

We can then make the following computation

$$\frac{1}{g(0)} \cdot f(x) = x^k \cdot \frac{g(x)}{g(0)} = \left(x \cdot \sqrt[k]{\frac{g(x)}{g(0)}} \right)^k = f_k \left(x \cdot \sqrt[k]{\frac{g(x)}{g(0)}} \right) .$$

Motivated by this computation we define the germs $\varphi : (\mathbb{R}, 0) \to (\mathbb{R}, 0)$ and $\psi : (\mathbb{R}, 0) \to (\mathbb{R}, 0)$ by

$$\varphi(x) = x \cdot \sqrt[k]{\frac{g(x)}{g(0)}} \quad \text{and} \quad \psi(y) = \frac{1}{g(0)} \cdot y .$$

Both φ and ψ are germs of diffeomorphisms, since

$$\varphi'(0) = \sqrt[k]{\frac{g(0)}{g(0)}} = 1 \neq 0 \quad \text{and} \quad \psi'(0) = \frac{1}{g(0)} \neq 0 .$$

If we insert φ and ψ in the above computation we see that $\psi \circ f = f_k \circ \varphi$, which proves that f and f_k are equivalent germs. This completes the proof of Theorem 8.3.2 $\square$

8.4 Morse germs

In this section we consider germs of smooth functions $f : (\mathbb{R}^n, 0) \to (\mathbb{R}, 0)$.

According to Theorem 8.2.3, a regular germ $f : (\mathbb{R}^n, 0) \to (\mathbb{R}, 0)$, i.e. a germ with $Df(0) \neq 0$, is equivalent to the germ $p_1 : (\mathbb{R}^n, 0) \to (\mathbb{R}, 0)$, defined by $p_1(x_1, \ldots, x_n) = x_1$. For these germs we have in other words already found a normal form.

We therefore now turn to the singular germs $f : (\mathbb{R}^n, 0) \to (\mathbb{R}, 0)$, i.e. germs for which $Df(0) = 0$, or, equivalently, germs for which all the partial derivatives

$$\frac{\partial f}{\partial x_1}(0) = \cdots = \frac{\partial f}{\partial x_1}(0) = 0 .$$

Among the singular germs, the Morse germs occupy a prominent place. Normal forms for the Morse germs were found by Marston Morse in 1925.

Definition 8.4.1. A singular germ $f : (\mathbb{R}^n, 0) \to (\mathbb{R}, 0)$ is called a *Morse germ* if the matrix of partial derivatives of second order, the so-called *Hesse matrix*,

$$\mathbf{H}(f) = \left[\frac{\partial^2 f}{\partial x_i \partial x_j}(0) \right]$$

is non-degenerate, i.e. a matrix of rank n.

The determinant of the matrix $\mathbf{H}(f)$, denoted by $\det \mathbf{H}(f)$, is often called the *Hessian* of f at 0. Then it is well known that the Hesse matrix $\mathbf{H}(f)$ is non-degenerate precisely when the Hessian $\det \mathbf{H}(f) \neq 0$.

Lemma 8.4.2. *A smooth germ* $g : (\mathbb{R}^n, 0) \to (\mathbb{R}, 0)$, *which is right equivalent to a Morse germ* $f : (\mathbb{R}^n, 0) \to (\mathbb{R}, 0)$, *is itself a Morse germ.*

Proof. Let $\Phi = (\varphi_1, \ldots, \varphi_n) : (\mathbb{R}^n, 0) \to (\mathbb{R}^n, 0)$ be a germ of a diffeomorphism such that $g = f \circ \Phi$. Since $Df(0) = 0$, the chain rule shows that $Dg(0) = Df(0) \circ D\Phi(0) = 0$, and thereby that g is a singular germ.

Using the chain rule, we first get

$$\frac{\partial g}{\partial x_j}(x) = \sum_{k=1}^{n} \frac{\partial f}{\partial x_k}(\Phi(x)) \cdot \frac{\partial \varphi_k}{\partial x_j}(x) ,$$

and next, by using the rule for differentiation of a product,

$$\frac{\partial^2 g}{\partial x_i \partial x_j}(x) = \sum_{k=1}^{n} \sum_{l=1}^{n} \frac{\partial^2 f}{\partial x_l \partial x_k}(\Phi(x)) \cdot \frac{\partial \varphi_l}{\partial x_i} \cdot \frac{\partial \varphi_k}{\partial x_j}(x)$$

$$+ \sum_{k=1}^{n} \frac{\partial f}{\partial x_k}(\Phi(x)) \cdot \frac{\partial^2 \varphi_k}{\partial x_i \partial x_j}(x) .$$

Since f is a singular germ, we get

$$\frac{\partial f}{\partial x_k}(\Phi(0)) = \frac{\partial f}{\partial x_k}(0) = 0 \ ,$$

and therefore the matrix equation

$$\mathbf{H}(g) = \mathbf{D}\Phi(0)^* \cdot \mathbf{H}(f) \cdot \mathbf{D}\Phi(0) \ ,$$

where $\mathbf{D}\Phi(0)^*$ denotes the transposed matrix of the Jacobi matrix $\mathbf{D}\Phi(0)$.

Since Φ is the germ of a diffeomorphism, the Jacobi matrix $\mathbf{D}\Phi(0)$ is non-degenerate, i.e. $\det \mathbf{D}\Phi(0) \neq 0$. Since also $\det \mathbf{H}(f) \neq 0$, it follows that $\det \mathbf{H}(g) \neq 0$. Hence $\mathbf{H}(g)$ is non-degenerate, and we have proved that g is a Morse germ. $\square$

We now define an important invariant of a Morse germ, called the index. For that purpose we remind the reader that a real, symmetric $n \times n$-matrix $\mathbf{A}$ has n real eigenvalues, when these are counted with multiplicity. The eigenvalues are the roots in the polynomial equation $\det (\mathbf{A} - \lambda \mathbf{I}) = 0$, where $\mathbf{I}$ denotes the identity matrix. When we factorize the polynomial

$$\det (\mathbf{A} - \lambda \mathbf{I}) = (-1)^n \cdot (\lambda - \lambda_1)^{m_1} \cdots \cdot (\lambda - \lambda_k)^{m_k}$$

into linear factors, we have $m_1 + \cdots + m_k = n$, and m_j is the multiplicity of the eigenvalue λ_j for $\mathbf{A}$. By a fundamental result in linear algebra (The Spectral Theorem), there exists an orthonormal basis $\{e_1, \ldots, e_n\}$ in $\mathbb{R}^n$, equipped with the canonical inner product, consisting of eigenvectors of $\mathbf{A}$.

The index of a real, symmetric matrix $\mathbf{A}$ is, by definition, the number of negative eigenvalues of $\mathbf{A}$ counted with multiplicity. Since the Hesse matrix of a smooth function is a symmetric matrix, we can then make the following definition:

Definition 8.4.3. The *index* of a Morse germ $f : (\mathbb{R}^n, 0) \to (\mathbb{R}, 0)$ is the index of the Hesse matrix $\mathbf{H}(f)$, i.e. the number of negative eigenvalues, counted with multiplicity, of the non-degenerate symmetric $n \times n$-matrix

$$\mathbf{H}(f) = \left[\frac{\partial^2 f}{\partial x_i \partial x_j}(0) \right] .$$

The index of f is an integer k in the interval $0 \leq k \leq n$.

The index is invariant under right equivalence.

Lemma 8.4.4. *Two right equivalent Morse germs* $f : (\mathbb{R}^n, 0) \to (\mathbb{R}, 0)$ *and* $g : (\mathbb{R}^n, 0) \to (\mathbb{R}, 0)$ *have the same index.*

Proof. Proceeding as in the proof of Lemma 8.4.2 we get to the matrix equation

$$\mathbf{H}(g) = \mathbf{D}\Phi(0)^* \cdot \mathbf{H}(f) \cdot \mathbf{D}\Phi(0).$$

Let $\langle \cdot, \cdot \rangle$ denote the canonical inner product in $\mathbb{R}^n$. For vectors $x, y \in \mathbb{R}^n$, we denote by x^*, y^* the vectors considered as column matrices with only one column.

For $x \in \mathbb{R}^n$, define $y \in \mathbb{R}^n$ by $y^* = \mathbf{D}\Phi(0) \cdot x^*$. Then by the definition of a transposed matrix, we have

$$\langle \mathbf{H}(g) \cdot x^*, x^* \rangle = \langle \mathbf{H}(f) \cdot y^*, y^* \rangle.$$

Subspaces of maximal dimension in $\mathbb{R}^n$ on which $\mathbf{H}(f)$ and $\mathbf{H}(g)$ are negative definite therefore correspond to each other under the regular matrix (isomorphism) $\mathbf{D}\Phi(0)$. From this follows that $\mathbf{H}(f)$ and $\mathbf{H}(g)$ have the same number of negative eigenvalues (counted with multiplicity), i.e. they have the same index. $\square$

We shall now prove the converse result to Lemma 8.4.4, namely that two Morse germs with the same index are right equivalent. We prove this by providing a normal form corresponding to each index. The result is due to Marston Morse around 1925 and is known as the Morse Lemma.

Theorem 8.4.5 (The Morse Lemma). *For every integer k in the interval $0 \le k \le n$ we define a smooth germ $f_k : (\mathbb{R}^n, 0) \to (\mathbb{R}, 0)$, for $k = 0$, $1 \le k < n$ and $k = n$, respectively, by*

$$f_0(x_1, \ldots, x_n) = x_1^2 + \cdots + x_n^2$$
$$f_k(x_1, \ldots, x_n) = -(x_1^2 + \cdots + x_k^2) + (x_{k+1}^2 + \ldots x_n^2)$$
$$f_n(x_1, \ldots, x_n) = -(x_1^2 + \cdots + x_n^2).$$

(1) $f_k : (\mathbb{R}^n, 0) \to (\mathbb{R}, 0)$ *is a Morse germ of index k.*

(2) *An arbitrary Morse germ $f : (\mathbb{R}^n, 0) \to (\mathbb{R}, 0)$ of index k is right equivalent to f_k.*

It is easy to prove that $f_k : (\mathbb{R}^n, 0) \to (\mathbb{R}, 0)$ is a Morse germ of index k as asserted in Theorem 8.4.5(1). In the proof of Theorem 8.4.5(2), we need the following lemma, which we have already met for $n = 1$ in Assertion 8.3.3.

Lemma 8.4.6. *Let U be an open convex neighbourhood of $0 \in \mathbb{R}^n$, and let $f : U \to \mathbb{R}$ be a smooth function. Then there exist smooth functions*

$g_i : U \to \mathbb{R}$, $i = 1, \ldots, n$, *such that*

$$f(x) = f(0) + \sum_{i=1}^{n} x_i \cdot g_i(x)$$

for all $x = (x_1, \ldots, x_n) \in U \subseteq \mathbb{R}^n$.

Proof. Consider a fixed point $x \in U$. Since U is convex, the line segment $\{t{\cdot}x \mid 0 \le t \le 1\}$ is completely contained in U. Since the mapping $t \mapsto f(t{\cdot}x)$ defines a smooth function, we get

$$f(x) = f(0) + \int_0^1 \frac{d}{dt} f(t \cdot x)\, dt \ ,$$

which by the chain rule can be rewritten as

$$f(x) = f(0) + \sum_{i=1}^{n} \left(\int_0^1 \frac{\partial f}{\partial x_i}(t \cdot x)\, dt \right) \cdot x_i \ .$$

By letting x vary we define the function $g_i : U \to \mathbb{R}$ by the formula

$$g_i(x) = \int_0^1 \frac{\partial f}{\partial x_i}(t \cdot x)\, dt \ .$$

Well known rules for differentiation of a function defined by an integral show that g_i is smooth. At the same time, g_i has been constructed exactly so that we get the formula in the lemma. $\square$

Proof of Theorem 8.4.5(2). We shall only do the case $n = 2$, but the proof given below can be generalized to arbitrary dimensions n.

For convenience, let $(x, y) \in \mathbb{R}^2$ denote the coordinates in $\mathbb{R}^2$.

Let $f : (\mathbb{R}^2, 0) \to (\mathbb{R}, 0)$ be a Morse germ of index k represented by a smooth function $f : U \to \mathbb{R}$ defined in an open convex neighbourhood U of $0 \in \mathbb{R}^2$, for example an open circular disc with centre $0 \in \mathbb{R}^2$.

Since $f(0) = 0$, we can choose smooth functions $g_1, g_2 : U \to \mathbb{R}$, by Lemma 8.4.6, such that

$$f(x, y) = g_1(x, y) \cdot x + g_2(x, y) \cdot y$$

for all $(x, y) \in U$.

It is easy to see that

$$g_1(0, 0) = \frac{\partial f}{\partial x}(0, 0) = 0 \quad \text{and} \quad g_2(0, 0) = \frac{\partial f}{\partial y}(0, 0) = 0,$$

since $(0, 0)$ is a singular point for f, and hence we can factorize further by Lemma 8.4.6 and write $f(x, y)$ in the form

$$f(x, y) = a(x, y) \cdot x^2 + b(x, y) \cdot xy + c(x, y) \cdot y^2,$$

where $a, b, c : U \to \mathbb{R}$ are smooth functions.

Now observe, that the determinant

$$\begin{vmatrix} 2a(0,0) & b(0,0) \\ b(0,0) & 2c(0,0) \end{vmatrix} = \begin{vmatrix} \frac{\partial^2 f}{\partial x^2}(0,0) & \frac{\partial^2 f}{\partial y \partial x}(0,0) \\ \frac{\partial^2 f}{\partial x \partial y}(0,0) & \frac{\partial^2 f}{\partial y^2}(0,0) \end{vmatrix} \neq 0,$$

because f is a Morse germ.

Since this determinant is nonzero, either $a(0,0) \neq 0$ or $b(0,0) \neq 0$.

(i) First assume that $a(0,0) \neq 0$.

We only have to worry about the germ of f, and hence we are free to shrink the neighbourhood U of $0 \in \mathbb{R}^2$ as we see fit. Since $a : U \to \mathbb{R}$ is continuous we can then assume that $a(x,y) \neq 0$, and with the same sign as $a(0,0)$ in all points $(x, y) \in U$. Let $\varepsilon(a) = \pm 1$ denote the sign of $a(0,0)$.

Then we can perform the following rewriting of f:

$$f(x,y) = \varepsilon(a) \cdot \left(\sqrt{|a(x,y)|} \cdot x + \frac{\varepsilon(a) \cdot b(x,y)}{2\sqrt{|a(x,y)|}} \cdot y \right)^2$$

$$+ \left(c(x,y) - \frac{b(x,y)^2}{4a(x,y)} \right) y^2.$$

For convenience we put

$$d(x,y) = c(x,y) - \frac{b(x,y)^2}{4a(x,y)} \quad \text{for} \quad (x,y) \in U.$$

The determinant $4a(0,0) \cdot c(0,0) - b(0,0)^2 \neq 0$, and hence $d(0,0) \neq 0$. Since $d : U \to \mathbb{R}$ is continuous, we can therefore assume that $d(x,y) \neq 0$, and with the same sign $\varepsilon(d) = \pm 1$ as $d(0,0)$, in all points $(x,y) \in U$.

Motivated by the above rewriting of f, we put

$$x_1 = \sqrt{|a(x,y)|} \cdot x + \frac{\varepsilon(a) \cdot b(x,y)}{2\sqrt{|a(x,y)|}} \cdot y$$

$$x_2 = \sqrt{|d(x,y)|} \cdot y.$$

Define the smooth germ $\Psi : (\mathbb{R}^2, 0) \to (\mathbb{R}^2, 0)$ by $\Psi(x,y) = (x_1, x_2)$.

The determinant of the Jacobian matrix

$$\begin{vmatrix} \frac{\partial x_1}{\partial x}(0,0) & \frac{\partial x_1}{\partial y}(0,0) \\ \frac{\partial x_2}{\partial x}(0,0) & \frac{\partial x_2}{\partial y}(0,0) \end{vmatrix} = \sqrt{|a(0,0)|} \cdot \sqrt{|d(0,0)|} \neq 0,$$

and hence by Theorem 8.1.2, the germ Ψ is the germ of a diffeomorphism.

Let $\Phi = \Psi^{-1}$ be the inverse germ.

Since $\Phi(x_1, x_2) = (x, y)$ we get immediately

$$f \circ \Phi(x_1, x_2) = \varepsilon(a)x_1^2 + \varepsilon(d)x_2^2 \ .$$

Possibly after an interchange of x_1 and x_2, this shows exactly that f is right equivalent to the Morse germ f_k.

(ii) Next assume that $a(0,0) = 0$, $c(0,0) = 0$, but $b(0,0) \neq 0$.

By performing the change of coordinates

$$(x, y) = (\bar{x} + \bar{y}, \bar{x} - \bar{y})$$

we get

$$f(\bar{x}, \bar{y}) = \bar{a}(\bar{x}, \bar{y}) \cdot \bar{x}^2 + \bar{b}(\bar{x}, \bar{y}) \cdot \bar{x}\bar{y} + \bar{c}(\bar{x}, \bar{y}) \cdot \bar{y}^2 \ ,$$

where $\bar{a}(0,0) \neq 0$.

Then we are back to case (i).

(iii) Finally assume, that $a(0,0) = 0$, but $c(0,0) \neq 0$.

Just by interchanging the coordinates x and y we immediately get back to case (i).

Thereby we have considered all possible cases, and we have completed the proof that f is right equivalent to the Morse germ f_k. $\square$

8.5 Whitney C^k-topology on function spaces

In singularity theory one often uses the phrase: "for 'almost all' functions in a function space ... " . This refers to a topology on the function space in question. In this section we shall illustrate these topologies by considering the function space of smooth functions of $\mathbb{R}$ into $\mathbb{R}$.

Let $C^\infty(\mathbb{R}, \mathbb{R})$ denote the set of smooth functions $f : \mathbb{R} \to \mathbb{R}$, and let k be a fixed non-negative integer, i.e. $k \in \mathbb{N}_0 = \mathbb{N} \cup \{0\}$.

We start out by considering a topology on $C^\infty(\mathbb{R}, \mathbb{R})$, which is metrizable if we allow an extended notion of metrics, taking values in *the extended real numbers* $\mathbb{R}^* = \mathbb{R} \cup \{-\infty, +\infty\}$.

For $f, g \in C^\infty(\mathbb{R}, \mathbb{R})$ we define

$$d^k(f, g) = \sup_{x \in \mathbb{R}} \left(\sum_{i=0}^{k} |f^{(i)}(x) - g^{(i)}(x)| \right)$$

$$= \sup_{x \in \mathbb{R}} \left(|f(x) - g(x)| + \cdots + |f^{(k)}(x) - g^{(k)}(x)| \right) \ .$$

Note that $d^k(f, g) \in \mathbb{R}^*$, since the supremum can assume the value $+\infty$ in the case considered.

It is easy to prove that for all $f, g, h \in C^\infty(\mathbb{R}, \mathbb{R})$ it holds that

MET 1 $d^k(f, g) \geq 0$, $d^k(f, g) = 0 \iff f = g$.

MET 2 $d^k(f, g) = d^k(g, f)$.

MET 3 $d^k(f, g) \leq d^k(f, h) + d^k(h, g)$.

This shows that $(C^\infty(\mathbb{R}, \mathbb{R}), d^k)$ is a *metric space in the extended sense*. Compared to the definition of a metric space in Definition 2.2.1, the only difference is that we allow the metric to assume the value $+\infty$. This difference will cause no serious problems.

Define the open ball with centre $f \in C^\infty(\mathbb{R}, \mathbb{R})$ and radius $r \in \mathbb{R}^+$ by

$$B_r(f) = \{g \in C^\infty(\mathbb{R}, \mathbb{R}) \mid d^k(f, g) < r\} .$$

Next we define a subset $U \subseteq C^\infty(\mathbb{R}, \mathbb{R})$ to be *open* if for every function $f \in U$ there exists an $r \in \mathbb{R}^+$ such that $B_r(f) \subseteq U$. The open sets in $C^\infty(\mathbb{R}, \mathbb{R})$ defined this way, form a collection of subsets $\mathcal{T}$ that satisfies the conditions TOP 1, TOP 2 and TOP 3 for a topology on $C^\infty(\mathbb{R}, \mathbb{R})$. This topology is called *the uniform C^k-topology* on $C^\infty(\mathbb{R}, \mathbb{R})$.

In an arbitrary topological space we can define the notion of convergence of sequences in the space. How does this notion work in $C^\infty(\mathbb{R}, \mathbb{R})$?

Let (f_n) be a sequence in $C^\infty(\mathbb{R}, \mathbb{R})$ and let $f \in C^\infty(\mathbb{R}, \mathbb{R})$. Then it is easy to prove that $f_n \to f$ for $n \to \infty$ in the uniform C^k-topology on $C^\infty(\mathbb{R}, \mathbb{R})$ if and only if

$$\forall \varepsilon > 0 \; \exists n_0 \in \mathbb{N} \; \forall n \in \mathbb{N} :$$
$$n \geq n_0 \implies \forall x \in \mathbb{R} : \sum_{i=0}^{k} |f^{(i)}(x) - f_n^{(i)}| < \varepsilon .$$

Convergence of a sequence (f_n) to f in the uniform C^k-topology on $C^\infty(\mathbb{R}, \mathbb{R})$ can in other words be described as uniform convergence of f_n, together with the first k derivatives of f_n, to f, and the corresponding derivatives of f.

In the uniform C^k-topology on $C^\infty(\mathbb{R}, \mathbb{R})$, uniform attention is paid to the function values at all points $x \in \mathbb{R}$. In many situations, however, one wishes to put stronger conditions (have more control) on the function values at '∞'. For that purpose, the Whitney C^k-topology on $C^\infty(\mathbb{R}, \mathbb{R})$, which we are about to introduce, has turned out to be convenient.

For an arbitrary continuous function $\varepsilon : \mathbb{R} \to \mathbb{R}^+$ with values in the positive real numbers, we put

$$U(\varepsilon, k) = \left\{ g \in C^\infty(\mathbb{R}, \mathbb{R}) \mid \sum_{i=0}^{k} |g^{(i)}(x)| < \varepsilon(x) \, , \, x \in \mathbb{R} \right\} .$$

Next for $f \in C^\infty(\mathbb{R}, \mathbb{R})$ we put

$$U(f; \varepsilon, k) = \left\{ g \in C^\infty(\mathbb{R}, \mathbb{R}) \mid (g - f) \in U(\varepsilon, k) \right\} .$$

Let $\varepsilon_1 : \mathbb{R} \to \mathbb{R}^+$ and $\varepsilon_2 : \mathbb{R} \to \mathbb{R}^+$ be continuous functions and put $\varepsilon = \min\{\varepsilon_1, \varepsilon_2\}$, i.e. $\varepsilon(x) = \min\{\varepsilon_1(x), \varepsilon_2(x)\}$ for $x \in \mathbb{R}$. It is not difficult to prove that $\varepsilon : \mathbb{R} \to \mathbb{R}^+$ is continuous and that for every $f \in C^\infty(\mathbb{R}, \mathbb{R})$ it holds that

$$U(f; \varepsilon_1, k) \cap U(f; \varepsilon_2, k) = U(f; \varepsilon, k) .$$

In $C^\infty(\mathbb{R}, \mathbb{R})$ consider the system of subsets $\mathcal{T} = W_k$ consisting of the subsets $U \subseteq C^\infty(\mathbb{R}, \mathbb{R})$ with the property that for each function $f \in U$ there exists a continuous function $\varepsilon : \mathbb{R} \to \mathbb{R}^+$ such that $U(f; \varepsilon, k) \subseteq U$.

Using the above property of the sets $U(f; \varepsilon, k)$, it is an easy exercise to prove that the system of subsets W_k in $C^\infty(\mathbb{R}, \mathbb{R})$ satisfies the conditions TOP 1, TOP 2 and TOP 3. In other words, W_k is a topology on $C^\infty(\mathbb{R}, \mathbb{R})$. This topology is called the *Whitney C^k-topology* on $C^\infty(\mathbb{R}, \mathbb{R})$.

It is easy to prove that the topologies W_k, for $k = 0, 1, 2, \ldots$, are related by the following tower of inclusions

$$W_0 \subseteq W_1 \subseteq \cdots \subseteq W_k \subseteq \cdots .$$

The union

$$W = \bigcup_{k=0}^{\infty} W_k$$

is again a system of subsets in $C^\infty(\mathbb{R}, \mathbb{R})$ that satisfies the requirements for a topology. This topology is called the *Whitney C^∞-topology* on $C^\infty(\mathbb{R}, \mathbb{R})$. Usually it is this topology which appears in statements in singularity theory like: "for 'almost all' functions ..." .

The Whitney C^k-topology contains enormously many open sets (the sets in W_k). In fact, already the Whitney C^0-topology contains so many open sets that it does not satisfy the first axiom of countability (Definition 2.5.8). In contrast, it is easy to see that a metric space (also in the extended sense) satisfies the first axiom of countability.

Theorem 8.5.1. *The Whitney C^0-topology on $C^\infty(\mathbb{R}, \mathbb{R})$ does not satisfy the first axiom of countability. In particular, it is not a metrizable topology.*

Proof. The proof is an indirect proof. With that in mind, we assume therefore that $\{B_1, B_2, \ldots, B_l, \ldots\}$ is a countable basis of open neighbourhoods of $f \in C^\infty(\mathbb{R}, \mathbb{R})$ in the Whitney C^0-topology. For every $l \in \mathbb{N}$, choose a continuous function $\varepsilon_l : \mathbb{R} \to \mathbb{R}^+$ such that $U(f; \varepsilon_l, 0) \subseteq B_l$. Furthermore, let $x_1, x_2, \ldots, x_l, \ldots$ be a sequence in $\mathbb{R}$ without any accumulation points, for example $1, 2, \ldots, l, \ldots$. Next construct a continuous function $\varepsilon : \mathbb{R} \to \mathbb{R}^+$ such that $\varepsilon(x_l) < \varepsilon_l(x_l)$ for all $l \in \mathbb{N}$; cf. Figure 8.1. (We leave this as an exercise for the reader!)

$\{B_1, B_2, \ldots, B_l, \ldots\}$ is a basis of open neighbourhoods of f in $C^\infty(\mathbb{R}, \mathbb{R})$ with the Whitney C^0-topology, and hence there exists an $m \in \mathbb{N}$ such that $B_m \subseteq U(f; \varepsilon, 0)$. But then $U(f; \varepsilon_m, 0) \subseteq U(f; \varepsilon, 0)$. This leads to a contradiction since it is possible to find a smooth function $g : \mathbb{R} \to \mathbb{R}$ where $|g(x) - f(x)| < \varepsilon_m(x)$ for all $x \in \mathbb{R}$, but $|g(x_m) - f(x_m)| > \varepsilon(x_m)$. For such a function $g \in C^\infty(\mathbb{R}, \mathbb{R})$, clearly $g \in U(f; \varepsilon_m, 0)$, whereas $g \notin U(f; \varepsilon, 0)$. This is a contradiction. Hence we conclude that the Whitney C^0-topology on $C^\infty(\mathbb{R}, \mathbb{R})$ does not satisfy the first axiom of countability. $\square$

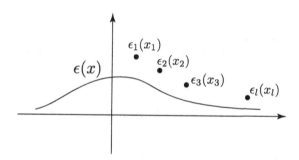

Fig. 8.1 The Whitney C^0-topology on $C^\infty(\mathbb{R}, \mathbb{R})$ is not metrizable (Theorem 8.5.1).

Without proof we now describe how the Whitney C^k-topology on $C^\infty(\mathbb{R}, \mathbb{R})$ works in connection with convergence of sequences.

A sequence (f_n) in $C^\infty(\mathbb{R}, \mathbb{R})$ converges to the function $f \in C^\infty(\mathbb{R}, \mathbb{R})$ in the Whitney C^k-topology on $C^\infty(\mathbb{R}, \mathbb{R})$ if and only if there exists a closed and bounded interval $[a, b] \subseteq \mathbb{R}$, which depends both on the sequence (f_n) and the function f, such that

(i) $f_n \to f$ for $n \to \infty$ uniformly on $[a, b]$ with respect to d^k, i.e.

$\forall \varepsilon > 0 \; \exists n_0 \in \mathbb{N} \; \forall n \in \mathbb{N} :$

$$n \geq n_0 \implies \forall x \in [a, b] : \sum_{i=0}^{k} |f^{(i)}(x) - f_n^{(i)}(x)| < \varepsilon .$$

(ii) $f_n = f$ on $\mathbb{R} \setminus [a, b]$, except for at most finitely many n.

In a completely analogous manner as for $C^\infty(\mathbb{R}, \mathbb{R})$, it is possible to describe a Whitney C^k-topology on the space of smooth mappings $C^\infty(M, \mathbb{R}^m)$ from an open set $M \subseteq \mathbb{R}^n$ into $\mathbb{R}^m$. In a slightly more technical manner, but using the same ideas, it is also possible to define a Whitney C^k-topology on the space $C^\infty(M^n, N^m)$ of smooth mappings of an n-dimensional manifold M^n into an m-dimensional manifold N^m.

Finally in this section, we add a few remarks on Baire spaces, a notion from general topology, which is of particular importance in the context of dynamical systems and singularity theory. We have already defined the notion of a dense subset in Definition 2.8.8, but repeat it here.

Definition 8.5.2. Let $(S, \mathcal{T})$ be a topological space.

(1) A subset W in S is said to be *dense* in S if $\overline{W} = S$.

(2) A subset W in S is said to be *residual*, or a *Baire set*, if it is a countable intersection of open and dense subsets in S.

(3) The topological space $(S, \mathcal{T})$ is said to be a *Baire space* if every residual subset in S is dense in S.

The notion of a Baire space emerged from work of the French mathematician René-Louis Baire (1874–1932).

There are a number of conditions under which a topological space is a Baire space. The most important result is the following theorem, which we state without proof.

Theorem 8.5.3. *If S is a compact Hausdorff space, or a complete metric space, then S is a Baire space.*

Example 8.5.4. As a special case of Theorem 8.5.3, the sets of real numbers $\mathbb{R}$ equipped with the usual topology is a Baire space, since by Theorem 3.5.5 it is a complete metric space.

Let $\mathbb{Q}$ and $\mathbb{I}$ denote the sets of rational and irrational numbers, respectively. The set of rational numbers is countable, and let $q_1, q_2, \ldots, q_n, \ldots$ be a list of all the rational numbers. Define for each $n \in \mathbb{N}$, the set $W_n = \mathbb{R} \setminus \{q_n\}$. Clearly, W_n is open and dense in $\mathbb{R}$. Since

$$\mathbb{I} = \bigcap_{n=1}^\infty W_n \,,$$

the set of irrational numbers $\mathbb{I}$ is a residual subset in $\mathbb{R}$. The set of rational numbers $\mathbb{Q}$ is also dense in $\mathbb{R}$, but it is not a residual subset in $\mathbb{R}$. Assume for a moment that $\mathbb{Q}$ is a residual subset in $\mathbb{R}$. Then the intersection $\mathbb{Q} \cap \mathbb{I} = \emptyset$ would also be a residual subset in $\mathbb{R}$, in particular a dense subset, since $\mathbb{R}$ is a Baire space, which is obviously not the case. ◀

In Chapter 2, page 39, we have defined a property of points in a topological space to be generic if it is valid for a dense subset of the points in the topological space. When dealing with points in Baire spaces it is often convenient to require a little more and to say that a property of points in the space is *generic* if it holds for the points in a residual subset of the Baire space. With this definition, we have seen in Example 8.5.4 that the property of being irrational is generic in the set of real numbers, whereas the property of being rational is not.

The following theorem, which we shall not prove here, is useful in singularity theory and in the theory of dynamical systems in connection with attempts to single out classes of objects that can be classified and still can be described as 'typical' objects for the situation. We state the theorem in a very general form.

Theorem 8.5.5. *A function space of smooth mappings equipped with the Whitney C^∞-topology is a Baire space.*

A residual subset in a function space of smooth mappings with the Whitney C^∞-topology is in particular dense in the function space. It is, however, 'denser' than just a dense subset, as we have seen in the case of the subsets of rational numbers and irrational numbers in the set of real numbers. When one uses the phrase "for 'almost all' functions in a function space" it relates to a residual subset of functions. When the function space is equipped with the Whitney C^∞-topology this implies in particular that an arbitrary smooth function can be approximated as closely as we wish with a smooth function from the residual subset, i.e. with a *generic* function.

8.6 How to prove results about genericity?

In this section we shall give three examples of how questions about the existence and the type of singular points for smooth mappings can be reformulated as geometrical (topological) questions. The geometrical questions take the form of how so-called jet-prolongations naturally associated to the mappings avoid certain 'bad' subsets in their image spaces. It is a generic property of the original mapping that the jet-prolongations avoid the 'bad' subsets in their image spaces, and hence one can in many situations obtain results about genericity of mappings with singularities of special type, by making use of the general notion of transversality of mappings.

8.6.1   *The space of smooth mappings of* $\mathbb{R}$ *into* $\mathbb{R}$

Let $f : \mathbb{R} \to \mathbb{R}$ be a smooth function. A point $p \in \mathbb{R}$ is a singular point of f if $f'(p) = 0$.

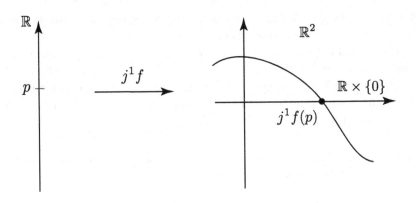

Fig. 8.2   The 1-jet-prolongation $j^1 f : \mathbb{R} \to \mathbb{R}^2$ of a smooth function $f : \mathbb{R} \to \mathbb{R}$.

Consider the smooth mapping

$$j^1 f = (f, f') : \mathbb{R} \to \mathbb{R}^2 .$$

The mapping $j^1 f$ is called the *1-jet-prolongation* of f. Note that $p \in \mathbb{R}$ is a singular point of f exactly when $j^1 f(p) \in \mathbb{R} \times \{0\} \subset \mathbb{R}^2$; cf. Figure 8.2.

Geometrically it is pretty clear that in general one cannot get rid of a singular point by a small perturbation of f in the Whitney C^1-topology.

Next consider the smooth mapping

$$j^2 f = (f, f', f'') : \mathbb{R} \to \mathbb{R}^3 .$$

The mapping $j^2 f$ is called the *2-jet-prolongation* of f. Obviously, both of the two derivatives $f'(p) = f''(p) = 0$ at a point $p \in \mathbb{R}$ if and only if $j^2 f(p) \in \mathbb{R} \times \{0\} \times \{0\} \subset \mathbb{R}^3$; cf. Figure 8.3.

Geometrically, $j^2 f$ describes a curve in the 3-dimensional space $\mathbb{R}^3$. The 1-dimensional curve $j^2 f$, will not 'typically' intersect the 1-dimensional axis

$\mathbb{R} \times \{0\} \times \{0\}$. Then it is quite clear that by an arbitrary small perturbation of f in the Whitney C^2-topology it is possible to push $j^2 f$ off $\mathbb{R} \times \{0\} \times \{0\}$. Thereby we get the following theorem.

Theorem 8.6.1. *There is an open and dense subset $\mathcal{F}$ of functions in $C^\infty(\mathbb{R}, \mathbb{R})$ equipped with the Whitney C^2-topology, such that every function $f \in \mathcal{F}$ satisfies that if $f'(p) = 0$ in a point $p \in \mathbb{R}$, then $f''(p) \neq 0$.*

Furthermore, the germs of the functions in $\mathcal{F}$ have the following normal forms:

(i) *If $f'(p) \neq 0$, then the germ of $f : \mathbb{R} \to \mathbb{R}$ at $p \in \mathbb{R}$ is equivalent to the germ $f_1 : (\mathbb{R}, 0) \to (\mathbb{R}, 0)$ defined by $f_1(x) = x$.*

(ii) *If $f'(p) = 0$ and $f''(p) \neq 0$, then the germ of $f : \mathbb{R} \to \mathbb{R}$ at $p \in \mathbb{R}$ is equivalent to the germ $f_2 : (\mathbb{R}, 0) \to (\mathbb{R}, 0)$ defined by $f_2(x) = x^2$.*

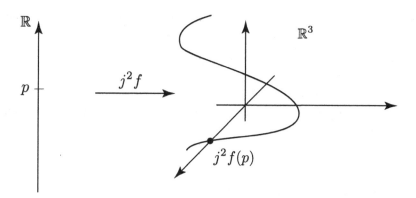

Fig. 8.3   The 2-jet-prolongation $j^2 f : \mathbb{R} \to \mathbb{R}^3$ of a smooth function $f : \mathbb{R} \to \mathbb{R}$.

The smooth functions in $\mathcal{F}$ are said to be generic in $C^\infty(\mathbb{R}, \mathbb{R})$. That we get the normal forms (i) and (ii) in Theorem 8.6.1 is a direct consequence of the classification of smooth germs $(\mathbb{R}, 0) \to (\mathbb{R}, 0)$ of finite order given by Theorem 8.3.2.

As an illustration of Theorem 8.6.1, one can think of the 'practical' task consisting in placing a narrow runner in a long hall. If we just roll out the runner quite arbitrarily, there will genericly only be smooth points (corresponding to $x \mapsto x$) or fold points (corresponding to $x \mapsto x^2$).

8.6.2 *Spaces of immersions*

Let $f = (f_1, f_2) : \mathbb{R} \to \mathbb{R}^2$ be a smooth mapping.

In agreement with the earlier Definition 7.8.1 of an immersion, the mapping f is called an *immersion*, or, a *regular curve*, if $f'(x) = (f_1'(x), f_2'(x)) \neq 0$ for all $x \in \mathbb{R}$.

In order to investigate whether $f : \mathbb{R} \to \mathbb{R}^2$ is an immersion or not, we consider the smooth mapping

$$j^1 = (f, f') : \mathbb{R} \to \mathbb{R}^2 \times \mathbb{R}^2 \ .$$

Again the mapping $j^1 f$ is called the 1-jet-prolongation of f. Note that $f'(x) \neq 0$ exactly when $j^1 f(x) \notin \mathbb{R}^2 \times \{0\}$.

Since $j^1 f$ describes a 1-dimensional curve in the 4-dimensional space $\mathbb{R}^4 = \mathbb{R}^2 \times \mathbb{R}^2$, it will 'almost always' avoid the 2-dimensional plane $\mathbb{R}^2 \times \{0\}$. Geometrically, it is then clear that the following theorem must be true.

Theorem 8.6.2. *The set of immersions* $\mathrm{Imm}(\mathbb{R}, \mathbb{R}^2)$ *is an open and dense subset in* $C^\infty(\mathbb{R}, \mathbb{R}^2)$ *equipped with the Whitney C^1-topology.*

This theorem is a special case of an immersion theorem due to Whitney for smooth mappings of an n-dimensional smooth manifold M^n into an m-dimensional smooth manifold N^m. The Whitney Immersion Theorem states that for $2n \leq m$, the space of immersions $\mathrm{Imm}(M^n, N^m)$ is an open and dense subset in $C^\infty(M^n, N^m)$ equipped with the Whitney C^1-topology.

8.6.3 *A space of Morse functions*

In this final example we shall consider Morse functions in two variables.

A smooth function $f : \mathbb{R}^2 \to \mathbb{R}$ in two variables $(x, y) \in \mathbb{R}^2$ is called a *Morse function* if all critical points of f are non-degenerate. Here a point $p \in \mathbb{R}^2$ is called a *critical point*, or a *singular point*, of f if

$$\frac{\partial f}{\partial x}(p) = \frac{\partial f}{\partial y}(p) = 0 \ ,$$

and a critical point $p \in \mathbb{R}^2$ of f is said to be *non-degenerate* if the determinant

$$\begin{vmatrix} \dfrac{\partial^2 f}{\partial x^2}(p) & \dfrac{\partial^2 f}{\partial y \partial x}(p) \\[2mm] \dfrac{\partial^2 f}{\partial x \partial y}(p) & \dfrac{\partial^2 f}{\partial y^2}(p) \end{vmatrix} \neq 0 \ .$$

Together with the smooth function $f : \mathbb{R}^2 \to \mathbb{R}$, we shall consider the 2-jet-prolongation of f,

$$j^2 f : \mathbb{R}^2 \to \mathbb{R} \times \mathbb{R}^2 \times \mathbb{R}^3 ,$$

with the coordinate functions

$$j^2 f = \left(f, \frac{\partial f}{\partial x}, \frac{\partial f}{\partial y}, \frac{\partial^2 f}{\partial x^2}, \frac{\partial^2 f}{\partial x \partial y}, \frac{\partial^2 f}{\partial y^2} \right) .$$

Let $(a, b, c) \in \mathbb{R}^3$ denote the coordinates in $\mathbb{R}^3$, and consider the conical surface of second order T in $\mathbb{R}^3$ given by

$$T = \left\{ (a, b, c) \in \mathbb{R}^3 \ \middle| \ \begin{vmatrix} a & b \\ b & c \end{vmatrix} = ac - b^2 = 0 \right\} .$$

The condition to be satisfied in order that $f : \mathbb{R}^2 \to \mathbb{R}$ is a Morse function is now that $j^2 f$ avoids the set

$$S = \mathbb{R} \times \{0\} \times \{0\} \times T \subset \mathbb{R} \times \mathbb{R}^2 \times \mathbb{R}^3 = \mathbb{R}^6 .$$

The set S is a '3-dimensional' subset in $\mathbb{R}^6$ and thus has 3 complementary dimensions. Then it is quite clear that by an arbitrary small perturbation of f in the Whitney C^2-topology we can push $j^2 f$ off the set S. (The proper tool to use is Thom's Transversality Theorem; cf. Remark 7.9.6, page 223.) Thereby we have the following theorem.

Theorem 8.6.3. *The set of Morse functions is an open and dense subset in $C^\infty(\mathbb{R}^2, \mathbb{R})$ equipped with the Whitney C^2-topology.*

At a critical point of a Morse function $f : \mathbb{R}^2 \to \mathbb{R}$, the germ of f is equivalent to exactly one of the Morse germs $f_0, f_1, f_2 : (\mathbb{R}^2, 0) \to (\mathbb{R}, 0)$, where $f_0(x, y) = x^2 + y^2$, $f_1(x, y) = -x^2 + y^2$ and $f_2(x, y) = -x^2 - y^2$; cf. Theorem 8.4.5.

Exercises and Further Results

Exercise 8.1. Let $F : (\mathbb{R}^n, x_0) \to (\mathbb{R}^m, y_0)$ and $G : (\mathbb{R}^m, y_0) \to (\mathbb{R}^k, z_0)$ be smooth germs, where the germ of G is taken at the point $y_0 = F(x_0)$.

1) Argue that we can define a *composed germ* $G \circ F : (\mathbb{R}^n, x_0) \to (\mathbb{R}^k, z_0)$ without any ambiguities.

2) Argue that if a germ $F : (\mathbb{R}^n, x_0) \to (\mathbb{R}^n, y_0)$ has a smooth representative $F : U \to V$ that maps an open neighbourhood U of $x_0 \in \mathbb{R}^n$ diffeomorphically onto an open neighbourhood V of $y_0 \in \mathbb{R}^n$, then we can, without any ambiguities, define an *inverse germ* $F^{-1} : (\mathbb{R}^n, y_0) \to (\mathbb{R}^n, x_0)$.

Exercise 8.2. Let L be an $m \times n$ matrix written as a block matrix

$$L = \begin{bmatrix} A & B \\ C & D \end{bmatrix},$$

where A is an invertible quadratic $k \times k$ matrix.
Prove that L has rank k if and only if $D = C \cdot A^{-1} \cdot B$.

Exercise 8.3. Let $F_0, F_1 : (\mathbb{R}^n, 0) \to (\mathbb{R}^m, 0)$ be equivalent smooth germs. Prove that either both of them are regular, or both of them are singular.

Exercise 8.4. Let $f : (\mathbb{R}^n, 0) \to (\mathbb{R}, 0)$ be a regular smooth germ. Prove in detail that f is right-equivalent to the germ $p_1 : (\mathbb{R}^n, 0) \to (\mathbb{R}, 0)$, represented by the projection of $\mathbb{R}^n$ onto the first coordinate axis.

Exercise 8.5. Prove that two smooth germs $F, G : (\mathbb{R}^n, 0) \to (\mathbb{R}^m, 0)$ are equivalent smooth germs if and only if they have the same rank.

Hint: Find a normal form for the smooth germs $F : (\mathbb{R}^n, 0) \to (\mathbb{R}^m, 0)$ of rank $rk_0 F = k$ for any $0 \leq k \leq \min\{n, m\}$.

Exercise 8.6. Let $f, g : (\mathbb{R}, 0) \to (\mathbb{R}, 0)$ be smooth germs of finite order. Prove that f and g are equivalent if and only if they have the same order.

Exercise 8.7. Let $f : I \to \mathbb{R}$ be an arbitrary smooth function defined in an open interval I about $0 \in \mathbb{R}$. Prove that for any integer $n \geq 1$ there exists a smooth function $g : I \to \mathbb{R}$ such that

$$f(x) = f(0) + f'(0)x + \frac{1}{2!}f^{(2)}(0)x^2 + \cdots + \frac{1}{n!}f^{(n)}(0)x^n + g(x)x^{n+1}.$$

Exercise 8.8. Let $U = U(x, \lambda)$ be the potential function for a mechanical system described by a real parameter $x \in \mathbb{R}$, and which depends on an adjustable parameter $\lambda \in \mathbb{R}$.

The equilibrium points for the system is the set of points

$$S = \{(x, \lambda) \in \mathbb{R}^2 \mid \frac{\partial U}{\partial x}(x, \lambda) = \frac{\partial U}{\partial \lambda}(x, \lambda) = 0\}.$$

1) Determine the set S of equilibrium points for the system described by the potential function

$$U = U(x, \lambda) = x^3 - \lambda x.$$

2) Draw the set S of equilibrium points for the potential $U = U(x, \lambda)$ in a plane $\mathbb{R}^2$ with a (λ, x)-coordinate system.

3) Now consider the potential function
$$V = V(x, \lambda) = \frac{x^2}{2} + 2\lambda \sin x.$$
Show that the germs $V : (\mathbb{R}^2, 0) \to (\mathbb{R}, 0)$ and $U : (\mathbb{R}^2, 0) \to (\mathbb{R}, 0)$ are equivalent Morse germs.

The smooth germ $U : (\mathbb{R}^2, 0) \to (\mathbb{R}, 0)$ defined by the potential function $U = U(x, \lambda)$ is, for good reasons, known as the *pitchfork bifurcation* in Bifurcation Theory.

4) Try to design a mechanical system that has the potential function $U = U(x, \lambda)$, or, maybe rather $V = V(x, \lambda)$.

<u>Idea:</u> A planar pendulum driven by an adjustable torque force might work.

Exercise 8.9. Let $U \subseteq \mathbb{R}^n$ be an open set in $\mathbb{R}^n$, and let $f : U \to \mathbb{R}^n$ be a differentiable mapping of class C^1. (One may think of f as a vector field in U with the vector $f(x) \in \mathbb{R}^n$ as the field vector at the point $x \in U$.) A zero of f, i.e. a point $x_0 \in U$, such that $f(x_0) = 0$, is called *non-degenerate* if the differential $Df(x_0) : \mathbb{R}^n \to \mathbb{R}^n$ of f at x_0 is an isomorphism.

Prove that f has at most finitely many zeros in any compact subset $K \subseteq U$, if we assume that all zeros of f are non-degenerate.

Exercise 8.10. Let $g = g(x) : \mathbb{R} \to \mathbb{R}$ be a polynomial function of degree $d \geq 2$, and define the smooth function $f = f(x, y) : \mathbb{R}^2 \to \mathbb{R}$ in two variables by $f(x, y) = y^2 - g(x)$.

Let $\mathcal{C} = \{(x, y) \in \mathbb{R}^2 \mid f(x, y) = 0\}$ be the *algebraic curve* defined by the set of zeros of f in $\mathbb{R}^2$.

The set S of singular points of f on $\mathcal{C}$ is called the *singular set* for $\mathcal{C}$.

1) Prove that $S = \{(x, 0) \in \mathbb{R}^2 \mid g \text{ has a multiple root at } x\}$.

2) Draw the algebraic curve $\mathcal{C}$ for each of the polynomials $g(x) = x^2 - 1$ and $g(x) = x^2 \cdot (1 - x^2)^3$.

Exercise 8.11. Consider $\mathbb{R}^2$ with coordinates $(u, x) \in \mathbb{R}^2$.

1) Determine the rank of the mapping $F : \mathbb{R}^2 \to \mathbb{R}^2$ at all points $(u, x) \in \mathbb{R}^2$, when F is defined by (i) $F(u, x) = (u, x)$; (ii) $F(u, x) = (u, x^2)$; (iii) $F(u, x) = (u, ux - x^3)$, respectively.

2) In the cases (ii) and (iii) we set
$$K = \{(u, x) \in \mathbb{R}^2 \mid \mathrm{rk}_{(u,x)} F = 1\} .$$
Show that K is a smooth curve in both cases. Find a simple parametrization $\alpha : \mathbb{R} \to \mathbb{R}^2$ of K, and determine the rank of $F \circ \alpha$ corresponding to all parameter values.

3) Show that the germs $(\mathbb{R}^2, 0) \to (\mathbb{R}^2, 0)$ defined by the above three smooth mappings are not pairwise equivalent.

Hint: Make use, for example, of a geometrical interpretation of the germs.

Exercise 8.12. Show that $f : (\mathbb{R}^2, 0) \to (\mathbb{R}, 0)$ defined by $f(x, y) = x^2 - xy$ is a Morse germ.

Determine the index of f, and provide a change of coordinates that will bring f into normal form.

Exercise 8.13. For every $l \in \mathbb{N}$, let $\varepsilon_l : \mathbb{R} \to \mathbb{R}^+$ be a continuous function. Furthermore, let $x_1, x_2, \ldots, x_l, \ldots$ be a sequence in $\mathbb{R}$ without any accumulation points, for example $1, 2, \ldots, l, \ldots$. Construct a continuous function $\varepsilon : \mathbb{R} \to \mathbb{R}^+$ such that $\varepsilon(x_l) < \varepsilon_l(x_l)$ for all $l \in \mathbb{N}$.

Exercise 8.14. Let $C^\infty(\mathbb{R}, \mathbb{R})$ be the set of smooth functions $f : \mathbb{R} \to \mathbb{R}$ equipped with the Whitney C^∞-topology.

Let $\mathcal{F}$ be the set of smooth functions $f : \mathbb{R} \to \mathbb{R}$ for which at least one derivative $f^{(k)}(p) \neq 0$, $k \geq 1$, at every point $p \in \mathbb{R}$.

Argue that $\mathcal{F}$ is generic in the Whitney C^∞-topology on $C^\infty(\mathbb{R}, \mathbb{R})$ and that all functions $f \in \mathcal{F}$ have the property that the germ of f at any point $p \in \mathbb{R}$ is equivalent to a germ of finite order.

Exercise 8.15. Introduce (with proofs) the necessary ingredients to define the Whitney C^k-topology on the space $C^\infty(\mathbb{R}^n, \mathbb{R}^m)$ of smooth mappings of $\mathbb{R}^n$ into $\mathbb{R}^m$.

Chapter 9

An Introduction to Geometric Variational Problems

The shapes of nature are uncompromisingly effective, and a general study of optimal properties of geometric objects is therefore both interesting and important. Optimal geometry lies at the roots of the calculus of variations.

In this chapter we present some examples of geometric variational problems, starting with a few problems that can be solved by geometric reasoning alone and progressing to a problem that requires analysis in infinite dimensional spaces, and can be considered as an introduction to Morse theory in infinite dimension.

9.1 Fermat's principle for light propagation

Optics was studied already by the old Greeks and they knew for example the basic law of reflection of light: *The angle of incidence* (the angle of the incoming light ray to the normal of the reflecting surface) *is equal to the angle of reflection* (the angle of the outgoing light ray to the normal of the reflecting surface).

Around 1660, the French mathematician Pierre de Fermat (1601–1665) formulated light propagation as a variational principle:

Principle of Least Time. *Light follows the path requiring least time.*

In a homogeneous medium, the path requiring least time corresponds exactly to the *shortest path*.

It is easy to deduce the law of reflection from Fermat's principle of light propagation by purely geometrical reasoning. The argument goes like this.

Since the reflecting surface can be substituted by the tangent plane at the reflection point and since (by Fermat's principle) a light ray must follow a normal plane to the surface, it is sufficient to consider the situation in a plane where the reflecting surface is represented by a line tangent to the surface; cf. Figure 9.1. Now consider a light ray that follows the path from P to Q after reflection at the point R on the reflecting line. Let Q^* be the symmetrical point to Q with respect to the reflecting line. The shortest path from P to Q, and hence the path that the light ray must follow according to Fermat's principle, will then be the broken line that corresponds to the straight line from P to Q^*, i.e. such that the line segments RQ and RQ^* are symmetrical with respect to the reflecting line. Elementary geometry then shows that the angle i of incidence is equal to the angle r of reflection.

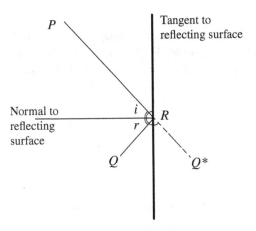

Fig. 9.1   The law of reflection: Angle of incidence $i =$ Angle of reflection r.

Fermat's principle is an archetypical example of how a physical theory can be fruitfully based on a variational principle and illustrates why physicists look for variational principles as the basis for physical theories. Thereby the calculus of variations becomes one of the important subjects of mathematical physics.

9.2   Triangles as optimal figures

Consider a triangle ABC in which the perimeter has the fixed length L and in which we keep the length g of the base AC fixed. We shall examine how to design this triangle so that it has the largest possible area.

The sum of the length of the sides AB and BC must be constant:

$$|AB| + |BC| = L - g.$$

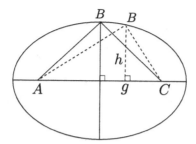

Fig. 9.2 Among all triangles with a given perimeter and one side kept fixed, the isosceles triangle over the fixed side has the largest area.

Points B with this property lie on an ellipse with the points A and C as foci; cf. Figure 9.2. Let the altitude of the triangle have length h. Then the area of the triangle is given by $\frac{1}{2}h \cdot g$. It is now obvious that the altitude h in the triangle is largest possible, whereby the area is largest possible, when B falls in the vertex of the ellipse. Hence the triangle with the largest area is the isosceles triangle in which the two sides AB and BC each have the length $\frac{1}{2}(L - g)$. Thereby we have proved the following theorem.

Theorem 9.2.1. *Among all triangles with a given base and a given perimeter the isosceles triangle has the largest area.*

We now try our hand with a triangle ABC in which we only keep the length L of the perimeter of the triangle fixed. How shall the triangle be designed, then, to obtain the largest possible area? This problem is known as *the isoperimetric problem* for triangles.

The problem can be tackled as follows. Start out with an arbitrary triangle with prescribed perimeter L. By an iterative process, we can construct a sequence of triangles with perimeter L and increasing area by making each triangle isosceles over one of the sides in turn. This (infinite) sequence of triangles in the limit approaches the equilateral triangle with perimeter L; cf. Figure 9.3. By continuity of the area of a triangle on its shape (Heron's formula) it follows that the equilateral triangle is the triangle enclosing the largest area among all triangles with the prescribed perimeter L.

Thereby we have proved the following theorem.

Theorem 9.2.2. *Among all triangles with a prescribed perimeter the equilateral triangle has the largest area.*

We can base an alternative proof of Theorem 9.2.2 on the method of

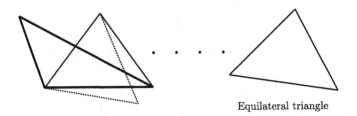

Equilateral triangle

Fig. 9.3   The equilateral triangle maximizes the area of isoperimetric triangles.

Lagrange multipliers described in Theorem 7.1.5.

Proof. Consider an arbitrary triangle ABC with sides a, b and c and perimeter L. Let s be half the perimeter, i.e. $L = 2s$.

According to Heron's formula, the area of the triangle is then given by

$$A = \sqrt{s(s-a)(s-b)(s-c)}.$$

We therefore have to maximize the function

$$f(a,b,c) = s(s-a)(s-b)(s-c)$$

given the constraint

$$g(a,b,c) = a + b + c - 2s = 0,$$

together with the restrictions $0 \le a \le s$, $0 \le b \le s$, $0 \le c \le s$ on the variables a, b, c.

In this case, the vector equation for Lagrange multipliers has the form $\nabla f(a,b,c) = \lambda \nabla g(a,b,c)$, which leads to the scalar equations

$$-s(s-b)(s-c) = \lambda$$
$$-s(s-a)(s-c) = \lambda$$
$$-s(s-a)(s-b) = \lambda,$$

to be solved together with the equation

$$a + b + c = 2s.$$

If one of the variables $a, b, c = 0$, we get a degenerate triangle, namely a line segment. This will correspond to area 0 and to a minimum point for the area function in the 3-dimensional cube defined by the restrictions $0 \leq a \leq s$, $0 \leq b \leq s$, $0 \leq c \leq s$ on the variables a, b, c.

If all the variables $a, b, c \neq 0$, we get the unique solution

$$a = b = c = \frac{2s}{3} = \frac{L}{3}.$$

This solution defines the maximum point for the area function in the relevant range of the variables and corresponds to the equilateral triangle. This completes the proof. $\square$

9.3 The isoperimetric problem for closed polygons

The isoperimetric problem for closed polygons can be formulated as follows:

> *Among all n-gons without self-intersections and with a prescribed perimeter find the one that encloses the largest area.*

The answer is, as one would expect, the regular n-gon.

Theorem 9.3.1. *Among all n-gons without self-intersections and with a prescribed perimeter, the regular n-gon encloses the largest area.*

Proof. First we prove the existence of an object with maximal area in the class of n-gons without self-intersections and with a prescribed perimeter L. For this purpose, choose a rectangular coordinate system in the Euclidean plane. An n-gon can then be described by the $2n$ coordinates of its corners. Now note, that all possible shapes of an n-gon without self-intersections and fixed length L can be described by the coordinates in a subset K of a closed ball of radius $\sqrt{2n}\,L$ in $2n$-dimensional Euclidean space $\mathbb{R}^{2n}$, e.g. by fixing one of the corners of an n-gon of length L at the origin of the Euclidean plane making all coordinates of the corners numerically less than L. The points in the closure $\overline{K}$ in $\mathbb{R}^{2n}$ of such a set K will represent the class of all n-gons without self-intersections with the fixed prescribed length L and degenerate cases where parts of edges may come together. An n-gon in $\overline{K}$ has an area which depends continuously on the $2n$ coordinates of its corners. Hence the area function attains a maximum value in the closed and bounded set $\overline{K}$ according to Theorems 3.2.4 and 3.6.3. The maximal value of the area function must clearly be attained in a non-degenerate case, i.e. in an n-gon without self-intersections. This proves the existence of a solution to the isoperimetric problem for n-gons.

To prove that the regular n-gon is the solution to the isoperimetric problem for n-gons, granted the existence of a solution, we employ a method used by the Swiss mathematician Jacob Steiner (1796-1863).

Consider a closed n-gon Γ_n without self-intersections and of a fixed prescribed length L with the property that among all such n-gons it encloses the largest plane area.

It is clear that Γ_n must be convex, since if this were not the case we could create a closed n-gon of the same length L as Γ_n, but enclosing a larger area, by reflecting an inbuckling to an outbuckling. It is also clear that Γ_n must be equilateral, i.e. all edges have the same length, since otherwise by Theorem 9.2.1, we could create an n-gon with a larger area than Γ_n by turning the triangle determined by a neighbouring pair of edges of different lengths into an isosceles triangle.

For technical reasons, we first consider the case $n = 2m$. Then we can choose two corners A and B on Γ_{2m} that divides the $2m$-gon into two polygonal arcs of equal length $\frac{L}{2}$. The chord AB divides the enclosed figure into two pieces, each of which must have the same area, since otherwise, we could construct a figure of larger area by replacing the smaller piece with the mirror image in AB of the larger piece; cf. Figure 9.4.

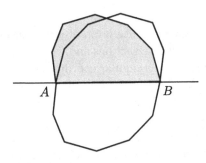

Fig. 9.4   Determining the shape of closed polygons maximizing the area.

Consider now the 'half' figure bounded by one of the polygonal arcs from A to B of length $\frac{L}{2}$ and the line segment AB. Let P be an arbitrary corner on the arc, and consider the triangle APB. Imagine that the arc is made of steel with a hinge at P. Without changing the length of the arc we can then bend the triangle at the corner P. Thereby it is easily seen that the 'half' figure has maximal area precisely when the triangle has a right angle at P; cf. Figure 9.5. We conclude that P lies on the semicircle with diameter AB, and hence that all corners of the original $2m$-gon, Γ_{2m}, lie on that circle. This proves that Γ_{2m} is the regular polygon with length L.

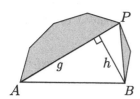 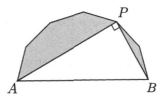

Fig. 9.5 The regular n-gon maximizes the area of isoperimetric n-gons.

If we connect only every other corner in a regular $2m$-gon, we get a regular m-gon. If this regular m-gon does not have maximum area in relation to its perimeter, it can be replaced by the 'maximal' equilateral m-gon corresponding to this perimeter, thereby creating a new $2m$-gon with the same perimeter as the original regular $2m$-gon, but with a larger area. This would be a contradiction, and hence the 'maximal' equilateral m-gon must be the regular m-gon. Thereby, we also solve the isoperimetric problem for n-gons with an odd number of corners.

This completes the proof of the theorem. $\square$

9.4 The general isoperimetric problem

For closed curves in the plane with finite length (rectifiable Jordan curves) and without self-intersections, one has the general isoperimetric problem. In its simplest form the problem can be stated as follows: Find the closed plane curve without self-intersections of a fixed prescribed length (rectifiable Jordan curves) that encloses the maximal plane area.

The ancient Greeks took it for granted that the solution to the problem is what they regarded as the most perfect of all closed curves, namely the circle. However, not until the 19th century was a complete proof given, and by that time it had emerged that in variational problems it is often the bare question of the *existence* of an optimal object that is the problem. Jacob Steiner suggested several ingenious proofs of the isoperimetric problem in the 1840s, but he did not prove the existence of a maximal object. This was only clarified in famous lectures by Weierstrass at the University of

Berlin in the 1870s, where he developed the general tools employed in the existence part of the proof of Theorem 9.3.1.

The existence of a solution to the general isoperimetric problem can be proved by approximating an arbitrary closed (rectifiable) curve with polygonal curves. The proof of the following theorem can then be completed in the manner, suggested by Steiner, that was employed in the proof of Theorem 9.3.1.

Theorem 9.4.1. *Among all closed curves in the plane without self-intersections and with a prescribed length (rectifiable Jordan curves), the circle encloses the largest area.*

9.5 Elements of the history of calculus of variations

The isoperimetric problem is one of the classical problems of the early history of the calculus of variations. The development in this field took place especially after Sir Isaac Newton (1642-1727) and Gottfried Wilhelm Leibnitz (1646-1716) developed the calculus with infinitesimal quantities (differential and integral calculus) at the end of the seventeenth century. Among the pioneers of the calculus of variations must be singled out the brothers Jakob Bernoulli (1655-1705) and Johann Bernoulli (1667-1748) plus, not least, the prolific mathematician Leonhard Euler (1707-1783).

Even though the powerful tools from infinitesimal calculus are decisive for the systematic use of calculus of variations, it has always been considered to be especially attractive to be able to carry through purely geometical arguments. It is rare, however, to succeed so well in this as with the isoperimetric problem.

The first method of solving the isoperimetric problem for triangles is a good example of a method in the calculus of variations that is known under the name *the direct method*. By this method you demonstrate in a problem (here the isoperimetric problem for triangles), the existence of a solution with a required extremal property and, at the same time, lay down the solution (here the equilateral triangle) by constructing an appropriate approximating sequence of objects (here isosceles triangles).

The isoperimetric problem can be generalized in innumerable ways, and the term *an isoperimetric problem* is now used for a whole family of problems, in which a quantity (usually an integral of some parameters) must be maximized while preserving another quantity. The immediate generalization to higher dimensional Euclidean spaces is as expected. For example, the spherical form is the figure in 3-space that encloses the largest volume with a prescribed surface area. The first to prove this was H.A. Schwarz in an article from 1884.

Among the mathematicians from the nineteenth century that have contributed significantly to the calculus of variations, we shall here content ourselves with singling out Weierstrass, who perfected the basis of the theory in the 1870s, in the earlier mentioned lectures at the University of Berlin. In the twentieth century, the calculus of variations has developed tremendously, and we shall only mention Marston Morse, after whom, as earlier mentioned, a fruitful topological theory within the field is named.

9.6   Minima for rubber bands on rigid cylinders

The final problem to be considered in this chapter serves as an introduction to Morse theory. The exposition is based on the author's article *Minima for elastics on rigid cylinders*, SIAM Review Vol. 31, No. 2, 1989, 310–317.

9.6.1   *The problem*

If you take a rubber band and wind it around a circular frictionless rigid cylinder, physical experience tells you that it will contract into a state of equilibrium, where the rubber band winds an integral number of times homogeneously around the rigid cylinder, either preserving or reversing a prescribed sense of orientation.

We shall explain these observations using methods from global calculus of variations in spaces of maps. The rubber band and the rigid cylinder are both modelled by circles, and winding the rubber band around the cylinder then corresponds mathematically to mapping a circle into itself. The number of times the elasic band winds around the cylinder is the so-called degree of the associated map of the circle into itself, and the equilibrium positions of the rubber band are identified as minima for a suitable energy function on the space of all smooth maps of the circle into itself. It is perhaps the only application of global calculus of variations where one can work entirely with smooth maps. But it does involve all the main ideas.

9.6.2   *Energy and degree of a circle map*

Let S^1 be the unit circle in the plane, and let

$$C^2(S^1, S^1) = \{f : S^1 \to S^1 \mid f \text{ of class } C^2\}$$

denote the space of maps $f : S^1 \to S^1$ of class C^2.

Put

$$x = (\cos t, \sin t), \quad t \in \mathbb{R},$$

and consider $f : S^1 \to S^1$ as a function of $t \in \mathbb{R}$ by the definition

$$f(t) = f(\cos t, \sin t), \quad t \in \mathbb{R}.$$

Define the *energy* of f by

$$E(f) = \frac{1}{2} \int_{S^1} \|f'(x)\|^2 dx = \frac{1}{2} \int_0^{2\pi} \|f'(t)\|^2 dt.$$

Corresponding to $f : S^1 \to S^1$, there is a so-called *angular function* for f, i.e. a continuous function $\theta : \mathbb{R} \to \mathbb{R}$ such that

$$f(\cos t, \sin t) = (\cos \theta(t), \sin \theta(t)), \quad t \in \mathbb{R};$$

cf. Figure 9.6.

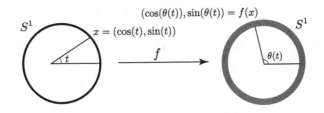

Fig. 9.6 Angular function $\theta(t)$ for a circle map $f(\cos t, \sin t) = (\cos \theta(t), \sin \theta(t))$.

Lemma 9.6.1. *Angular functions do exist and are unique up to addition of an integral multiple of 2π.*

Furthermore, if f is differentiable of class C^r, then any angular function θ for f is also differentiable of class C^r.

Proof. For the existence part, the function $\theta : \mathbb{R} \to \mathbb{R}$ can be pieced together from functions defined as compositions of the inverse trigonometric functions Arccos and Arcsin with the coordinate functions of $f(t)$. Differentiability of f is clearly transferred to θ.

For the uniqueness part, note that if

$$(\cos \theta(t), \sin \theta(t)) = (\cos \tilde{\theta}(t), \sin \tilde{\theta}(t)), \quad t \in \mathbb{R},$$

then $\tilde{\theta}(t) - \theta(t) = n_t \cdot 2\pi$ for an integer n_t for every $t \in \mathbb{R}$. For reasons of continuity, n_t has to be constant, i.e. $n_t = n$ is independent of $t \in \mathbb{R}$. This completes the proof. $\square$

If θ is an angular function for f, it follows by periodicity of the involved trigonometric functions, that there exists a unique integer $\deg(f)$ satisfying

$$\theta(2\pi) - \theta(0) = \deg(f) \cdot 2\pi.$$

This integer is called the *degree* of $f : S^1 \to S^1$; a notion introduced in 1910 by the Dutch mathematician Luitzen Egbertus Jan Brouwer (1881–1966). The degree counts how many complete windings, sense of direction taken into account, the map $f : S^1 \to S^1$ has completed in mapping the domain circle into the image circle.

If θ is an angular function for f we get

$$f'(t) = (-\sin\theta(t), \cos\theta(t))\frac{d\theta}{dt},$$

and therefore

$$E(f) = \frac{1}{2}\int_0^{2\pi}\left(\frac{d\theta}{dt}\right)^2 dt.$$

9.6.3 Estimate of the energy of maps

By the Cauchy-Schwarz inequality (page 21) in the vector space of continuous functions on the closed interval $[0, 2\pi]$ with the inner product defined in Example 4.4.6, page 97, we get for any two continuous functions u and v the inequality

$$\left|\int_0^{2\pi} uv\, dt\right| \le \sqrt{\int_0^{2\pi} u^2\, dt} \cdot \sqrt{\int_0^{2\pi} v^2\, dt}.$$

From this we get the following lower bound on the energy of f:

$$E(f) = \frac{1}{2}\int_0^{2\pi}\left(\frac{d\theta}{dt}\right)^2 dt = \frac{1}{2}\int_0^{2\pi}\left(\frac{d\theta}{dt}\right)^2 dt \cdot \frac{1}{2\pi}\int_0^{2\pi} 1^2\, dt$$

$$\ge \frac{1}{4\pi}\left|\int_0^{2\pi}\frac{d\theta}{dt}\, dt\right|^2 = \frac{1}{4\pi}(\theta(2\pi) - \theta(0))^2 = \frac{1}{4\pi}4\pi^2(\deg(f))^2$$

$$= \pi \cdot (\deg(f))^2.$$

9.6.4   *Manifolds of maps*

We can define a metric for the C^2-topology on $C^2(S^1, S^1)$ by

$$d^2(f, g) = \sup_{t \in [0, 2\pi]} \left\{ \rho(f(t), g(t)) + \left| \frac{d\theta}{dt} - \frac{d\psi}{dt} \right| + \left| \frac{d^2\theta}{dt^2} - \frac{d^2\psi}{dt^2} \right| \right\},$$

where $\rho(\cdot, \cdot)$ denotes arclength on S^1, and θ and ψ are angular functions for f and g, respectively.

Denote by

$$C^2(S^1, \mathbb{R}) = \{v : S^1 \to \mathbb{R} \mid v \text{ of class } C^2\}$$

the space of maps $v : S^1 \to \mathbb{R}$ of class C^2. This space is a vector space with norm

$$\|v\|_{(2)} = \sup_{t \in [0, 2\pi]} \{|v(t)| + |v'(t)| + |v''(t)|\},$$

where we consider $v : S^1 \to \mathbb{R}$ as a function of $t \in \mathbb{R}$ by

$$v(t) = v(\cos t, \sin t), \quad t \in \mathbb{R}.$$

Define the map:

$$p_f : C^2(S^1, \mathbb{R}) \to C^2(S^1, S^1)$$

by

$$p_f(v)(\cos t, \sin t) = \big(\cos \left(\theta(t) + v(t) \right), \sin \left(\theta(t) + v(t) \right) \big), \quad t \in \mathbb{R}.$$

In other words: If f has the angular function θ, then $p_f(v)$ has the angular function $\theta + v$.

Note, that since $v(2\pi) = v(0)$, it follows that $\deg(p_f(v)) = \deg(f)$. It can also be proved (but this is not completely trivial) that the space of maps $f : S^1 \to S^1$ of a fixed degree k, denoted $C_k^2(S^1, S^1)$, is a *path-component* in $C^2(S^1, S^1)$, i.e. it is a pathwise connected subset of $C^2(S^1, S^1)$, which is maximal with this property.

It is relatively easy to verify that p_f maps a neighbourhood of 0 (the vector function with constant value 0) in $C^2(S^1, \mathbb{R})$ homeomorphically onto a neighbourhood of f in $C^2(S^1, S^1)$. This proves the following theorem.

Theorem 9.6.2. *The space* $C^2(S^1, S^1)$ *is a manifold modelled on the Banach space* $C^2(S^1, \mathbb{R})$.

The manifold is, in fact, of class C^∞ and the energy function E is a smooth function. More precisely we have the theorem.

Theorem 9.6.3. $E \circ p_f : C^2(S^1, \mathbb{R}) \to \mathbb{R}$ *is differentiable at* $0 \in C^2(S^1, \mathbb{R})$ *and the differential is given by:*

$$D(E \circ p_f)(0)(v) = \int_0^{2\pi} \theta' v' \, dt = - \int_0^{2\pi} \frac{d^2\theta}{dt^2} v \, dt \ .$$

Proof. The proof is a consequence of the computations below, since the function $\varepsilon(v)$ defined in the computations is an ε-function:

$$E \circ p_f(v) - E \circ p_f(0) = \frac{1}{2} \int_0^{2\pi} [(\theta' + v')^2 - (\theta')^2] dt$$

$$= \int_0^{2\pi} \theta' v' \, dt + \left(\frac{1}{2} \int_0^{2\pi} \frac{v'}{||v||_{(2)}} v' \, dt \right) ||v||_{(2)}$$

$$= [\theta' v]_0^{2\pi} - \int_0^{2\pi} \frac{d^2\theta}{dt^2} v \, dt + \varepsilon(v) ||v||_{(2)}$$

$$= - \int_0^{2\pi} \frac{d^2\theta}{dt^2} v \, dt + \varepsilon(v) ||v||_{(2)} \ .$$

In the computation we have used that $\theta'(0) = \theta'(2\pi)$, since $f'(0) = f'(2\pi)$. Also note that $v(0) = v(2\pi)$.

This completes the proof. $\square$

In a local extremum for the energy function E at the function f, the differential $D(E \circ p_f)(0)$ must satisfy

$$D(E \circ p_f)(0)(v) = 0, \quad \text{for all} \quad v \in C^2(S^1, \mathbb{R}),$$

or equivalently,

$$\int_0^{2\pi} \frac{d^2\theta}{dt^2} v \, dt = 0, \quad \text{for all} \quad v \in C^2(S^1, \mathbb{R}).$$

By a standard argument, essentially setting $v = \frac{d^2\theta}{dt^2}$, this implies that

$$\frac{d^2\theta}{dt^2} = 0,$$

and therefore, that $\theta = kt + \theta_0$ for two constants k and θ_0. By the definition of the degree of a map, obviously, $k = \deg(f)$, whereas θ_0 can be chosen arbitrarily.

For any integer $k \in \mathbb{Z}$, let $f_k : S^1 \to S^1$ be the map with angular function $\theta_k(t) = kt$. An easy calculation shows that the map f_k, and similarly the map f with angular function $\theta(t) = kt + \theta_0$, has the energy $E(f_k) = \pi k^2$.

Note, that the map f with angular function $\theta(t) = kt + \theta_0$ corresponds to winding S^1 around itself k times followed by a fixed rotation θ_0.

It follows from the above considerations that a necessary condition for the energy function E to have a local extremum at f is that the angular function for f has the form $\theta(t) = \deg(f) \cdot t + \theta_0$. On the other hand, it also appears that at such a function, the energy has a global minimum in the path-component $C_k^2(S^1, S^1)$ of $C^2(S^1, S^1)$ containing the maps of constant degree $k = \deg(f)$.

Collecting the above results, we have proved the following theorem, which explains the observations made when a rubber band is wound around a rigid cylinder.

Theorem 9.6.4. *The only local extrema for the energy function*

$$E : C^2(S^1, S^1) \to \mathbb{R}$$

are the local minima f_k composed with a rotation, for all integers $k \in \mathbb{Z}$. On the path-component $C_k^2(S^1, S^1)$ in $C^2(S^1, S^1)$ consisting of the C^2-maps $f : S^1 \to S^1$ of degree k, the energy function E has the minimum value $\pi \cdot k^2$, which is attained at the map f_k, or, at f_k composed with a rotation, and only at these maps.

9.6.5 *Final comments*

There are many other geometric problems in the calculus of variations, many of which are nowadays tackled by Morse theory. Among other subjects, we shall mention the study of properties of curves in curved surfaces that locally minimize distances (geodesic curves). A particularly fascinating subject is the study of the so-called minimal surfaces, i.e., (curved) surfaces that locally minimize the area. Such surfaces can be found, among other places, as membranes in living organisms, or as the soap films that are formed when a closed wire curve of arbitrary shape is dipped into a suitably strong soap solution.

Exercises and Further Results

Exercise 9.1. Consider two media separated by a plane and suppose that the velocity of light is c_1 in the first medium and c_2 in the second. A light ray passes from a point P in the first medium to the point Q in the second medium. Let i be the angle of incidence (the angle of the incoming light ray to the normal of the plane of separation) and r the angle of refraction (the angle of the refracted light ray to the normal of the plane of separation). Deduce the law of refraction from Fermat's principle of light propagation (the Principle of Least Time):

$$\frac{\sin i}{\sin r} = \frac{c_1}{c_2}.$$

This law is called *Snell's law* after the Dutch mathematician Willebrord Snell (1580–1626) who established it in 1621.

Exercise 9.2. Consider a right triangle with the constant perimeter L. How shall the triangle be designed, in the class of right triangles, in order to enclose the maximal possible area?

Exercise 9.3. A *Reuleaux triangle* is constructed from an equilateral triangle by substituting each of its sides with the circular arc centred at the opposite corner. A Reuleaux triangle has the interesting property of being a closed curve of constant width. Constant width is therefore not a property restricted to a circle, as one might immediately believe.

Now consider the class of plane triangle-like figures with three corners connected by circular arcs. Discuss the isoperimetric problem for this class of figures and in particular the role of the Reuleaux triangle in relation to its solution.

Exercise 9.4. In this exercise we consider a class of plane figures which all have a well-defined perimeter L and a well-defined area A.

1) Show that in the class of triangles with a prescribed area A, the equilateral triangle with area A has the smallest perimeter.

2) Show that in the class of n-gons with a prescribed area A, the regular n-gon with area A has the smallest perimeter.

3) Prove that the following two questions are equivalent:

(a) which figure with a prescribed perimeter L maximizes the area?

(b) which figure with a prescribed area A minimizes the perimeter?

Exercise 9.5. Show that every continuous map $f : S^1 \to S^1$ has a unique integer $\deg(f)$, called the *degree*, associated with it. It can be defined making use of angular functions.

Let $f, g : S^1 \to S^1$ be continuous maps and consider the composite map $g \circ f : S^1 \to S^1$. Prove that

$$\deg(g \circ f) = \deg(g) \cdot \deg(f).$$

Exercise 9.6. Let $C^0(S^1, S^1)$ be the space of continuous maps $f : S^1 \to S^1$ equipped with the metric

$$d(f, g) = \sup_{t \in [0, 2\pi]} \{\rho(f(\cos t, \sin t), g(\cos t, \sin t))\},$$

where $\rho(\cdot, \cdot)$ denotes arclength on S^1.

Let $C^0(S^1, \mathbb{R})$ be the Banach space of continuous functions $v : S^1 \to \mathbb{R}$ with norm

$$\|v\| = \sup_{t \in [0, 2\pi]} \{|v(\cos t, \sin t)|\}.$$

For every fixed $f \in C^0(S^1, S^1)$ define the continuous map

$$p_f : C^0(S^1, \mathbb{R}) \to C^0(S^1, S^1), \text{ by}$$

$$p_f(v)(\cos t, \sin t) = \big(\cos\big(\theta(t) + v(t)\big), \sin\big(\theta(t) + v(t)\big)\big), \quad t \in \mathbb{R},$$

where θ is an angular function for f. Note, that $p_f(v)$ then has the angular function $\theta + v$.

1) Prove that $C^0(S^1, S^1)$ is a topological manifold modelled on the Banach space $C^0(S^1, \mathbb{R})$.

2) Let $C^0_k(S^1, S^1)$ be the set of continuous maps $f : S^1 \to S^1$ of degree $k \in \mathbb{Z}$. Prove that $C^0_k(S^1, S^1)$ is an open set in $C^0(S^1, S^1)$.

3) Prove that $C^0_k(S^1, S^1)$ is a path-connected set for any degree $k \in \mathbb{Z}$.

Exercise 9.7. Make a study of the minima for a suitable energy function on the space of maps $C^2(S^1, S^1 \times [0, h])$ consisting of maps $f : S^1 \to S^1 \times [0, h]$ of class C^2 of the circle S^1 into the circular cylinder $S^1 \times [0, h]$ of height $h > 0$. Relate the result obtained to the problem of winding a rubber band around a circular cylinder of height $h > 0$.

Bibliography

In the bibliography below, every book is listed under the subject of special relevance to the present book. The bibliography is by no means exhaustive.

General topology

J.D. Dugundji: "Topology," Allyn and Bacon, Boston, 1966.

G.F. Simmons: "Introduction to Topology and Modern Analysis," McGraw-Hill, New York, 1963.

W.A. Sutherland: "Introduction to Metric and Topological Spaces," Oxford University Press, 1975.

Analysis in Banach spaces

H. Cartan: "Calcul différentiel," Herman, Paris, 1967.

J. Dieudonné: "Foundations of Modern Analysis," Academic Press, New York, 1960.

V.L. Hansen: "Functional Analysis - Entering Hilbert Space," World Scientific, Singapore, 2nd edition, 2016.

S. Lang: "Analysis II," Addison-Wesley, Reading Massachusetts, 1969.

Analysis on manifolds, foundations of global analysis

R. Abraham, J.E. Marsden, T. Ratiu: "Manifolds, Tensor Analysis, and Applications," Addison-Wesley, Reading Massachusetts, 1983.

D.D. Bleecker, B. Booss-Bavnbek: "Index Theory with Applications to Mathematics and Physics," International Press of Boston, Inc., 2013.

N. Bourbaki: "Varietés différentielles et analytiques," Herman, Paris, 1967.

Y. Choquet-Bruhat, C. de Witt-Morette, M. Dillard-Bleick: "Analysis, Manifolds and Physics," North-Holland, Amsterdam, 1977.

J. Dieudonné: "Treatise on Analysis," Vol. III, Academic Press, New York, 1972.

S. Lang: "Introduction to differentiable manifolds," (Second edition) Universitext, Springer-Verlag, New York, 2002.

Differential geometry and topology

T. Bröcker, K. Jänich: "Introduction to differential topology," Cambridge University Press, 1982.

B.A. Dubrovin, A.T. Fomenko, S.P. Novikov: "Modern Geometry-Methods and Applications," Vol. I, II, Springer, Berlin-Heidelberg-New York, Vol. I 1984, Vol. II 1985.

V. Guillemin, A. Pollack: "Differential Topology," Prentice-Hall, New Jersey, 1974.

M.W. Hirsch: "Differential Topology," Springer, Berlin-Heidelberg-New York, 1976.

J. Margalef Roig, E. Outerelo Dominguez: "Differential Topology," North-Holland Mathematics Studies, 173. North-Holland Publishing Co., Amsterdam, 1992.

P. W. Michor: "Topics in differential geometry," Graduate Studies in Mathematics, Vol. 93, American Mathematical Society, Providence, RI, 2008.

J. Milnor: "Topology from the differentiable viewpoint," (Revised reprint of the 1965 original) Princeton University Press, Princeton, NJ, 1997.

F.W. Warner: "Foundations of Differentiable Manifolds and Lie Groups," Scott-Foresman, Dallas Texas, 1971.

Dynamical systems

V.I. Arnold: "Geometrical methods in the theory of ordinary differential equations," Springer, Berlin-Heidelberg-New York, 1983.

D.R.J. Chillingworth: "Differential topology with a view to applications," Pitman, London, 1976.

R.L. Devaney: "An introduction to chaotic dynamical systems," (Second edition) Addison-Wesley, Redwood City, CA, 1989.

J. Guckenheimer, P. Holmes: "Nonlinear oscillations, dynamical systems, and bifurcations of vector fields," Springer, Berlin-Heidelberg-New York, 1983.

M.C. Irwin: "Smooth dynamical systems," Academic Press, New York, 1980.

D.J. Luo, L.B. Teng: "Qualitative theory of dynamical systems," World Scientific, Singapore, 1993.

J. Palis, W. de Melo: "Geometric theory of dynamical systems, an introduction," Springer, Berlin-Heidelberg-New York, 1982.

L. Perko: "Differential Equations and Dynamical Systems," Springer, Berlin-Heidelberg-New York, 3rd edition, 2001.

D. Ruelle: "Elements of differentiable dynamics and bifurcation theory," Academic Press, Inc., Boston, MA, 1989.

S. Wiggins: "Introduction to applied nonlinear dynamical systems and Chaos," Springer, Berlin-Heidelberg-New York, 2nd edition, 2003.

Singularity theory and catastrophe theory

V.I. Arnold: "Catastrophe Theory," Springer, Berlin-Heidelberg-New York, 1984.

V.I. Arnold, A.N. Varchenko, S.M. Gusein-Zade: "Singularities of Differentiable Maps," Vol. I, II, Birkhäuser Boston Inc., Cambridge Massachusetts, Vol. I 1985, Vol. II 1988.

M. Golubitsky, V. Guillemin: "Stable Mappings and Their Singularities," Springer, Heidelberg-New York, 1973.

M. Golubitsky, D.G. Schaeffer: "Singularities and Groups in Bifurcation Theory," Vol. I, Springer, Berlin-Heidelberg-New York, 1985.

V.L. Hansen: "Geometry in Nature," AK Peters Ltd., Wellesley, Massachusetts, 1993.

M. Golubitsky, I. Stewart, D.G. Schaeffer: "Singularities and Groups in Bifurcation Theory," Vol. II, Springer, Berlin-Heidelberg-New York, 1988.

T. Poston, I. Stewart: "Catastrophe Theory and its Applications," Dover Publications, Inc., Mineola, New York, 1996 (reprint of the 1978 original).

Calculus of variations

J. Eells, L. Lemaire: "Selected topics in harmonic maps," CBMS Regional Conference Series in Mathematics, No. 50, American Mathematical Society, Providence, Rhode Island, 1983.

V.L. Hansen: "Shadows of the Circle - Conic Sections, Optimal Figures and Non-Euclidean Geometry," World Scientific, Singapore, 1998.

W. Klingenberg: "Lectures on closed geodesics," Springer, Berlin-Heidelberg-New York, 1978.

J.W. Milnor: "Morse Theory," Princeton University Press, 1963.

R.S. Palais: "Foundations of global non-linear analysis," Benjamin, New York, 1968.

H. Seifert, W. Threlfall: "Variationsrechnung im Grossen," Teubner, Leipzig, 1938.

L.C. Young: "Lectures on the Calculus of Variations and Optimal Control Theory," Saunders, Philadelphia, 1969.

Theoretical mechanics

R. Abraham, J. Marsden: " Foundations of Mechanics," (Second edition) Benjamin/Cummings, Reading Massachusetts, 1978.

V.I. Arnold: "Mathematical methods of classical mechanics," Springer, Berlin-Heidelberg-New York, 1978.

D.D. Holm: "Geometric mechanics. Part I & II," (Second edition) Imperial College Press, London, 2011.

S. Johannesen: "Smooth manifolds and fibre bundles with applications to theoretical physics," CRC Press, Boca Raton, Florida, 2017.

J.M. Knudsen, P.G. Hjorth: "Elements of Newtonian mechanics. Including nonlinear dynamics," Advanced Texts in Physics. Springer-Verlag, Berlin, 3rd edition (revised and enlarged), 2000.

J. M. Leinaas: "Classical Mechanics and Electrodynamics," World Scientific, Singapore, 2019.

List of Symbols

Symbol	Explanation	Page
$Gl(E)$	the general linear group in Banach space E	136
$\sum_{n=1}^{\infty} x_n$	sum of series in a Banach space E	137
$L^r(E, F)$	space of r-linear continuous mappings between normed vector spaces	159
$D^r f$	r^{th} derivative of mapping f	162
$D_{ij}f(x)$	second order partial derivative of f at x	163
$\mathcal{A}$	atlas for differentiable structure on manifold	189
$T_p M$	tangent space to manifold M at point p	196
TM	tangent bundle of manifold M	202
$T_p f$	differential (tangential mapping) of mapping $f : M \to N$ between manifolds at point $p \in M$	210
$\mathcal{E}(M, p)$	germs of functions on manifold M at point p	199
$\mathcal{D}er(M, p)$	the space of derivations in $\mathcal{E}(M, p)$	231
$C^\infty(M)$	the algebra of smooth real-valued functions on a smooth manifold M	207
$v_p[f]$	directional derivative of f along tangent v_p	199
$X[f]$	directional derivative of f along vector field X	206
$[X, Y]$	Lie product of vector fields X and Y	208
$f \pitchfork_p W$	f transversal to W at point p	220
$\mathbf{H}(f)$	Hesse matrix for germ $f : (\mathbb{R}^n, 0) \to (\mathbb{R}, 0)$	241
$\mathbb{R}^n$	Euclidean n-dimensional real number space	22
S^n	the unit sphere in $\mathbb{R}^{n+1}$	192
$\mathbb{RP}^n$	the real projective space of dimension n	228
$\mathbf{O}(n)$	the orthogonal group in $\mathbb{R}^n$	229

Index

Printed in the United States
By Bookmasters